Geography of Kerala

About the Author

Dr. Srikumar Chattopadhyay (b. 1953) retired from CESS as Scientist G and Head, Resources Analysis Division. He obtained his M. Sc. and Ph. D. degree in Geography from the University of Calcutta and Post Graduate diploma in Applied Geomorphology and Environmental Survey from the International Institute of Aerospace Survey and Earth Sciences (ITC), The Netherlands. He was a Post-Doctoral Fulbright Fellow in the Environment and Policy Institute, East West Centre, Hawaii, USA. Recipient of Dr. S. Vasudev award (1994) instituted by the Kerala State Council for Science, Technology and Environment, Government of Kerala, and Prof. R.N. Dubey memorial award (2001) instituted by the Boovigyan Vikas Foundation, New Delhi. After retirement from CESS in 2013 he was awarded Senior Fellow position by ICSSR hosted in the Centre for Development Studies (CDS), Thiruvananthapuram. In 2016, he served as a Fellow in the HWK Institute for Advanced Studies, Delmenhorst, Germany. He also worked as a Senior Consultant, Kerala Development Innovation Strategic Council (K-Disc), Government of Kerala, Thiruvananthapuram. At present he is a National Fellow awarded by the ICSSR, housed in the Gulati Institute for Finance and Taxation Thiruvananthapuram.

Dr. Chattopadhyay's work area covers natural and environmental resource management, terrain analysis, human dimension of environmental change, participatory local level planning, and sustainable development. He contributed substantially in evolving Participatory Panchayat Resource Mapping and River Bank Mapping programmes, which were widely acclaimed. He served as visiting faculty in the Department of Geography and Environmental Studies, Montclair State University, New Jersey, USA; Leibnitz Centre for Tropical Marine Ecology (ZMT), Bremen, Germany and several other Indian Universities including Delhi, Calcutta, Mumbai, Burdwan, and Vidyasagar. He has produced more than 100 research papers, authored/edited eight books, one Atlas and completed 40 research projects funded by the state, national and international agencies.

Several Ph. D. and MS students from India and abroad completed their theses under his guidance. Dr. Chattopadhyay served in various committees, academic bodies and professional organisations both at the national and state level. He was the President of Institute of Indian Geographers, Pune during 2014-15.

GEOGRAPHY OF KERALA

Srikumar Chattopadhyay

CONCEPT PUBLISHING COMPANY PVT. LTD.
NEW DELHI-110059

ISBN: 978-93-5439-001-2 (HB)
978-93-5439-009-8 (PB)

First Published 2021

Published and Printed by

Concept Publishing Company Pvt. Ltd.
A/15-16, Commercial Block, Mohan Garden,
New Delhi-110059 (INDIA)
T : 91 11 41101460, 91 11 25351460, 91 11 25351794
E : publishing@conceptpub.com; *W* : www.conceptpub.com

Cataloging in Publication Data--Courtesy: D.K. Agencies (P) Ltd. <docinfo@dkagencies.com>

Chattopadhyay, Srikumar, author.
pages cm
Includes bibliographical references and index.
ISBN 9789354390012 (HB)
ISBN 9789354390098 (PB)

1. Kerala (India)—Geography. 2. Kerala (India)—Economic conditions. I. Title.

LCC DS485.K4C43 2021 | DDC 915.48 23

DEDICATION

Rekha Chatterjee

(Born on 15 August, 1929 — Died on 31 July, 2010)

This book is dedicated to the memory of my mother Late Rekha Chatterjee who during her first visit to Kerala in 1986 exclaimed *"Prakirtir ki Asim Abadan"* (What an unbound gift of nature!)

Preface and Acknowledgements

I came to Kerala in September 1979 and joined the Resources Analysis Division (RAD) of Centre for Earth Science Studies (CESS), now rechristened as the National Centre for Earth Science Studies (NCESS) after the institute was taken over by the Government of India in January 2014. One of the projects undertaken by CESS from its inception was 'Resources Atlas of Kerala'. I got the opportunity to work in this project from the beginning. During the course of our work, we gradually realised about the bountiful natural endowment and the role of geography in Kerala's development process. We also noted that there was hardly any book on the geography of Kerala, barring few articles dealing with Kerala as part of Indian geography. The idea of a book on 'Geography of Kerala' was discussed among colleagues in CESS at the behest of late Dr. N.G.K. Nair, Head of Resources Analysis Division (RAD) during that time, however, the project could not take off. There were several other assignments and always a deadline to meet for completion of projects. Nevertheless, the idea of a book on the geography of Kerala was always in the back of my mind. I was hoping to start writing the book at an opportune moment.

I retired in 2013 and therefore, could devote all my time to the book, although some of the chapters were written well before my retirement. During the 34 years of my service in CESS, I executed several projects related to various branches of geography covering both physical and human aspects, conducted field studies across Kerala, and led multidisciplinary teams for the execution of various projects. This experience and interaction with scholars of other disciplines both in CESS and other institutes, particularly, Centre for Development Studies (CDS), and also the people of Kerala, outside the institutional sphere, helped me know the state intimately, understand its development trajectory, and sharpen my viewpoint. Over the years, I learnt about the role of geography in shaping Kerala ecologically, socially, economically, and politically.

I am indebted to a large number of people both in my office and outside to complete this book, a long-awaited project finally materialised. I am grateful to (late) Prof. C. Karunakaran, Dr. Harsh K. Gupta, (late) Subrata Sinha, (late) Dr. K.M. Nair, Dr. M. Baba, and Dr. N.P. Kurien all former directors of CESS for shaping my ideas and providing all facilities. I am thankful to Dr. N. Purnachandra Rao, Director of NCESS for according permission to affiliate myself to NCESS for my ICSSR National Fellowship and extending facilities. Subsequently, I moved to the Gulati Institute of Finance and Taxation (GIFT) Thiruvananthapuram. I take this opportunity to thank Prof. K.J. Joseph, Director, GIFT to host me for availing my National Fellowship. My affiliation with NCESS and GIFT has been useful for finalising the manuscript.

I have drawn from the discussions with Prof. M.K. Prasad and Dr. M.P. Parameswaran, both of them are associated with the Kerala Sasthra Sahithya Parishat (KSSP), a premier organisation to popularise science in Kerala, Dr. T.M. Thomas Isaac, Hon'ble Minister for Finance and Coir, Government of Kerala, Dr. K. Soman, former Scientist, and Head, Resources Analysis Division, CESS, and Dr. K. Narayanan Nair, former Director, CDS. I take this opportunity to thank all of them. I also thank Dr. R. Anilkumar, (Retd.) Professor in Geography, University College, Thiruvananthapuram for many discussions over various occasions and raising the question of geography education in Kerala. I acknowledge the support rendered by Prof. Anu Kapur, Department of Geography, Delhi School of Economics, Delhi, while finalising the book.

I am grateful to Prof. Dr. Richard W. Franke, Emeritus Professor, Montclair State University, New Jersey, United States of America (USA), Prof. Dr. Tim Jennerjahn, Head, Working Group on Ecological Biogeochemistry, Leibnitz Centre for Tropical Marine Research (ZMT), Bremen, Germany and Dr. Susan Fuchs, Society Programme, Hanse-Wissenschaftskolleg (Institute for Advanced Studies), Delmenhorst, Germany for facilitating my visit to their respective institutes and academic exchanges that helped organise my ideas. For arrangement and treatment of subject matters in some of the chapters, I drew inspiration from a book titled 'A Geography of New Jersey— the City in the Garden' by Charles A. Stansfield Jr. presented to me by Prof. Harbans Singh, emeritus professor in the Department of Earth and Environmental Sciences, Montclair State University, USA. Idea of the chapter 'Changing Geography of Kerala' germinated from a book 'The Changing Geography of Asia' edited by Graham P. Chapman and Kathleen M. Baker. Late Prof. Graham P. Chapman, personally presented this book to me. I fondly remember my association with him.

Maps are integral parts of geography. This book contains some maps which are prepared specifically for this book. There are also some other maps, included in this book, which were prepared for various research projects/papers executed by me over the years. S. Suresh Kumar, Dr. Shaji Johnson, Dr. V. Dhanya, C.S. Mereena, S.S. Jayalakshmi, Dr. M. Vandana, K. Eldhose, and Rafeeque Chembrasseri assisted me by processing the maps and photographs. I thankfully acknowledge their services.

My wife, Dr. Mahamaya Chattopadhyay, (Retd.) Scientist and Head, Resources Analysis Division, NCESS, helped me on various occasions. I thank her for all the support. I am also grateful to her for letting me convert our dining table as my temporary office for writing and making it untidy. My daughter Suchismita deserves special mention for helping in organising the book in final stage. I am also thankful to my other family members, son Suvojit, daughter-in-law Sanjana, and grandsons Ayaan and Dhruv for sparing their demands on my time and allow me to concentrate on this book.

I thank the Concept Publishing Company Pvt. Ltd., New Delhi for undertaking the task of publishing this book.

Srikumar Chattopadhyay

Thiruvananthapuram
Dated: February, 2021

Contents

List of Tables and Figures

LIST OF TABLES

LIST OF FIGURES

Abbreviations and Glossary

Adivasi	'Original'; member of a tribal group or Scheduled Tribe
Ala	Alappuzha district
amsl	Above mean sea level
Ayacut	The command or distribution area of an irrigation project
ARC	Administrative Reforms Committee
Base flow	River water flow during dry period/non-monsoon season without any addition of rain water
BFA	Beneficiary Farmers' Association
bgl	Below Ground Level
Biomass	Total vegetative matter in an area
Black soil	Type of soil, black in colour with high amount of clay particles, mostly developed in areas of lava flow, found in parts of Palakkad district
BOD	Biological Oxygen Demand
CADA	Command Area Development Authority
CAPEX	Kerala State Cashew Workers Apex Co-operative Society
CBR	Crude Birth Rate
CBSE	Central Board of Secondary Education
CBNP	Community Based Nutrition and Poverty Alleviation Programme
CDS	Centre for Development Studies
CE	Christian Era
Cent	1/100 of an acre
CESS	Centre for Earth Science Studies
CGWB	Central Ground Water Board
Chagara	Mud bank/Suspended colloidal mud in sea water; also the great shrimp catch along the Alappuzha coast
Crore	Ten million
COIRFED	Kerala State Co-Operative Coir Marketing Federation Limited
CPCB	Central Pollution Control Board
CWC	Central Water Commission
CWRDM	Centre for Water Resources Development and Management
DO	Dissolved Oxygen
DT	Demographic Transition
DTM	Digital Terrain Model
DTCP	Department of Town and Country Planning
E	East
EDC	Eco-Development Committee
ENE	East North East

Ern	Ernakulam district
FAO	Food and Agriculture Organisation
EFA	Ecologically Fragile Area
GCA	Gross Cropped Area
GDP	Gross Domestic Product
GOI	Government of India
GOK	Government of Kerala
Gram Panachayat	Lowest administrative unit under three tier Panchayati Raj System
GSVA	Gross State Value Added
GWS	Global Water System
Ha/ha	Hectare—one hundredth of one square kilometre; 2.47 acres
HANTEX	Kerala State Handloom Weaver's Co-operative Societies
HANVEEV	Kerala State Handloom Development Corporation
HDI	Human Development Index
HYV	High Yielding Varieties
Brown Hydromorphic	Type of soil developed under wet condition in an enclosed basin
ICAR	Indian Council for Agricultural Research
ICSE	Indian Certificate of Secondary Education
Idu	Idukki district
IMD	India Meteorological Department
IMR	Infant Mortality Rate
INR	Indian National Rupees
IPCC	Intergovernmental Panel on Climate Change
IRE	Indian Rare Earths
IWT	Inland Water Transport
Jalanidhi	Programme for rural local water supply scheme in Kerala
JFM	Joint Forest Management
JFMC	Joint Forest Management Committee
JNV	Jawahar Navodaya Vidyalaya
Kan	Kannur district
Karapadam	Local term of paddy field in relatively elevated ground in Kuttanad
Kari	Local term for black peaty soil, acidic with high organic matter, found in Kuttanad
Kayal	Local term of low lying land with or without water. The term is also used for lakes/estuaries
Kas	Kasaragod district
Kavu	Small grove of natural vegetation associated with a religious place
KERI	Kerala Engineering Research Institute
KINFRA	Kerala Industrial Infrastructure Development Corporation
KITCO	Kerala Industrial and Technical Consultancy Organisation
Km	Kilometer
Km^2	Square kilometre
Kol	Kollam district
Kole land	Low lying waterlogged rice fields in Trissur district
Kot	Kottayam district
Koz	Kozhikode district
KMML	Kerala Minerals and Metals Limited
KSCDC	Kerala State Cashew Development Corporation
KSCSTE	Kerala State Council for Science Technology and Environment
KSEB	Kerala State Electricity Board
KSI	Kerala Statistical Institute
KSIDC	Kerala State Industrial Development Corporation

KSPCB	Kerala State Pollution Control Board
KSRTC	Kerala State Road Transport Corporation
KSSP	Kerala Sasthra Sahithya Parishat; Kerala People's Science Movement
KSTP	Kerala State Transport Project
KSWTD	Kerala State Water Transport Department
KURTC	Kerala Urban Road Transport Corporation
KV	Kendriya Vidyalaya
KWA	Kerala Water Authority
lpcd	Litre per capita per day
Lakh	Hundred thousand
Lineament	Deep seated fracture, a geological feature
LSG	Local Self Government
LUZ	Land Use Zone
M	Magnitude of earthquake in Richter scale. Richter scale measures magnitude of earthquake in a 10 point scale. Magnitude 0-2M—not felt by people, 2-3M felt little by people. With 4-5M cracks develops in buildings. Magnitude increases in logarithmic scale. 2M releases energy equivalent of 56 kg of explosives, 3M releases energy equivalent of 1800 kg of explosives and 4M releases energy equivalent of 56,000 kg of explosives.
Mal	Malappuram district
MAR	Maximum Assured Rainfall
MCM	Million Cubic Metre
MC Road	Main Central Road
MDG	Millennium Development Goal
mm	Millimeter
MSME	Micro, Small and Medium Enterprises
MW	Mega Watt
N	North
NATPAC	National Transport Planning and Research Centre
NCD	Non Communicable Disease
NE	North East
NEP	New Education Policy
NH	National Highway
NW	National Waterway
NNE	North North-East
NNMR	Neo Natal Mortality Rate
NNW	North North-West
NRSA	National Remote Sensing Agency, now known as National Remote Sensing Centre (NRSC)
Padasekharam	Local term for a group of paddy fields, mostly low lying
Pal	Palakkad district
Pat	Pathanamthitta district
PCB	Pollution Control Board
PHED	Public Health Engineering Department
PHC	Primary Health Centre
PPC	People's Plan Campaign
PPP	Public Private Partnership
PRA	Participatory Rural Appraisal
PRI	Panchayat Raj Institutions
PRM	Panchayat Resource Mapping
PWD	Public Works Department

PSU	Public Sector Undertaking
PSM	People's Science Movement
PTR	Pupil Teacher Ratio
RTE	Right to Education; Right of Children to Free and Compulsory Education Act
S	South
Sarpa kavu	Sacred groves (Kavu) associated with snake god
SC	Scheduled Caste
SE	South East
SH	State Highway
SHP	Small Hydro Power Project
SPB	State Planning Board
ST	Scheduled Tribe
SW	South West
TDS	Total Dissolved Solid
TFR	Total Fertility Rate
Thr	Thrissur district
TSEM	Total Energy Security Mission
Tvm	Thiruvananthapuram district
UA	Urban Agglomeration
UAM	Udyog Aadhar Memorandums
USDA	United States Department of Agriculture
VSS	Vana Samrakshna Samiti
W	West
Way	Wayanad district
WNW	West North West
WSW	West South West
YBP	Year Before Present

CHAPTER 1

Introduction

GEOGRAPHY MATTERS

Geography matters. It contributes to shaping various aspects of civilisation and human lives including the development process, political strategy, culture, and language (Marshall, 2015). This is well evident across the scale from local, regional to global. However, there is a tendency to overlook geography's role. The primary reason for such a trend may be the presumption that subscribing to the idea of geographical control on human activities is tantamount to accept determinism. The other reason for underplaying geography's role is, perhaps, that it is too simplistic an approach to explain the phenomena over the earth's surface. Besides, there is a strong belief that technology can help overcome geographical constraints during the course of the development process. Subscribing to this view is quite natural as technology has pervaded all walks of life, and rapid technological development has brought spectacular changes in the economic, social, and demographic characteristics of society. However, it is now being well realised that there is a limit to which natural systems can be stretched and or manipulated. The environmental drawdown and the impending impact of climate change is too heavy for the earth to carry forward the ongoing development process, perpetually, following business as usual approach.

Anthropogenic activities are principal drivers of accelerated environmental changes. It has been so intense, penetrating, and materially voluminous that recent years, more particularly, the period beginning with the industrial revolution in the latter part of the eighteenth-century is considered as 'Anthropocene' (Crutzen and Stoermer, 2000). Clearly, the landscape that we see today was not the same that people knew even fifty years back, and it is changing faster than what earth system can assimilate resulting in increasing ecological footprint, overshoot, and endangered earth. The Human dimension of environmental change has emerged as an important subject matter of scientific enquiry. It is important to understand the process triggering this change, so that, appropriate planning initiatives can be taken up to the internalisation of environment in the development process, which is no more an academic exercise rather it is an imperative, a necessary condition for sustainable development. Our actions can be best designed by learning nature's law and applying them correctly. This calls for multidisciplinary, holistic, process based and non-linear approaches, knowledge-based planning, and stewardship. Geographical knowledge is one of the essential elements to tread this path. Nobel Laureate Nelson Mandela in one of his speeches underscored importance of geography for sustaining the future of humanity while observing "....A devastated geography makes for a devastated people. It renders people vulnerable, and the traditionally vulnerable—women, children, the aged and disabled—will always be bearing the brunt of the suffering. Let us stand together to make of our world a sustainable source for our future as humanity on this planet" (Nelson Mandela as quoted in Wood, 2004:2,4).

There is a renewed thrust on geographical studies in recent years across the world. The report of the US National Research Council (1997) on "Rediscovering Geography: New Relevance for Science and Society" reflected on critical issues and challenges for geography to flourish further as an intellectual enterprise, to transcend disciplinary divides, and emerge as true multidisciplinary integrating science and contribute in the decision-making process, so very important in today's context of intensive human impact, environmental degradation and associated changes including climate change. Geographers' focus on spatial dimension and emphasis on location have equipped them to view the processes and phenomena quite differently from other disciplines (NSA, 1997). Geography helps us to connect with the surroundings and provides an integrated and co-evolutionary perspective to our understanding of human society and human-environment relationships. With interfacing nature and porous boundaries, the discipline of geography is potential to serve as an intellectual dynamic subject with a normative focus for integrating different viewpoints about the world around us. The geographic reasoning is capable of addressing the question of diversity of phenomena, heterogeneity, and behaviour in different scales, changing trend, the evolution of spatial/non-spatial hierarchy, and finally visualization of a future trend. This book "Geography of Kerala" is conceived under this theoretical backdrop.

Geography contributes significantly to shaping Kerala's ecologic, economic, social, and cultural lives, and the emergence of Kerala mode of development, which is nationally and internationally well acclaimed. The devastation of geography, particularly the deterioration of environmental and natural resource base can impinge upon the ongoing development process and can even jeopardise the future scope of development. The symptoms are already evident. This book consisting of 16 chapters deliberates on issues related to physical and human geography including human development, poverty, decentralised planning, natural hazards, disaster management, changing geography of Kerala, and regions. It is suggested that there is a need to focus on place-based studies and understand and conserve local geography for the future development of Kerala.

KERALA: SOME BASIC ISSUES

Modern Kerala came into existence by 1st November 1956. However, the antiquity of Kerala is undisputed (Menon, 1967). Kerala's name was referred in ancient Sanskrit literature, Epics, Puranas, and also in the writings of ancient classical writers of Greece and Rome (*ibid*). The word Kerala, which evoked various images since early history, continues to raise widespread interest not only in India but also across the World in recent years. Travellers, historians, geographers, economists, social scientists, journalists, politicians, and even common people describe Kerala as a state perceptibly different from the rest of the states in India. We note here some of them. The discussion begins with the derivation of the name of Kerala itself.

How did Kerala get her name? There are two common versions. According to Menon (1997) " ...though in popular imagination Kerala is the land of *Kera* (coconut), one of the theories relating to the origin of Kerala focuses attention on the geographic aspect. *Cher* or *Chernta* means added and *Cheralam* (Sanskritised as Keralam) means the land that was added on to the already existing mountainous country. In other words, the traditional land of Kerala was a reclamation or addition to the land from the Arabian Sea" (Menon 1997:1). This second version finds its support from the popular legend that much of Kerala's present land emerged from the sea when Parasurama threw his axe across the sea. While the origin of this legend may be devoid of any factual basis, it is a well-established fact that there are several phases of sea transgression and regression in geological recent years. As a consequence, there were emergence and submergence of landmass along the sea-front from time to time. Geological evidence indicates that many parts of the present day coastal tract emerged during the last couple of thousand years. The name Kerala is, therefore, justified on both counts.

Kerala was well known for spices since the early days. Historical evidence suggests spices trade even in the 3rd millennium BC (Menon, 1967). Merchants from Phoenicia and Greece visited Malabar Coast for buying spices in the 8th century BC. Spices trade with Europe took place through Arab countries. Vasco de Gama's visit to Kozhikode in 1498 was in connection with the spices trade. The export of spices through sea routes was the main activity in those days. Kerala enjoyed trade links with various Asian and European countries. Black pepper constituted major share of spices trade. Europeans used to refer to Kerala as a 'land of black pepper'. At present, Kerala accounts for 98 per cent of black pepper production in India. The biophysical set up of Kerala is conducive for the luxurious growth of condiments and spices like black pepper, ginger, cardamom, and turmeric. In 2016-17, export earnings from Kerala's spices industry were worth of INR 38,579 million.

A traditional image of Kerala is that of a land of Ayurveda and medicinal plants. Inspired by the local people's use of plants for various therapeutic use, Van Rheede, the Dutch Councillor of erstwhile Malabar, compiled *'Hortus Indicus Malabaricus'* (1687-1693)—the first comprehensive printed book written in a European language with able assistance from a native physician (Nicolson *et al.*, 1988). This is the first published classical work in medicinal plant botany. It is now reported that out of 3700 species of vascular flora in Kerala about 900 are known for medicinal properties (Pushpangadan and Nair, 1997). There has been a resurgence of people's interest in traditional medicine throughout the globe, at present, due to harmful side effects of the allopathic system of medicines.

Kerala is a type locality of laterite for the geologists. In fact, laterite and laterite derived materials cover 60 per cent of the total area of Kerala. Laterite is a product of chemical weathering in tropics. It was Francis Hamilton Buchanan of East India Company who reported about laterite from Angadipuram, a village in Malappuram district of Kerala for the first time in 1800-01 (Buchanan, 1807). Geological Survey of India has erected a monument in Angadipuram in the year 1980 in recognition of Buchanan's observation. Kerala is the gateway of monsoon for mainland India. Monsoon hits Kerala on 1st June in most of the years.

Kerala is globally acclaimed for her achievement in the social sector and human development in recent years. She is well ahead of other states in India in the matter of literacy, health care, life expectancy, birth control, reduction of infant mortality rates, and women's empowerment. Kerala model or Kerala mode of development is a subject matter of deliberations in various fora. More often than not, Kerala is cited as a model to be emulated by not only other Indian states but also by other developing countries. The successful introduction of *Panchayati Raj* system for local-level governance and people's participation in the planning process are two recent phenomena often discussed in planning parlance across the country. The welfare approach of the successive governments, irrespective of political formations, underpins this remarkable social development.

Kerala is well known for emigration and remittance of money. There were 16 lakh people from Kerala living abroad by 2013. Migration to Gulf countries is an all Kerala phenomenon.

Kerala's economy is heavily dependent on foreign remittance. According to a survey report of the Reserve Bank of India, Kerala topped among the states in India by receiving 19 per cent of total foreign remittances credited to the country by 2016-17.

Culturally, Kerala epitomises peaceful coexistence of different religions and social groups from the early days. Hardly, there had been any social tension reported from Kerala. Among people in Kerala, there is a common cultural identity, which is composite, cosmopolitan and evolved through the contribution of several people and races; as a result, there is vitality, strength, tolerance, and catholicity (Menon, 1997). According to legends the Syro-Malabar Catholic Church was built by St. Thomas, one of the 12 Apostles of Jesus Christ in 52 AD at Palayoor in Thrissur district. The first synagogue in India was built in Kodungallur, Thrissur district in 4th Century AD. Legends also claim that the first Muslim mosque in India was built in 629 AD at Methala, Kodungallur, Thrissur district. There were interactions, exchange, and assimilations of cultures. A common tradition evolved throughout Kerala, which is well evident even in *Vasthusastra*. A study on vernacular houses distributed throughout the state and inhabited by different social groups brought out Kerala's unique tradition of synthesising regional *vasthusastra* with that of local content (Widiastuti, 2013).

Environmentalists fondly refer to Kerala for the Silent valley movement in late 1970s and early 1980s to save a type locality of tropical rain forest, and Plachimada struggle of local people against multinational to save precious groundwater in 2002. There are several such initiatives undertaken by people's science movement, voluntary organisations, individuals, and communities. The activism of people in Kerala is a much talked about subject.

To the political scientists, and social observers, Kerala is known for setting the example of electing the first communist government in a democratic set up way back in 1957. This first Government is credited for introducing several milestone Bills including the Land Reform Bill and the Administrative Reforms Bill. Kerala is also noteworthy for electing alternatively Left dominated and Congress dominated governments for administering the state every five years.

FIVE BROAD THEMATIC AREAS CHARACTERISING KERALA

Perusing all these issues we can identify, perhaps, five broad thematic areas that characterise today's Kerala: (i) the first and foremost is the physical landscape. Placement with respect to the Lakshadweep sea and the Western Ghats, latitudinal position, geological history, geomorphic evolution, climate, topography, and their interactions as apparent through soil formation, growth of natural vegetation, and landuse development have provided Kerala a unique biophysical foundation and endowed it with rich natural resource base, (ii) the second theme refers to human development, particularly demographic characteristics and social infrastructure. Civil society movement, social activism, proactive nature of people, and consciousness form part of this trait, (iii) the third observation is about long tradition of interaction with outer World mainly through the spices route and subsequent sea travels. Spice trade, particularly black pepper, and the emergence of commercial agriculture facilitated the interaction of the people of Kerala with the outer world. There was assimilation of cultures and exchange, (iv) the fourth one is of recent origin related to foreign remittance through emigration which significantly influences the state's economy and landuse, and finally (v) it is the overarching political power relations, and social democracy that have facilitated many of the initiatives with long term implications in human-environment relationships and development processes.

KERALA: A VERNACULAR REGION

A vernacular region is an area that has been identified based on people's perception of culture. Vernacular or regional language highlights the cultural connection. Modern Kerala is a linguistic state. In 1956, reorganisation of states in India was based on language. Malayalam speaking princely states of Travancore-Kochi were amalgamated with Malayali dominated Malabar areas of Madras state and Kasaragod taluk of South Kanada. As domination of Malayalam language was the sole criterion, the Tamil speaking Nagercoil-Kanya Kumari part of Travancore princely state was carved out, and added to Madras state (now Tamil Nadu) during reorganisation exercise in 1956. Geopolitically, areas under different administrative set up were brought together to form a state purely based on language. Or language is the sole criterion to draw Kerala's present boundary. Although Kerala is a linguistics state of Malayalam, the language is spoken in different ways in different parts of the state. The mannerism of people also differs from one locality to other, which is quite natural.

KERALA: A GEOGRAPHICAL REGION

The present boundary of Kerala also represents a distinct physical entity between the Lakshadweep Sea and the Western Ghats. The eastern boundary of Kerala broadly coincides with the crest line of the Western Ghats, while the western boundary is the coastline. Kerala is, perhaps, one of the few states in India where administrative boundary largely coincides with physical (watershed) boundary. Thus, a vernacular region takes the shape of a geographical region.

ADMINISTRATIVE DIVISIONS OF KERALA

At the time of the formation of the state of Kerala in 1956, there were only nine districts, which increased to 11 in the year 1978, and to 14 in the year 1984 (Fig. 1.1). By 2018, there were 14 districts, 77 talukas, 152 community development blocks, 1647 revenue villages, 941 panchayats, 6 municipal corporations, 87 municipalities, and one township. Idukki is the largest district with an area of 4358 km^2 and Malappuram is the most populous district with 41.13 lakh people (2011 Census). However, among the districts in Kerala, Thiruvananthapuram district has the highest density of population (1508 persons per km^2). List of districts, area, population, and population density are given in Table 1.1.

KERALA IN INDIA AND AMONG OTHER SELECTED STATES

Why does Kerala generate so much interest? The answer lies in the fact that as a constituent province of India, Kerala has achieved significantly in social and human development outperforming all other states, some of which can boost of strong agricultural, industrial and financial bases. There are widespread interest to understand— how can a relatively small state perform so well? Steady performance of Kerala in social and human development, vibrant civil society, urge of people in Kerala to participate in governance and people centric approach of government to address the problems have generated interest among other states in our country, and also among other countries in the world.

It is important to understand Kerala's position in India and amidst the neighbouring two

Table 1.1: Area, Total Population and Population Density by Districts in Kerala

Name of the District/State	*Area (in Km^2)*	*Total Population (in lakh), 2011*	*Population density (persons/km^2)*
Thiruvnanthapuram	2192	33.01	1508
Kollam	2491	26.35	1061
Pathanamthitta	2637	11.97	452
Alappuzha	1414	21.28	1504
Kottayam	2208	19.75	895
Idukki	4358	11.09	255
Ernakulam	3068	32.82	1072
Thrissur	3032	31.21	1031
Palakkad	4480	28.1	627
Malappuram	3550	41.13	1157
Kozhikode	2344	30.86	1316
Wayanad	213	8.17	384
Kannur	2966	25.23	852
Kasaragod	1992	13.07	657
Kerala	38863	334.04	860

Source: Government of Kerala, 2018; Census of India, Kerala, 2011.

states of Karnataka and Tamil Nadu, and few other selected states in the country. The geographical area, population, economy, nature of development and culture of the constituent provinces of India vary considerably. Land in Kerala is intensely used and dominated by non-food crops. Hardly, any part of the state barring rock out crops, and bare sands is devoid of vegetation. Net Sown Area of Kerala is 58 per cent compared to the national average of 47 per cent, however, area under food crops in Kerala is little over 44 per cent, whereas the national average is over 62 per cent. Kerala ranks 13th in population size among all the states and union territories in India. It covers 1.2 per cent of total area of India but accommodates 2.76 per cent of India's population (Table 1.2). According to 2011 census, population density of Kerala was 860 persons per km^2, next to Bihar (1106), and West Bengal (1028). Kerala's population density was 2.25 times higher than that of the all India average of 382 persons/km^2, and 1.55 times, and 2.7 times higher than that of Tamil Nadu, and Karnataka respectively. The decadal (2001-2011) growth rate of population in Kerala was 4.9 per cent against the national average of 17.64 per cent.

By 2016, when Kerala completed 60 years of its existence in the present form as a constituent province of Indian union, it outshone all other states in India in crucial human development indicators, like sex ratio, literacy, infant mortality rate (IMR), life expectancy, human development index (HDI), and other associated parameters. Even in per capita income, Kerala is now placed well above the national average, and is counted as one among the first few major states in the country. The final country report of India on Millennium Development Goals (MDG) has noted that Kerala has achieved all MDG 2015 targets well in advance (Government of India, 2017a). Due to all these accomplishments, Kerala is often considered as a front runner state among Indian states having the potential to transit to sustainable development, a professed goal under the UN charter.

KERALA'S UNIQUE HUMAN SETTLEMENT SYSTEM

Human settlement in Kerala is unique, and differs from the rest of India. Dwelling houses are intermingled with the natural environment. In fact, human habitations are so well gelled with plants and trees that a land use category of 'settlement with mixed tree crops' has been especially

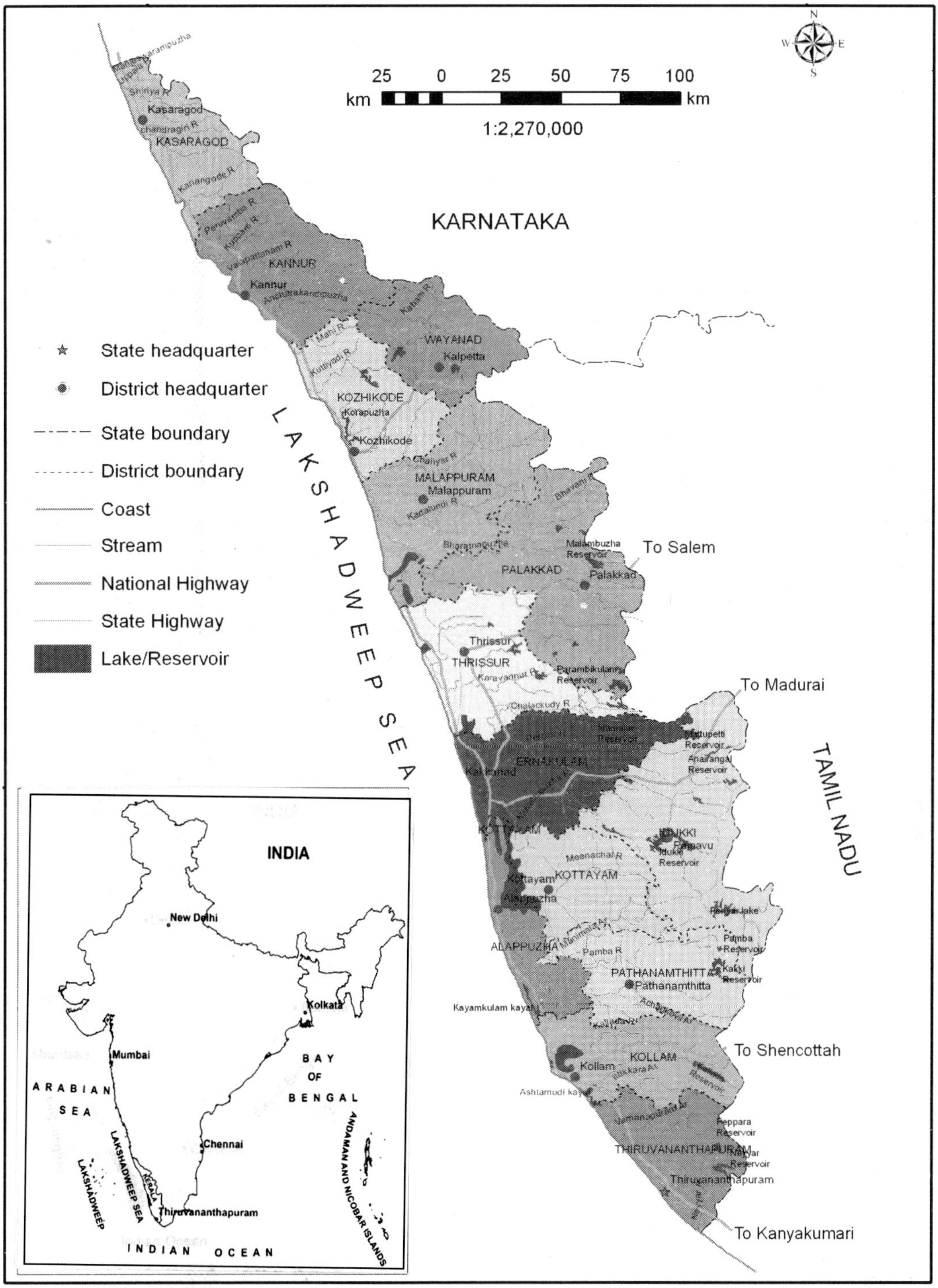

Fig. 1.1: Location of Kerala in India and Administrative Divisions of Kerala Showing Districts.

Table 1.2: Position of Kerala in India and among Selected States in Terms of Selected Parameters

Variables/units	*India*	*Kerala*	*Tamil Nadu*	*Karnataka*	*Uttar Pradesh*	*Bihar*	*West Bengal*	*Punjab*
Area (per cent)*	100	1.20	4.10	6.05	7.60	2.97	2.80	1.60
Net Sown Area (per cent) (2014-15)**	100	1.46	3.44	7.16	11.84	3.34	3.74	2.94
Population (per cent) (2011)*	100	2.76	5.96	5.05	16.49	8.58	7.55	2.30
Population density (2011)*	382	860	555	319	828	1102	1029	550
Population growth rate (2001-2011) average Annual*	1.64	0.48	1.46	1.47	1.85	2.26	1.31	1.30
Urban population 2011 (per cent)*	31.16	47.72	48.45	38.47	22.28	11.30	31.89	37.49
Sex ratio (Female per 1000 male)*	943	1084	996	973	943	918	950	895
Literacy (per cent)*	74.04	93.91	80.09	75.36	67.68	61.80	77.08	75.84
Infant Mortality Rate (2015)**	34	10	17	24	43	38	25	21
Life expectancy (years)**	68.3	75.2	71	69	64.5	68.4	70.5	72.1
Per capita income (PPP INT$)#	6965	9891	10258	11408	3380	2141	5265	8433
Human Development Index (rank, 2017)$	0.639 (rank)	0.784 (1)	0.708 (11)	0.682 (18)	0.583 (35)	0.566 (36)	0.637 (28)	0.721 (8)

Source: * Census of India, Provisional Tables, 2011, ** GOI, 2018: Ministry of Agriculture and Farmers Welfare, $ Sub national HDI—Area Data base, Global Data lab; # Ministry of Statistics and Programme Implementation, GOI, 2019.

identified for land use studies in Kerala. The whole of Kerala appears like a city in the garden with main towns as high density nodes within the city. The settlements in Kerala are continuous and have no distinct core, nor do the villages have any marked nodality. Along the coastal plain, and parts of midland from one end to the other, Kerala appears to be a single garden city or high density village admixture with urban functions. Hardly, there is any distinction between rural and urban settlements. Kerala's speciality of settlement system is the rural-urban continuum. This trend is not of recent origin. Famous traveller Ibn Batuta recorded this scenario in the 14th century. C.E. Logon (1887) noted that the coastal tract of Malabar (Kerala) was so densely populated that it was difficult to say where one municipal town begins and the other ended. Settlements along the coastal plain are in an unbroken chain. Urban centres can only be recognized by the presence of some more commercial activities and markets.

Kerala presents a hybrid landscape where human played a significant role in enhancing the garden character by constructing settlements within trees and moulding society accordingly. There are trees in every household even in urban areas. Settlements are found side by side of mixed tree crops, coconut gardens, and rubber plantations. Trees are grown in multi-tier with an upper canopy dominated by coconut plantations followed by other tree crops and plants in different layers including vegetables in the ground. On average, there are 3.2 households in one ha of garden land. Kerala is famous for homestead biodiversity.

Barring few pockets of indigenous settlements within the Western Ghats, Kerala is inhabited by people whose lifestyle, occupation and outlook largely resemble urban dwellers. Traditional city countryside dichotomy is rarely seen in the state. It is only recently that multi-storied buildings have mushroomed and people started living in flats. The city's functions and people have melded into the countryside and created a new type of rural-urban continuum. Underpinned by unique ecologic economic factors, geographical space formation takes a new turn in Kerala, realization of which is fundamental to understand the geography of Kerala.

CHAPTER

2

Physical Geography

THE PHYSICAL SETTING

Kerala is distinct from its adjoining states both in its physical geography as well as human geography. As a narrow strip of land between the Lakshadweep sea to the west and the Western Ghats to the east, Kerala occupies parts of the Western Ghats, and adjoining coastal plain along the south western corner of the Indian Peninsula. The physical setting of Kerala including its latitudinal position, orography, and placement with respect to sea and mountain has played important role in the patterns of human settlement, cultural, and economic development. Landscape ecological variations within short distances underscore the significance of landforms, underlying geology, climate, drainage, and interactions of geology, topography, and climate as manifested in the development of soil and natural vegetation.

Kerala extends from 8° 18′ N latitude to 12° 48′ N latitude, and from 74° 52′ E longitude to 77° 22′ E longitude. This latitudinal position controls receipt of solar insulation, length of day and night, and overall climate thereby impacts the productivity of land and water including the growth of plants and crops. The state of Karnataka bounds Kerala in the north and parts of the east, and Tamil Nadu borders Kerala for the rest parts of the east and south. The common boundary between Kerala and Tamil Nadu runs for a distance of around 600 km, and that between Kerala and Karnataka is for 200 km.

The coastline of Kerala is 560 km long and nearly straight in north north west-south south east direction indicating structural control. An average of 69 km^2 land area supports 1 km length of coastline. This has a profound impact on the productivity of coastal waters. The maximum width of Kerala is 120 km in a crow fly distance from the coastline to the eastern border in Kochi-Anamudi stretch. If a triangle is drawn considering the northern and southern endpoints of Kerala along the coast as the base and the farthest point (120 km from the coast) located to the east of Anamalai in Idukki district as the apex, it may be observed that around 80 per cent of the total geographical area of Kerala will fall within this obtuse-angled triangle.

RELIEF

Kerala's altitude ranges from sea level to 2695 m above mean sea level at the Anamudi, the highest point in Peninsular India. Some parts of Kerala around the Kuttanad area even lie at 1 to 2 m below mean sea level. Relief distribution is asymmetric as evident from the digital elevation model and area-altitude distribution (Fig. 2.1, Fig. 2.2). In a diagrammatic plan, the landscape of Kerala appears to rise in steps, where altitude increases gradually from the coastal plain to the foothills, and thereafter

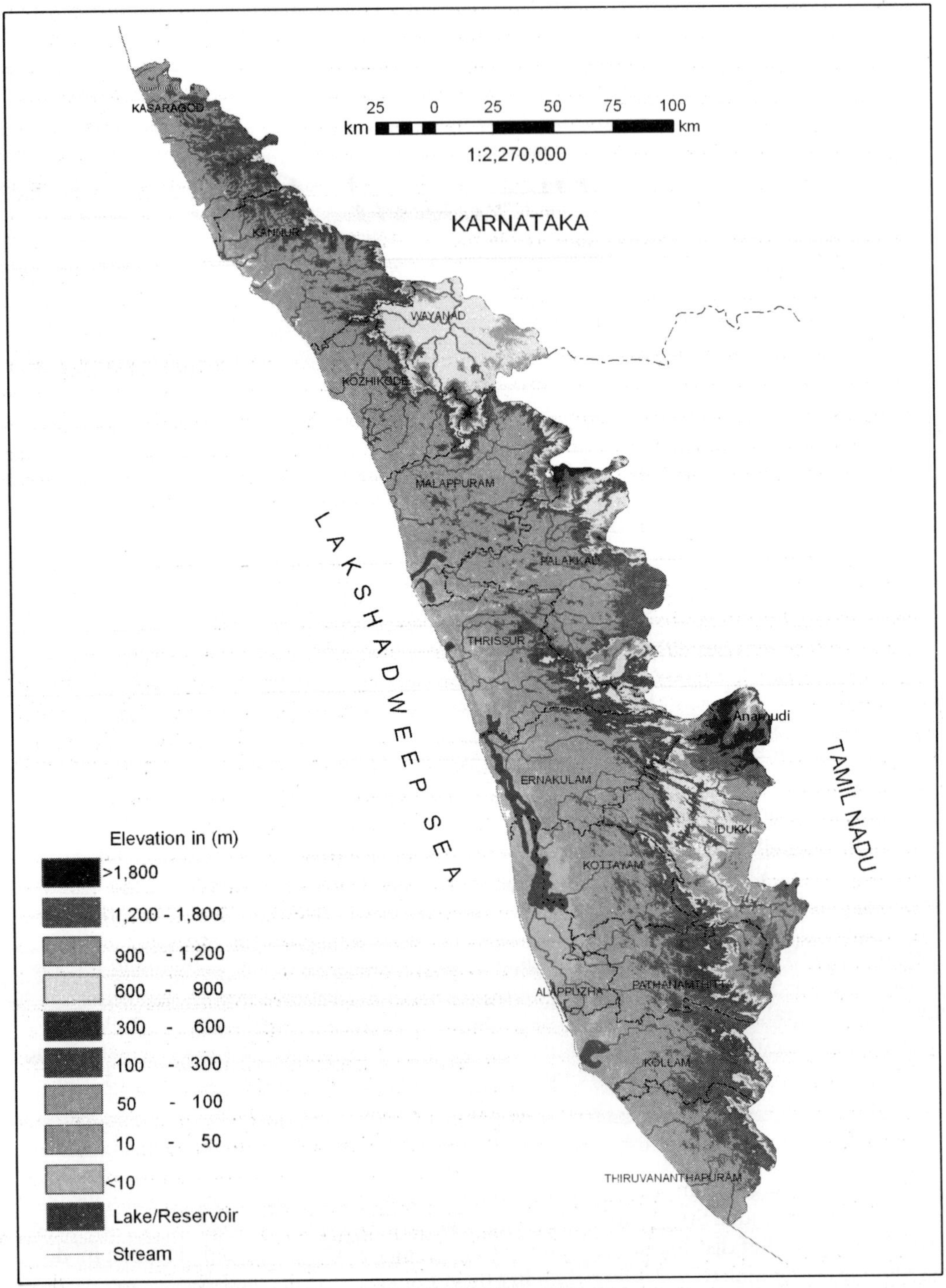

Fig 2.1: Digital Elevation Model Showing Altitudinal Zones in Kerala.

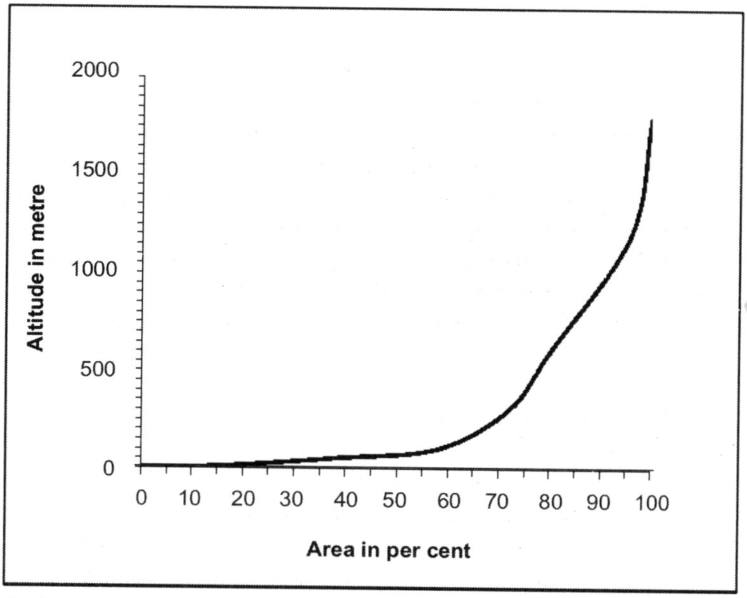

Fig 2.2: Area Altitude Distribution in Kerala.

the rise is abrupt through scarp face to the Western Ghats plateaus and the crest. The term 'Ghats' stands for steps like rising. The average rise of relief is 27 m for every one Kilometre from the coast to the Western Ghats crest. The rate of rising is low (1.4 m/km) in lower altitudes (up to 100 m), however, it increases incrementally towards the east along with rising in elevation. The Western Ghats scarp stands like a vertical wall. The average distance between the Western Ghats crest line and the coastline is around 40-45 km. This distance touches the maximum in the Anaimalai stretch (around 80 km), and it decreases both towards north and south.

Around 58 per cent of the total geographical area of Kerala lies below 100 m altitudes that include coastal plain and adjoining undulated lateritic landscape. The Western Ghats including the scarp slope and piedmont slope cover rest 42 per cent area. Strong relief is associated with hard rocks. Landmasses developed over the Recent and Tertiary sediments are characterized by low relief. The Palghat gap having an average width of 30 km breaks the continuity of the monolithic Western Ghats, and connects Kerala with Tamil Nadu plain and the rest of the Indian peninsula. This gap has a significant influence on the environment, economy, and social milieu of Kerala. In the southern end, Kerala opens to Tamil Nadu through Nagercoil valley, and in the northern tip, Kerala merges with Karnataka coastal plain over laterite surfaces. Apart from the Palghat gap, there are other gaps/passes like the Kambam pass, Bodinaykunnur pass, Arambadi pass, Aryankavu pass, Perambadi pass, etc. connecting Kerala with neighbouring states. Altitudes drop down significantly in these gaps/passes.

GEOLOGY

Geologically, the Kerala region is part of the south Indian Precambrian terrain. The bulk of the rocks of Kerala, especially the granulites and associated gneisses belong to the Precambrian. Sporadic late Precambrian—Early Palaeozoic granites and associated pegmatites, and Meso-Cenozoic dykes intrude these rocks (Soman, 1997; 2002). The Precambrian crystallines consist chiefly of garnet-sillimanite gneisses, charnockite, pyroxene granulites, hypersthene diopside gneisses, granites, garnetiferous leptynite, and garnet biotite geneiss. The on-land sedimentary formations are confined to the Neogene period only. It includes pebble beds, sandstone, grit, clay with shells, marl, and limestones. All the rock types (crystallines and sedimentaries) are lateritised at variable depth. Duricrust formations are marked in places. The Recent and sub-recent sediments cover low-lying areas, coastal plain, and river valleys. Rock types, their composition, and degree of weathering influence landform development. Rugged terrains are mostly on hard rocks and radial drainages characterise the areas affected by granite intrusion. Structural control on drainage patterns and morphology is well pronounced. River capture, presence of beheaded streams, and elbow shaped right-angled turning along with river courses, and unpaired terraces are probable evidence of neotectonic activities.

CLIMATE

Latitudinal and longitudinal position, placement with respect to ocean and continent, and orography, in combination, governs the climatic condition of Kerala. The temperature drops with altitude. Diurnal and seasonal variations in temperature are limited. The Palghat gap almost in the centre of Kerala modifies the pattern of wind movement. It acts a as zone of convergence and experiences effects of the land breeze and the sea breeze by connecting the dry Tamil Nadu plain with the wet Kerala plain.

Climate is perhaps the most widely used parameter to classify the earth's surface into biomass production zones. According to Koppen's scheme of classification, Kerala is grouped under Amw (tropical monsoon variety), and under Miller's scheme, Kerala's climate is Am1 (equatorial monsoon type) (Jarrett, 1977). Southern Kerala with two rainfall maxima shows the characteristics of equatorial climate. Based on macro-level data, Subramanyam and Murthy (1982) divided Kerala into two zones: the tropical monsoon, and the tropical savanna. Parts of Kannur and Kozhikode districts in north Kerala, more than half of Kottayam and Alappuzha districts, and the entire Kollam, Pathanamthitta, and Thiruvananthapuram districts are designated as the tropical savanna. The rest of Kerala falls under the tropical monsoon type. The difference between tropical monsoon, and tropical savanna is determined by the amount of rainfall; the savanna area receives comparatively less rainfall.

Rainfall

The long term annual average rainfall in Kerala (from 1871 to 2018) is 282 cm. Rainfall varies widely over the years and it experiences a decreasing trend (Fig. 2.3). Kerala received the highest annual rainfall of 394 cm in the year 1924 followed by 391 cm in 1961. Both these years witnessed the incidence of a devastating flood. The year of 2018 was also a year of high rainfall (352 cm) and witnessed unprecedented flood havoc. The long-term (1871-2018) average monsoon (June, July, August, and September) rainfall is 193 cm. Rainfall data since 1980 indicate that the annual average and the average monsoon rains are 269 cm and 182 cm respectively. During the last 10 years (2009-2018), monsoon rainfall varied from 131 cm in 2016 to 252 cm in 2018. Spatial distribution of rainfall shows an increasing trend from the coast to inland and the maximum rainfall is recorded along the foothills around Neriyamangalam (451 cm) in the south and Kuttiyadi (417 cm) in the north (Simon and Mohankumar, 2004). Rainfall decreases further east and the lowest rainfall (<100 cm) is noted in the rain shadow region along the leeward eastern slopes of the Anamudi around Chinnar and North Marayur (Fig. 2.4).

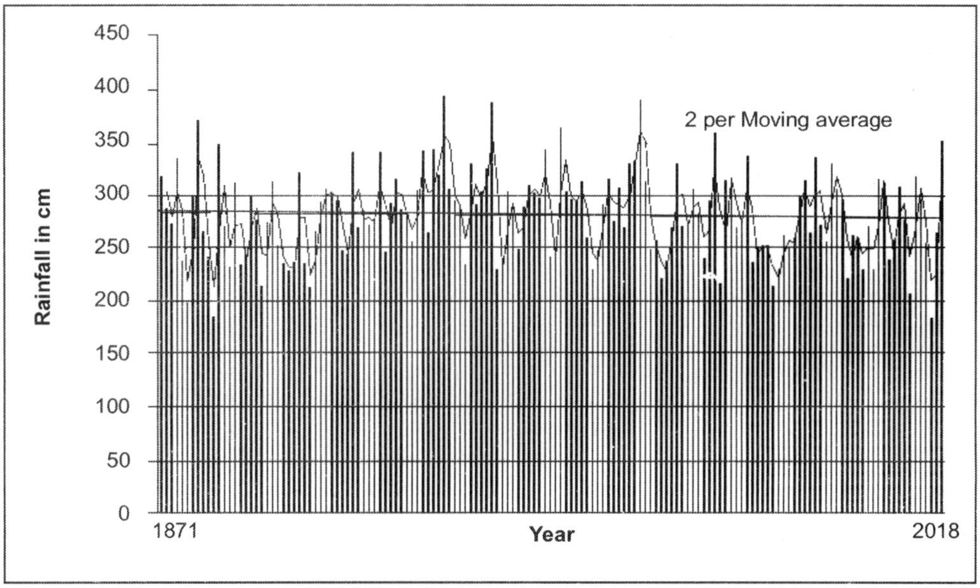

Fig. 2.3: Trend of Annual Rainfall in Kerala (1871-2018).

Note : Curved line represents 2 year period moving average and the inclined line is the trend line, which indicates declining trend of rainfall.

Source: Compiled from IMD data.

Due to a sudden reduction in relief, the Palghat Gap disrupts rainfall trends resulting in considerably low rainfall in and adjoining the gap area. Kerala receives rainfall almost every month. This year-round rainfall and high precipitation in the foothills have contributed to the perenniality of the rivers. The orographic influence on rainfall is well pronounced. Moisture-laden air condenses in the hills, and thus the foothills receive the maximum rainfall. As there is another level of high altitude over the plateau, rainfall is quite high even on plateau surfaces. Both Idukki and Wayanad districts receive high rainfall.

Seasons

Based on rainfall characteristics and cloud formation, four seasons can be identified in Kerala. These are: South-West monsoon (locally known as *Kalavarsham*) (June to September), North-East monsoon (locally known as *Thulavarsham*) (October and November), Winter (December-February) and Pre-monsoon (March-May). Total rainfall received during south-west monsoon increases steadily from Thiruvananthapuram in the south to Kasaragod in the north along the coast (Sampath and Vinayak, 1989). The contribution of the south-west monsoon rain to total annual rainfall also increases in the same direction. The south-west monsoon sets over the southern parts of Kerala normally by 1st June and extends over the entire state by 5th June. Contribution of the north-east monsoon is high in the southern part and it is considerably less in the northern part. In Thiruvananthapuram, around 33 per cent of annual rain is received

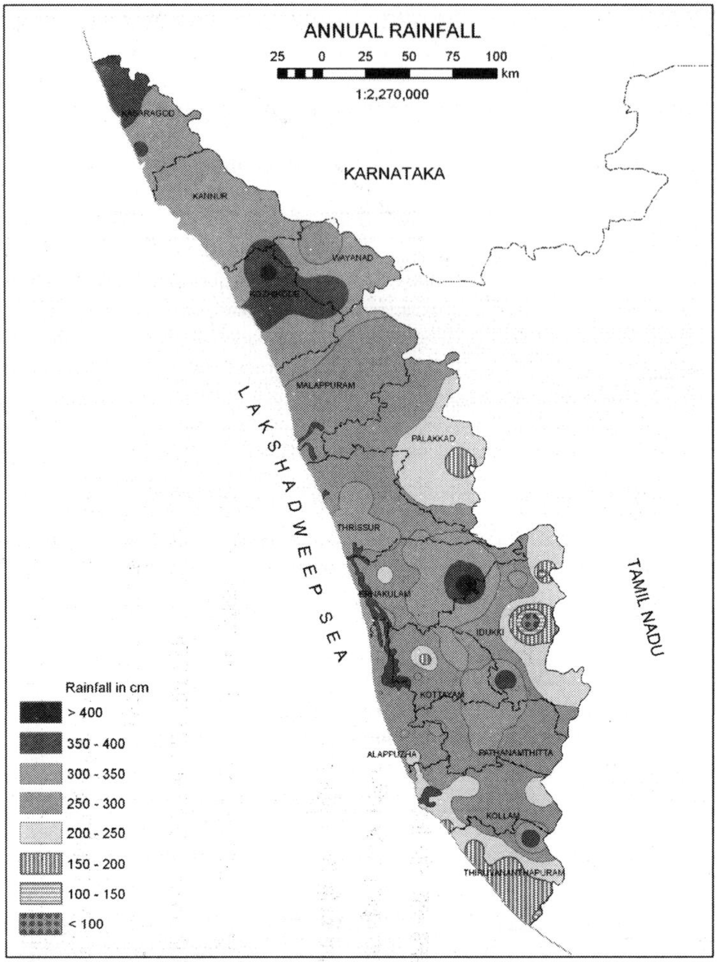

Fig 2.4: Distribution of Annual Rainfall in Kerala.

Data Source: Compiled from IMD Data and Resource Atlas of Kerala.

during the north-east monsoon and in Kasaragod, the north-east monsoon contributes only 9 per cent (*ibid*). The northern station, particularly Kannur is characterized by a single peak in the month of July. All other stations record two peaks, one either in the month of June or July coinciding with the south-west (SW) monsoon and the other peak during the month of October, corresponding to the north-east (NE) monsoon. Stations lying to the north of the Palghat gap like Kozhikode and Palakkad record the SW monsoon peak during the month of July, whereas the southern stations (south of Palghat gap) show the SW monsoon peak during the month of June (Fig. 2.5). The June-July peak is higher in all cases except for Chinnar in the Idukki district, where the highest peak is in the month of October denoting the predominance of the north-east monsoon (CESS, 1984; Shravankumar *et al.,* 1988).

It is generally observed that if rainfall is low during the SW monsoon, the NE monsoon season brings relatively higher rainfall. Monsoon failures for both SW and NE monsoon periods are very rare. Kerala, therefore, enjoys a certain amount of assured rainfall each year. The

Maximum Assured Rainfall (MAR) during the SW monsoon period is expected to be realized during the last week of July while during the NE monsoon, the MAR occurs during the third week of October (Prabhakaran *et al.*, 1994).

Winter months are characterized by the minimum clouding and rainfall, and the major thunderstorm activities take place during pre-monsoon months (Ananthakrishnan, *et al.*, 1979). The thunderstorm rain in the pre-monsoon months of April and May are locally known as *Edavappthi*. It may be noted here that Kerala does not experience winter season like other parts of the country. The minimum temperature in the winter months does not fall below 20°C in most parts of Kerala. Some parts of the state falling in the mountainous region, like Wayanad and Idukki experience low temperatures during winter months on account of high altitudes.

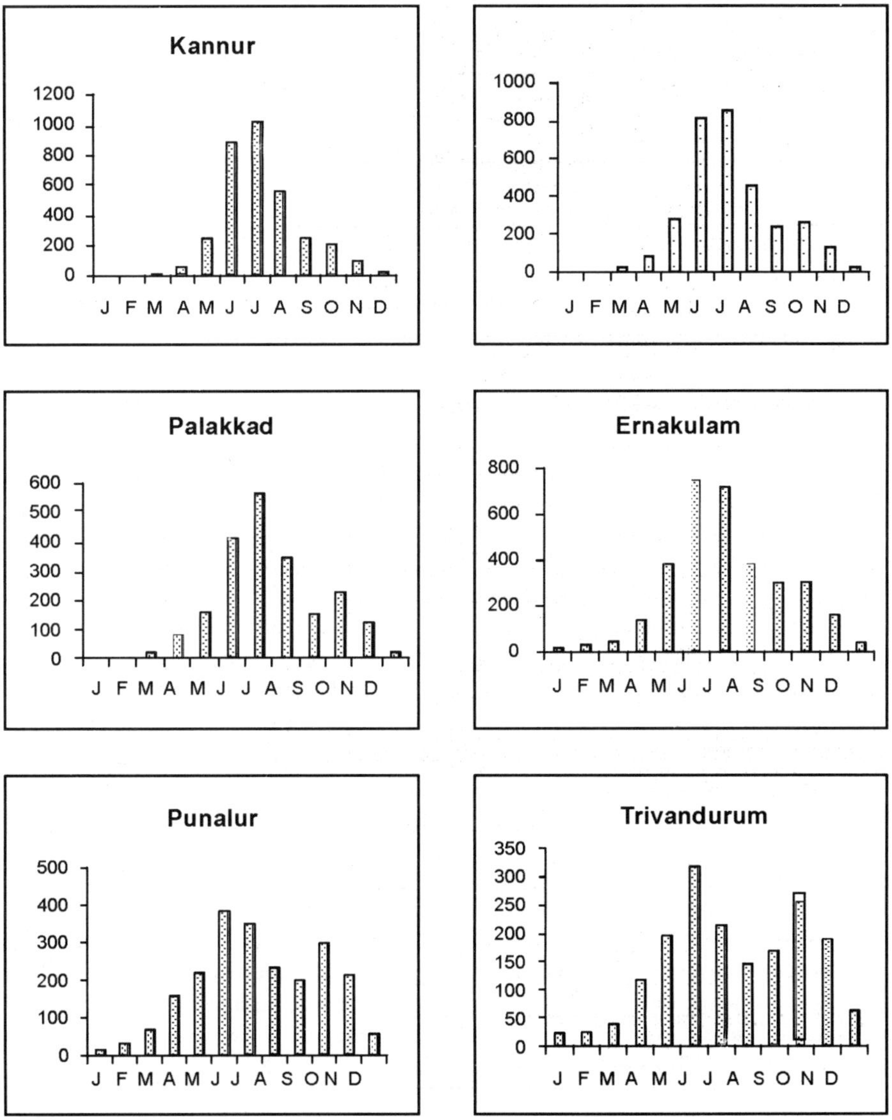

Fig. 2.5: Average Monthly Rainfall in Six Selected Stations in Kerala.

Temperature

Temperature is high (>32°C) during summer months—March to May. It gradually comes down from the month of June due to heavy monsoon rain. An increasing trend of temperature is also noticed in the months of October-November, followed by relatively lower temperatures (<27°C) in the months of December-January (CESS, 1984). Seasonal and diurnal variations of temperature are not uniform throughout Kerala. Stations located near the coast are influenced by land and sea breezes. Seasonal and diurnal variations of temperature in these stations are almost in the same range (5°C to 7°C). At Palakkad, the mean seasonal variation is less than the diurnal variation, but in the high ranges, which are typically sub-tropical, the diurnal variation is very high (>15°C in some months). This is a typical example of a par-humid area where the tropical climate has been remarkably modified by altitudes.

The zone of the highest mean annual temperature (>27.5°C) falls in the midland region of Kerala. High summer temperatures are recorded in Palakkad and Punalur due to their location along the Palghat gap and Shencotta pass respectively (Table 2.1). Temperature is moderate along the coast, whereas it is low towards east due to orographic effect. This type of temperature variation is due to the presence of the sea in the west and high relief in the east which have endowed Kerala with a unique agroclimatic condition favourable for the cultivation of large varieties of crops. The highland region (>500 m) with a relatively cool climate favours plantation crops like tea, coffee, and cardamom, but the lowland and midlands with warm humid climate sustain various seasonal, annual, and tree crops including rubber.

Table 2.1: Average Mean Monthly Temperature (in °C) of Selected Stations in Kerala

Month	*Kannur*	*Kozhikode*	*Palakkad*	*Kochi*	*Punalur*	*Trivandrum*
January	26.6	26.9	27.6	27.0	26.4	25.9
February	27.4	27.4	29.1	27.7	27.3	26.8
March	28.5	28.6	30.7	28.7	28.7	27.9
April	29.3	29.4	30.6	29.1	29.1	28.3
May	29.0	29.1	29.1	28.4	29.2	28.2
June	26.6	26.6	26.4	26.5	27.5	26.4
July	25.8	25.7	25.5	25.9	27.0	26.0
August	26.1	26.1	26.0	26.2	27.2	26.2
September	26.4	26.5	26.8	26.5	27.3	26.4
October	27.0	27.1	27.2	26.8	27.1	26.5
November	27.0	27.3	27.3	27.3	26.6	26.0
December	26.7	26.8	27.1	27.0	26.3	25.7

Source: http://en.climate-data.org/asia/India/Kerala accessed on 30.03.2018.

DRAINAGE

Rivers in Kerala are small by all-India standards. There are 44 rivers in Kerala based on a yard stick of 16 km length to designate a natural channel as a river. Out of these 44 rivers, 41 rivers flow westward to the Lakshadweep Sea and the rest three are east-flowing rivers. Some of the west-flowing rivers join together before their outfall; therefore, there are 34 river basins drained by 44 rivers (Fig. 2.6). Some of the west-flowing rivers have a certain portion of their catchments in adjoining states of Tamil Nadu and Karnataka. Altogether, 10 rivers originate outside Kerala. There are some more streams such as the Kumbala, the Kalnad, the Bekal, and the

Pooramparamba, with separate watersheds which drain into the Arabian Sea directly or through backwaters. However, all these rivers are small, each having a length of fewer than 15 km.

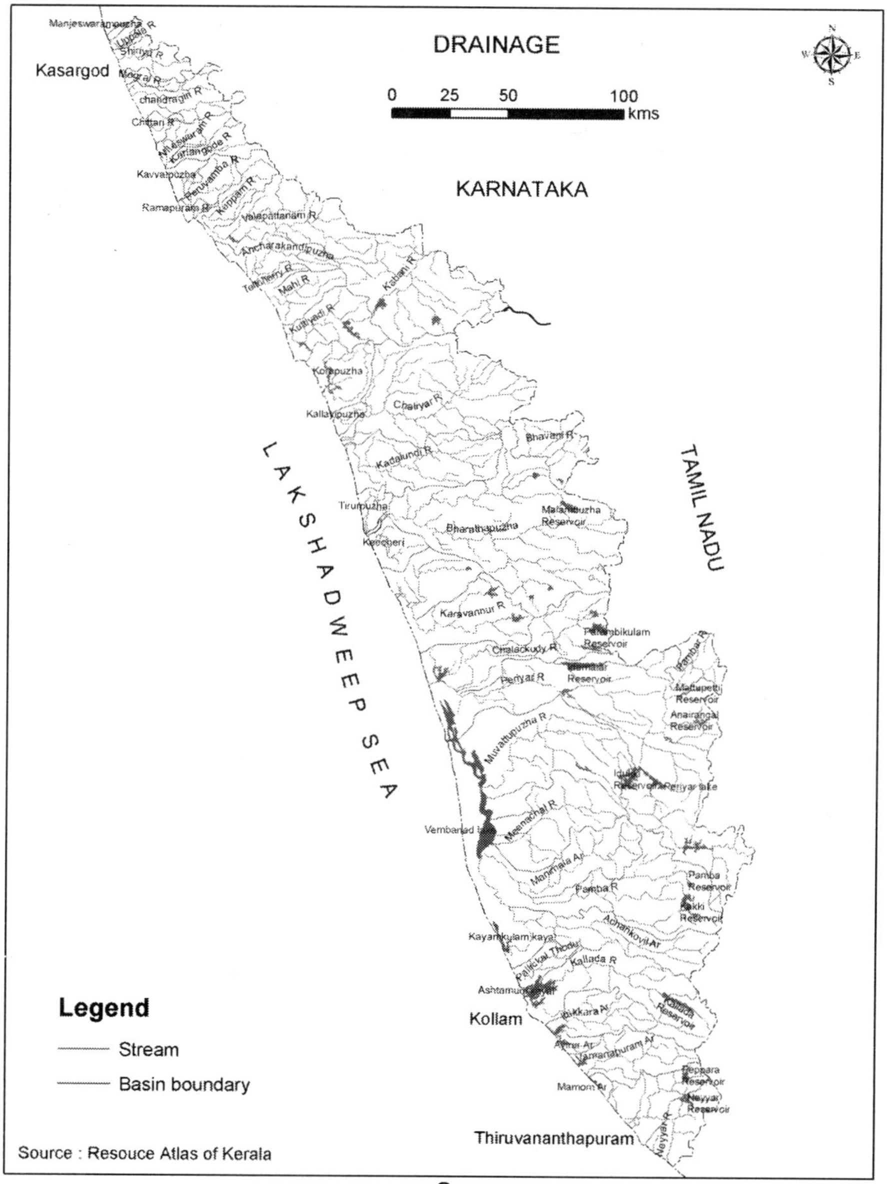

Fig. 2.6: Drainage Network along with Drainage Basin Boundary and Selected Reservoirs in Kerala.

River length varies from 16 km for the Manjeswar river to 244 km for the Periyar, the longest river in Kerala. There are only 11 rivers, each having more than 100 km length. These are, from the north, the Chandragiri (105 km), Valapattanam (110 km), Chaliyar (169 km), Kadalandi (130 km), Bharatapuzha (209 km), Chalakudy (130 km), Periyar (244 km),

Muvattupuzha (121 km), Pamba (176 km), Achancovil (129 km) and Kallada (121 km). The Chalakudy river is a tributary of the Periyar river. The Achancovil river joins the Pamba in its downstream. The Bharatapuzha river has the largest catchment area of 6186 km^2, of which 1786 km^2 falls in Tamil Nadu. The Periyar river drains a catchment area of 5398 km^2, of which 114 km^2 are in Tamil Nadu. The combined length of the trunk stream of all the major rivers in Kerala is around 3200 km. It has been estimated that, on average, one-kilometre length of the main channel is sustained by 12 km^2 of the catchment area in Kerala. This indicates a higher capacity of land in Kerala to nurture a river, and also a high rate of water yield compared to adjacent states. High rainfall, deep soil, and high biomass density have contributed to this high level of catchment efficiency.

Drainage Pattern

The general drainage pattern in Kerala is dendritic indicating the dominance of crystalline rocks. It is sub-parallel and radial at places. Most of the rivers are structurally controlled and follow lineaments, conspicuously, the general direction being NW–SE, and NE–SW. The radial drainages are associated with the granitic bodies. Drainages in the coastal plain are mostly coast-parallel. There are evidences of drainage rearrangements and river capture in the foothill region. One such capture is evident in the case of Ithikkara and Kallada rivers. While all the rivers have originated from the higher altitudes of the Western Ghats, the Ithikkara river is the only river that originated at an altitude of <600m. Perhaps, the Ithikkara river has lost its headwater to the Kallada river through river piracy. The orientation of main drainage courses also manifests structural control. The rivers to the north and south of the Palghat Gap show different evolutionary history. Rivers in the southern section (south of Palghat Gap) show northerly convex bend in their courses while traversing the lower midland segment. This convexity is not so evident in the case of the rivers draining the northern part of Kerala. The course of Bharatapuzha is largely straight in the east-west direction. The Achancovil river in south Kerala is an antecedent river maintaining a straight-line course even while flowing through the alluvial tract. It is draining one of the World's oldest troughs of Precambrian origin.

River gradients indicate that the coastal plain extends far more eastward in the central part of Kerala than in the northern, and southern parts. Flows of rivers are sluggish below 100 m altitude, and all rivers join the sea through lagoons. Some river mouths show the formation of spits. Rivers are swift-flowing with a steep thalweg gradient from the source to 100 m altitude. Extension of a high gradient to a considerable length from the source of some rivers suggests their youthful stage. For example, the Chalakudi, Periyar, and Pamba rivers have gradients of 1/250 or more for 3/4th of their courses from their origin.

Formations of river valleys are closely related to geology, climate, material, and relief. Valley incisions are marked at various levels associated with the main channels and also tributaries. Incised valleys are manifestations of change in the sea level. Most of the rivers are misfit in nature indicating complexity in valley formation. Rivers flow through broad valleys, which, the present rivers are not capable to carve out. This indicates that these valleys were carved out during high rainfall regime, prevalent in the mid-Holocene. There is evidence supporting occurrences of such phenomena from 10k to 4k years before the present (YBP) (Nair and Padmalal, 2004). The river valleys are also impacted by successive transgression/ regression events, which are reflected in the long profiles of the rivers and thalweg character. A series of knick points along with the river courses, draining flanks of the Western Ghats, indicate successive uplift of the Western Ghats and consequent channel adjustment. The lateritisation process impacts drainage development and valley formation.

LANDFORMS

Landforms in Kerala are polycyclic in nature and evolved through the interaction of various morphogenetic processes. East-west flowing rivers cut across the NNW-SSE altitudinal zones, thus general topography is characterized by alternate ridges and valleys in north-south traverse. Landform zones are placed accordingly. Kerala has attained her present altitude through successive stages of uplift with the hinge zone located somewhere along the coast, which is a fault coast. Three commonly referred to landform zones based on altitude, namely, the high land (> 80 m), midland (8 m to 80 m), and lowland (<8 m) follow longitudinal trend of the topographic grain of the state. This scheme of classification, proposed by the PWD, Government of Kerala, was primarily concerned with the extension of human activities and is still being widely used for various practical purposes. For Kerala as a whole, the area covered under lowland, midland and highland are 10 per cent, 42 per cent and 48 per cent respectively (Chattopadhyay and Mahamaya, 1995). Human activities have expanded further eastward from the midlands and questions are often raised about the sanctity of bounding altitude (80 m) between midland and highland zones.

Notwithstanding the altitudinal limit and administrative convenience, the three well identifiable landform zones with distinct geomorphic processes are the Western Ghats, the Coastal Plain, and the intermediate Undulating Lateritic Terrain (ULT) connecting these two units to the east and west respectively. Combining altitude with the landform assemblages, the ULT can be further divided into two units: low land and midland/upland. The low land of ULT borders the coastal plain and the upland part of the ULT merges with foothills. Similarly, the Western Ghats is composed of two units: Western Ghats slopes including foothills, and plateaus, and high hills. Altitudinally, and also based on landform features, there are seven landform zones that can be recognised in Kerala (Fig. 2.7). These are: (i) Coastal plain, (ii) Lateritic mesa and associated landform in Malabar area/lower piedmont plain (Malabar segment), (iii) Lower piedmont plain (southern segment), (iv) Palghat gap zone, (v) Northern foothill and the Western Ghats slopes, (vi) Southern foothills and the Western Ghats slopes, and (vii) the Western Ghats proper with plateaus

Western Ghats

The Western Ghats occupying the eastern part of Kerala forms the southern segment of the *Sahyadri* or the Great Indian Escarpment (Fig. 2.8). Composed mainly of the Precambrian crystalline rocks, and structurally controlled, this important geomorphic feature of the Peninsular India overwhelmingly impacts Kerala's environment, economy, and culture. The Western Ghats, provenance of all rivers, is a biodiversity hotspot, and houses plantations of cardamom, tea, coffee, and a host of other hill products. Dense tropical evergreen and semi-evergreen forests occupy the steep scarp slopes and portions of the plateau. The Western Ghats is composed of several structural blocks with varying lithology. It is most likely that the evolutionary process is not uniform throughout the Western Ghats, and therefore, the resultant landforms record variations.

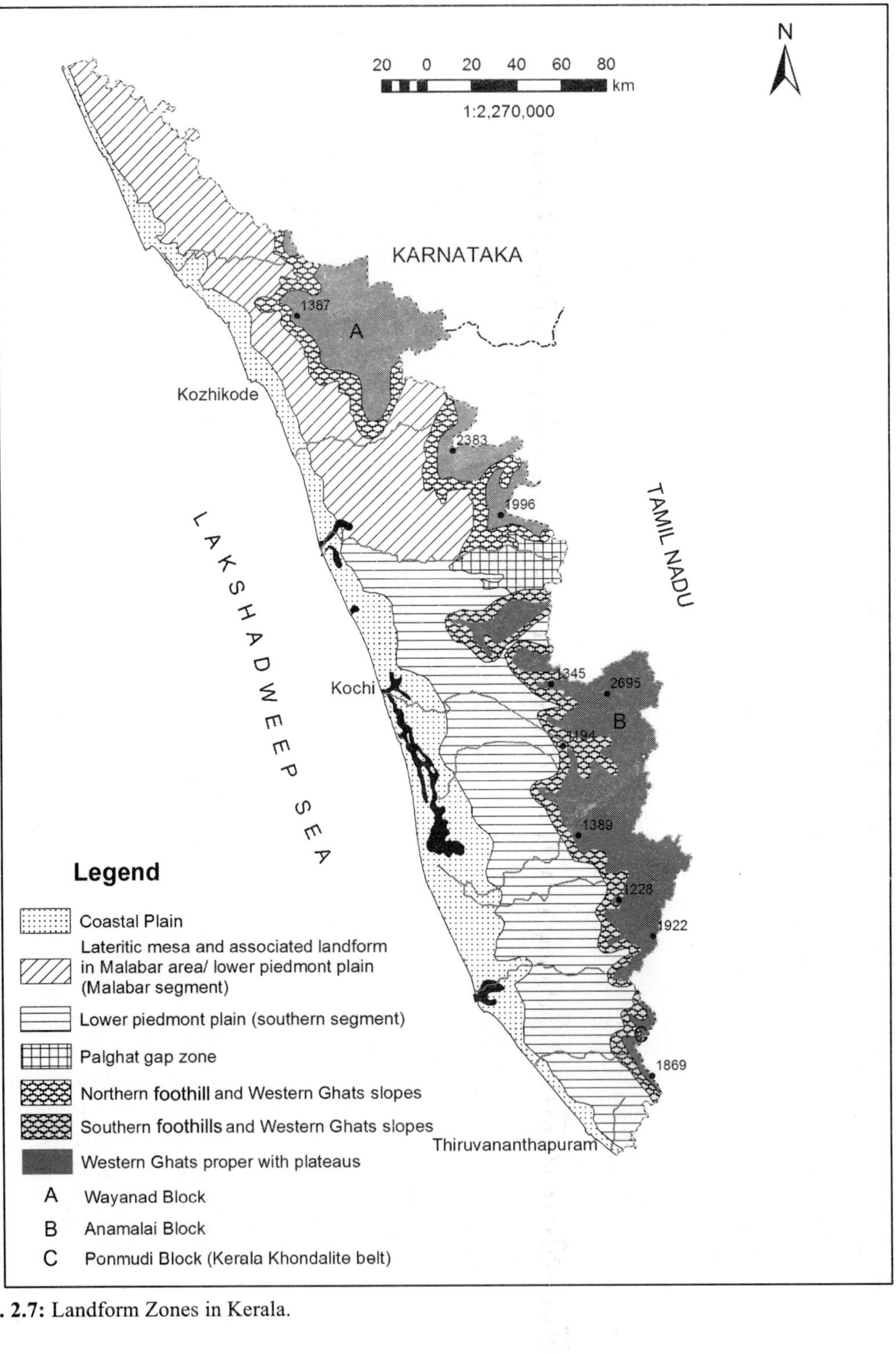

Fig. 2.7: Landform Zones in Kerala.

Fig. 2.8: Panoramic View of Parts of the Great Indian Escarpment (The Western Ghats) from Ponmudi, Thiruvananthapuram. Photo by Srikumar Chattopadhyay.

Three Terrain Blocks

Two prominent tectonic blocks of the Wayanad and the Anaimalai are located to the northern and southern segments of the Palghat gap respectively (Balakrishnanan, 2001). The north-easterly tilted Wayanad plateau, drained by the Kabini river, is well developed at an altitudinal range of 700 to 900 m and is characterised by half orange relief and broad valleys. The high altitudinal zone extends further south to the Kunda hills, Silent valley, and Attapadi valley adjoining the Nilgiris, where the Eastern Ghats merges with the Western Ghats to the north of the Palghat gap. The rivers of Valapattanam, Kuttiyadi, and Chaliyar are important drainage systems that drain the western flank of the Wayanad plateau.

Part of the Western Ghats extending from the south of the Palghat gap to Trivandrum-Nagercoil is sometimes considered as the Southern Ghats or the Southern Sahyadri whose evolutionary processes might have been different. The southern wall of the Palghat gap is steep with rocky scree slope bordering the Nelliampathy plateau, which merges with the Munnar-Periyar plateau further south. This segment is around the Anamudi (2695 m), the highest peak in Peninsular India. The general slope of this plateau complex is towards the west except for some parts of the Anamalai drained by the Pambar river in a north-easterly direction. This Pambar river is one of the three east flowing rivers in Kerala. Distinct scarp face like the one seen around the western face of the Wayanad plateau is not so evident in the case of the western face of the Nelliampathy-Periyar segment. Instead, the scarp face is quite remarkable along the eastern fringe of this segment falling in Tamil Nadu. The rivers of Chalakudi, Periyar, Muvattupuzha, and Pamba drain this segment of the Western Ghats, which houses all major hydroelectric projects of Kerala. Some patches of good virgin forests are found here. The Sabarimala shrine is located in the upstream of the Pamba catchment. The rivers originating from this block are considered as water surplus basins in Kerala.

The third Block is a litho terrain unit of the Kerala Khondalite belt designated as the Ponmudi block lying to the south of the Achankovil river (Fig. 2.7). The boundary between the khondalite (garnet-sillimanite metapelites) and the charnockite massif in south Kerala is the Achankovil river which follows a lineament similar in straightness and geometry to the Proterozoic shear belts further north in Tamil Nadu and Kerala (Hansen *et al.*, 1987). Due to the relatively high erosivity of khondalite rocks, the existing landmass to the south of the Achankovil shear is comparatively thinner than that of the Wayanad and Nelliayampathi-Munnar-Periyar blocks, which are wider and are characterised by gneisses and charnockites. The rivers of Kallada, Ithikkara, Vamanapuram, Karamana, and Neyyar drain this segment.

Undulated Laterite Terrain

Undulated Laterite Terrain (ULT) connecting the Western Ghats and the Coastal Plain is primarily a subdued terrain evolved through pedimentation, lateritisation and parallel slope retreat, valley formation, and expansion of aggradational plain. Laterite has developed as a cap on all types of rocks including recent sediments. Lateritic mesa at different elevations below 300 m altitude are conspicuous landform features in Kerala. Whether these surfaces are products of different erosion cycles and belong to different ages or they are part of the differentially eroded single planation surface is still being debated. Mesa landforms with duri crust formations are seen throughout Kerala in patches; however, these surfaces are most prominent in north Kerala particularly in Kannur and Kasaragod districts (Fig. 2.9).

Fig. 2.9: Lateritic Mesa in Kannur District of Kerala. Note the flatness of mesa surface and a metalled road cutting across the mesa. Photo by Mahamaya Chattopadhyay.

These lands are not suitable for regular cultivation due to hard crust cap and are left as wastelands in most cases. Kannur and Kasaragod districts possess maximum wastelands among all districts in Kerala. Narrow alluvial valleys indented into laterite landscapes give rise to undulated landform, where ridges and slopes are lateritic and valleys are alluvial. The alluvial valleys have developed along the east-west running rivers and streams cutting the lateritic plane mostly along weaker zones like lineaments. Around 60 per cent of Kerala's landmass is lateritic. Plantation crops like cashew, rubber, coconut, and various other tree crops, tapioca, and pepper grow well in laterite areas.

Coastal Plain

Kerala coast is a high-energy coast manifested through sudden steepness, irregularity, cliff formation, and wave erosion. The coastal plain is narrow in most places with low relief varying between 4 m to 6 m above the present-day sea level. The central part of the coast characterized by permeable recent sediments is wide (25 to 30 km) and it tapers on both sides. All three geological formations, the Crystalline, Tertiary, and Recent sediments characterise the coastal plain. Scattered hillocks and rocky elevated cliffs are associated with the Crystalline and Tertiary rocks. One of the Tertiary cliffs is well exposed at Varkala, around 35 km north of Thiruvananthapuram city (Fig. 2.10). Geological Survey of India identified this Tertiary cliff as one of the Geosites in India. The cliff section shows hard crust lateritic cap rock underlain by soft clay material. Cliff face is receding. Springs, developed along the cliff face, accelerates cliff erosion.

Fig. 2.10: Sea Cliff on Tertiary Sediments at Varkala in Kerala. Photo by Shaji Johnson.

In addition, the Kerala coast has certain distinct peculiarities, like backwaters, barrier bars, spits, and offshore mud banks. Three prominent barrier island systems and 28 spits of more than 1 km length can be identified in the Kerala coast (Kunte, 1995). The formation of the offshore bar and subsequent development of the lagoon due to sea-level change is a unique feature of the Kerala coast. All the rivers flow to the sea through backwaters/lagoons/estuaries (locally known as *Kayals*). Some of the estuaries are coast perpendicular. They are mostly foundered river mouth surrounded by lateritic landscape. The coast parallel estuaries/lagoons have developed due to offshore bar formation, and are surrounded by alluvium. Fractal dimension measuring crenulations of the estuarine shores is higher for the coast perpendicular estuaries compared to the coast parallel estuaries (Chattopadhyay and Suresh Kumar, 2007). The estuaries are mixing ground, where both fluvial and marine processes interact. They are nutrient-rich and highly productive. Sediments brought down by the rivers are initially settled in the estuaries before flowing down to thc sea. The presence of barrier islands in different coastal stretches of Kerala and associated palaeo beach ridges are manifestations of partial submergence of beach ridges formed in the earlier phase of geological events.

There are 34 *Kayals* (lagoon/estuary) characterising Kerala coastal plain. These *Kayals,* besides providing great scope for inland navigation, play important ecologic and economic functions benefiting not only the people living around but also the state as a whole. Besides fishing, coir retting is also practiced in these brackish water bodies. The Vembanad Lake, south of Kochi, is the largest (205 km^2) brackish water body in Kerala followed by Ashtmudi lake (61.4 km^2) in Kollam district further south.

In spite of so many rivers discharging into the sea at various points, no major delta has been formed as the coastal shelf off Kerala is not conducive for delta formation. However, there is an inland delta of the Pamba-Manimala-Achankovil river system occupying the southern fringe of the Vembanad lake. Around 250 km of Kerala coastline is affected by severe erosion. The continental shelf bordering Kerala coast varies in width and depth. The coastal plain is densely populated and houses all major urban centres. Coconut plantations, fisheries, and extensive paddy fields are the main land use types. The main transport lines, both, railways, and roadways have been developed along the coastal plain, besides, there are waterways. All airports are also located in the coastal plain.

The strip of the inner shelf between the coast and 10 m bathymetry is narrow in the southern part of Kerala, and it gradually widens towards the north. The bottom topography of the inner shelf in the central Kerala coastal plain is different from the other parts, which is also reflected in the fishing technology used by the fishermen.

SOIL

Soil is the topmost part of the land surface, which is used for all activities and the bulk of human food supply depends on the qualities of the top 15 to 30 cms of soil. Lying over the bedrock at different depths this complex mixture of animal, vegetal, and the mineral matter is one of the four prime requisites of life. Soil is primarily influenced by parent material (underlying rock), climate, relief, slope, drainage, vegetation cover, animal activities (from earthworms and insects to people and their domestic animals), and perhaps time. Time to develop balance among all these parameters is critical to soil formation.

Kerala's predominating Precambrian Crystalline rocks, tropical weathering, and formation of laterites, transgression, and regression in geological recent past, dense vegetation cover, long hours of sunshine, and abandoned rainfall all have contributed to the development of complex soil patterns. There is a general spatial correlation between landform and soil. Traditionally, soils in Kerala have been classified into ten broad groups based on morphological features and physico-chemical properties (Government of Kerala, 1978). These are coastal alluvium, red loam, riverine alluvium, greyish Onattukara soil, brown hydromorphic soil, hydromorphic saline soil, acid saline soil, lateritic soil, black soil and forest loam (Table 2.2).

The nature of the soil is changing due to prolonged use for agriculture, irrigation, fertilizer use, and reshaping of land particularly in the valleys and slopes. Human activities have become important modifiers of natural soil qualities. Some of the best quality soils are no longer available for cultivation in the areas where there is a high demand of land for housing and other non-farm activities as a consequence of urban expansion. Reclamation of wetlands and removal of top soils in many places have altered soil character.

Table 2.2: Broad Soil Types and Characteristics (selected) of Soils in Kerala

Soil type	*Textural class*	*Organic carbon*	*pH*	*Major available Nutrient status (kg/ha)*		
				N	*P*	*K*
Coastal alluvium	Sandy loam	0.25	5.2-6.5	102-164	9.6-14.8	56-83
Red loam	Loam	0.82	4.5-5.4	154-171	18-28.6	52.3-112.8
Riverine alluvium	Sandy loam	0.63	4.7-5.6	295-341	22-34	172-194
Acid saline	Silty clay	3.67	2.8-5.3	242-256	6.85-10.9	60-208
Hydromorphic saline	Clay	2.90	3.3-6.0	182-500	5.88-57.2	24-592
Brown hydromorphic	Clay	1.47	5.0-5.5	382-419	28-48	112-128
Grayish Onattukara	Loamy sand	1.00	5.2-6.5	90-152	8.2-14.5	58-78
Laterite	Sandy loam	1.25	4.6-5.5	78-175	14.6-17.8	55.3-101.1
Black soil	Loam	0.61	6.3-8.3	151-172	18.5-23.2	48.6-59.3
Forest loam	Clay loam	4.40	4.8-5.2	384-550	16.2-25.6	152-258

Source: KSCSTE 2007a, Vol. 1, pp. 57-65; Government of Kerala, 1978.

NATURAL VEGETATION AND WILDLIFE

Kerala is well endowed with natural vegetation and maintains a high level of productivity compared to the rest of India. Natural virgin forest occupies highly diverse lands ranging from valleys to steep rugged slopes and parts of plateaus. The climatic variability and physiographic diversity are well manifested in the vegetation types, which include wet evergreen and semi-evergreen, moist deciduous, dry deciduous, sub-tropical montane, and temperate shola forests and grasslands. The famous Silent Valley movement in the late 1970s and early 1980s was to protect a piece of virgin tropical rainforest in this part of India. The Western Ghats is a biodiversity hotspot and houses numerous valuable medicinal plants. Apart from timbers, forest produce includes various non-timber minor forest products, which provide livelihood sustenance to a large section of people, particularly tribal communities. Kerala forests provide home to a variety of wild animals like elephants, various types of deer, wild bears, tigers, panthers, Nilgiri taur, Nilgiri langur, and liontailed macaques, besides a large variety of snakes. Detailed discussions on natural vegetation, forestry, forest management, and wildlife sanctuaries are provided in Chapter 7 dealing with Animal husbandry, fishing, and forest resources.

CHAPTER

3

Geomorphic Evolution

GEOMORPHIC PROCESSES

Dominant geomorphic processes active in Kerala are marine, estuarine, fluvial, and denudational (Chattopadhyay and Mahamaya, 1995). Around 60 per cent of the total area of Kerala is under denudational processes. Therefore, Kerala is a region of net soil loss. Characteristics of sea waves, tidal range, and direction of littoral current are the main factors influencing marine, estuarine and coastal processes. Although, Kerala enjoys both the monsoons, the south-west monsoon is stronger. Sometimes, the wind speed during this period (June-September) exceeds 60 km/hr., correspondingly, wave heights reach as high as 6.5 m in some days (Baba *et al.*, 1997). However, wave height is even less than 1 m in a fair weather period. Wave energy reduces towards the north (*ibid*). The tidal range is the maximum (90 cm) at Kochi. In addition to these, the rivers directly debouching into the sea have a pronounced influence on coastline geomorphology, sediment distribution and beach character. During monsoon months huge quantities of water brought down by these rivers modify wave propagation. Formation of barrier bars in several stretches of the coast associated with river mouths manifests the marine-riverine interaction, and therefore, landforms developed in this zone are of mixed nature bearing imprints of both the processes. All these interactions are explicit in evolving coastal landforms of erosional, depositional and complex nature. The aeolian landforms, particularly windblown sand dunes, have developed in certain pockets of the Kerala coast.

The fluvial processes are active in major parts of Kerala, although large depositional units under these processes can be marked only in certain areas like greater Kuttanad, Ernakulam-Thrissur stretch, and Palakkad. Flood plains, fluvial terraces, and other depositional landforms associated with major rivers have developed in patches due to topographical constraints. River valleys of various configurations are seen in different parts of Kerala.

Lateritisation, a process of chemical weathering, typical of tropical regions with alternate dry wet climatic conditions, has a pronounced impact on the geomorphic character of Kerala. Laterite forms during the final stage of land surface reduction and provides a cap rock protecting the old surface (McFarlane, 1981). Besides preservation of the old landform or planation surface due to duricrust formation, lateritisation facilitates parallel slope retreat or pediplanation in tropical areas, where landform develops mainly due to peneplanation (Chattopadhyay and Mahamaya, 1995; Narayanswamy and Chattopadhyay, 1996). Lateritisation follows peneplanation, which facilitates the horizontal movement of water. All high altitude laterites are uplifted after their formation at the sea level. Break-in-slope, marked in the lateritic topography, is often described as morphological discontinuity and part of polycyclic landform.

The diffusive process or slope wash is an important geomorphic process. Apart from geology and relief, the slope processes are controlled by climatic factors and vegetation. Half orange relief with graded slope develops in a tropical rain forest area with high slope wash. Wayanad plateau and bordering parts of Kozhikode and Kannur districts exhibit this type of graded slope condition (Chattopadhyay and Mahamaya, 1989). The eastern part of the Palakkad district experiencing semi-arid climatic conditions shows the formation of inselbergs. Slopes are long under the sub-humid or temperate conditions in the rain shadow, high altitude Munnar-Anamudi stretch. Climatic fluctuations over time have also impacted landform development. There are relict landforms bearing imprints of palaeoclimate. Hill tops and adjoining scree slopes produce a considerable amount of loose materials, which are washed down the hills and are deposited in the form of alluvial fans over which ephemeral channels develop. The north-facing foothills along the southern wall of the Palghat gap is a typical area exhibiting the formation of this type of landform.

There are multiple denudational centres with gentle to moderate slopes in low relief areas (Mahamaya, 1998). Therefore, identification of break-in slope is significant at the micro-level (Chattopadhyay *et al.,* 1984). Soils are very deep in the tropical areas due to intense weathering processes. Even the steep slopes are characterised by very deep soils.

Finally, human interventions, particularly in the modification of cover types, have a significant impact on landform development as both sediment movement and surface runoff undergo modifications. Landforms in Kerala are extensively modified by various human activities like removal of rocks and soils from elevated areas, clay mining from paddy fields, sand mining from river beds and palaeo channels, filling up of wetlands, and terracing of slopes. This point has been further elaborated in Chapter 15 discussing 'Changing Geography of Kerala'

LINEAMENTS

Landforms in Kerala are structurally controlled. Disposition of major landforms, the orientation of rivers and stream, and the development of scarp face bear imprints of the contribution of geologic structure in their evolution. Lineaments, defined as deep-seated fractures, provide significant information on a tectonic activity that a region is subjected to. Lineaments in Kerala are oriented in the directions of NW-SE, WNW-ESE, NNW-SSE, N-S, NE-SW, and E-W (Katz, 1978; Varadarajan and Nair, 1978; Varadarajan and Balakrishnan, 1980). The NE-SW to ENE-WSW trending lineaments can be considered as the youngest (Nair, 1990).

One of the major NNW-SSE lineaments (N40°W-S40°E) can be traced from Mananthody in the Wayanad plateau through the Bavalipuzha river to the coastal belt north of the Ezhimala. The Mananthavady river in the Wayanad plateau and the Bavalipuzha river have developed along this lineament in the eastern and western segments respectively. Distribution of lateritic mesa following alignments of lineaments indicates structural control on topography. The Kuppom river is flowing along a lineament. Mesa surfaces developed on both sides of the river, at comparable altitudes, perhaps points to a pre-existing planation surface over which the river has developed along the lineament. Part of the scarp face of the Wayanad plateau coincides with a lineament more or less parallel to the coastline in the NW-SE direction.

The Palghat gap is placed between two WNW-ESE running lineaments. The Idamalayar lineament (NW-SE) can be traced from the vicinity of the Idukki reservoir to Ottappalam, where the course of the Bharatapuzha shows a structural meander. The Periyar river follows the Periyar lineament (WNW-ESE) for the most part of its course. A possible extension of this lineament can be traced up to the Kambam lineament (Singh *et al.*, 1992). The Idamalayar and Periyar lineaments represent shear zones (Rao, 1978). The NW-SE trending Kambam lineament skirting the eastern part of the Periyar plateau is also a major shear zone (Rao, 1978; Katz, 1982). The Munnar lineament is another major lineament in the NE-SW direction. The general lineament patterns in this Munnar-Periyar plateau are sub-parallel to the major lineaments. The Achankovil and Thenmala shear zone, further south, in WNW-ESE is an important structural feature that delimits the northern boundary of the Khnodalite belt in south Kerala. The Kallada lineament more or less runs parallel to this shear zone.

NEOTECTONICS

Change in sea level, river courses, and foundering of river mouths are noted in various parts of Kerala. Valdiya and Rajagopalan (2000) in their study at Mysore plateau pointed out that neotectonic movements along the N-S and NE-SW oriented faults had caused blockage of streams and the formation of lakes. Marshes developed in the Wayanad plateau falling more or less in a similar direction and the Pukkot lake are perhaps the manifestations of neotectonics. Raised old river terraces, pebble beds, the abrupt change in river direction found in the Palghat gap might have resulted from tectonic movements (Rajendran and Rajendran, 1996). Reactivation of structures in the Palghat gap during the Cretaceous-Tertiary period is indicated by the development of the NW-SE dyke system and fractures cutting across the older generation folds (Krishnaswamy, 1981). The presence of slicken-sided micacous planes in NW-SE trending dyke in the Kottayam-Kakkiar hill tracts and the intermittent occurrences of earth tremors in this area are suggestive of weak structural planes that are still active indicating continued tectonic activity along these faults since Cretaceous times (Soman, 2002). The foundering of Asthamudi mouth is related to neotectonic movements.

PLANATION SURFACES AND THEIR CHARACTERISTICS

Identification of the planation surface (sometimes referred to as erosional surface or geomorphic surface) is an attempt to recreate the denudation chronology of an area, as the planation surface is a non-relief forming denudational topographic surface. There were various attempts to identify planation surfaces in Kerala (Chattopadhyay, 2004). Besides, coastal plain and aggradational surfaces, there are primarily two main levels of planation surfaces developed in Kerala, and each of these two main levels can be divided into a number of sub-levels (Fig. 3.1, Table 3.1). The existence of these planation surfaces is well evident from long profiles of the rivers, which manifest distinct break-in slope/knick points at different altitudes, the prominent one being in the altitudinal range of 700-800 m (Fig. 3.2). The lower limit of high-level planation surface more or less follows this height. All the major plateaus in Kerala-Wayanad, Nelliampathi, and Periyar are located above this altitude. Further upstream, a prominent break can be marked at the altitudinal range of 1200 to 1300 m. Characteristics of these planation surfaces are briefly described here.

High-Level Planation Surfaces (HLPS) and Prominent Landforms

The formation of high-level plateaus is linked with uplift. The older peneplain of the Cretaceous period was uplifted during Miocene resulting in the formation of the Wayanad-Mysore plateau.

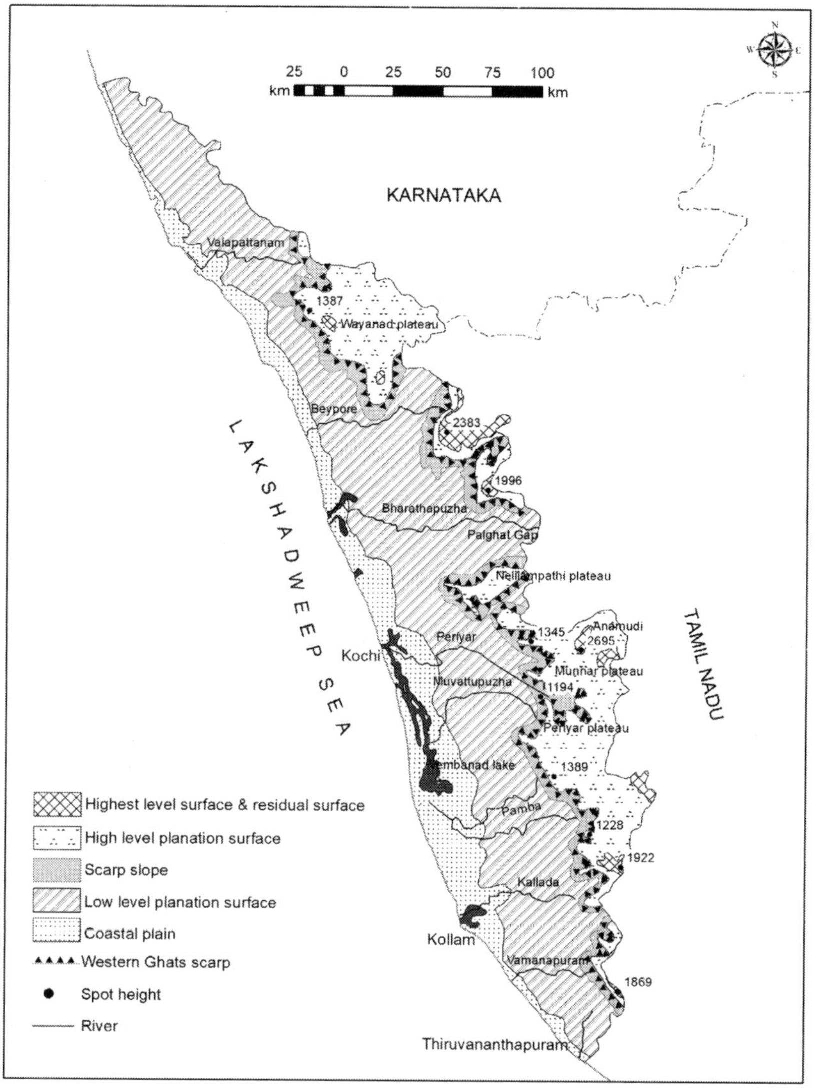

Fig 3.1: Planation Surfaces in Kerala.

As a continuation of these movements, it is possible that further block uplifts of segments of the old plateau took place resulting in the upraising of these previously uplifted plateaus to comparatively greater heights (Radhakrishna, 2001). High-level planation surfaces are found in two levels, at above 1500 m at the Munnar, Palghat hills/Bhavani tract, Kunda hills, south-western part of the Wayanad plateau, and at 700 m to 1000 m level covering the Periyar plateau, Nelliampathi plateau, and Wayanad plateau proper. The planation surface of > 1500 m level is well developed around Munnar, and it is only the remnants that are preserved in other areas.

Wayanad Plateau: Easterly tilt of the Wayanad plateau, an extension of the Mysore plateau, is related to crustal upwarping in the direction of NNW-SSE and continental distension even prior to the separation of the continents, probably in early Cretaceous (Vaidhyanadhan, 1967;

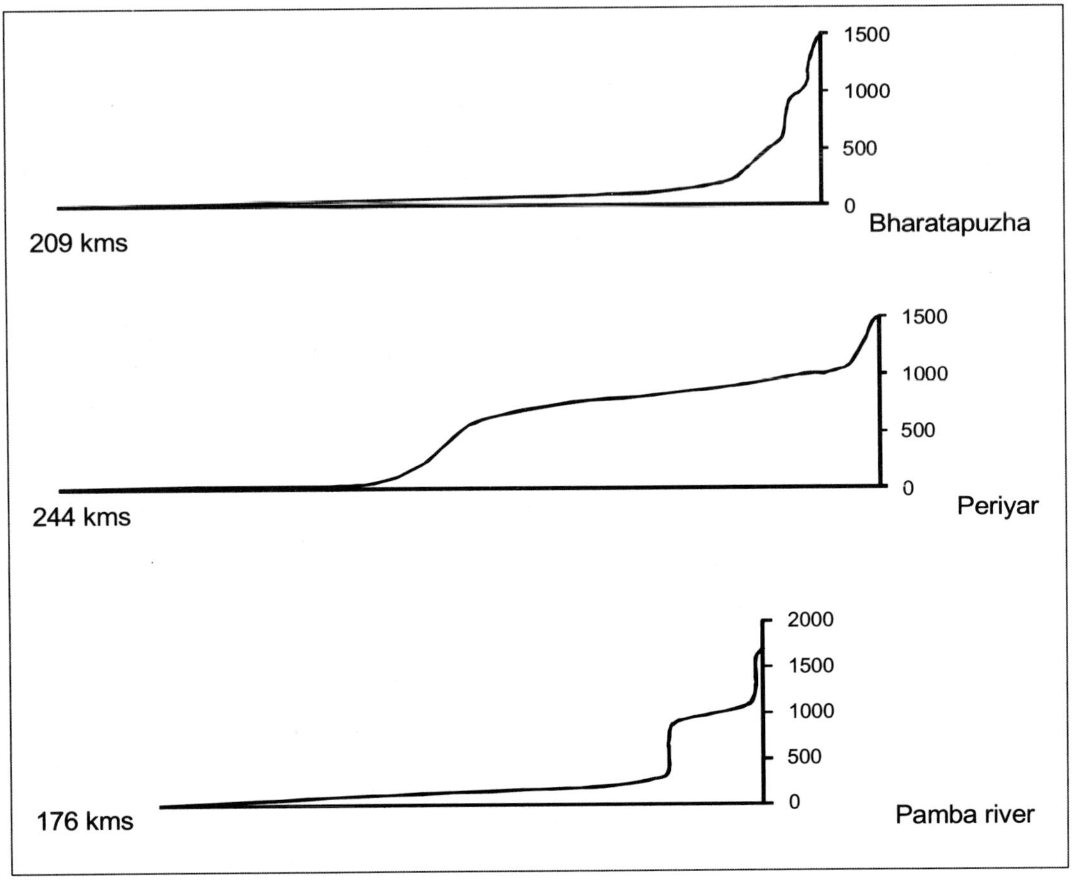

Fig 3.2: Long Profile of Selected Rivers in Kerala.

Subramanian, 1987). The main planation surface is found to have developed within the altitudinal range of 700 m to 900 m. The upper surface above the main plateau is traceable as shoulder slope of the Nadapuram Mudi (1395 m), Banasurama hill, and as residual hills like the Kalpatta and Ambalavayal. These residual hills are composed of granite or chanockite and the plateau proper is basically gneissic terrain. East flowing Kabini river traversing the plateau follows the Bhavali lineament, which in all probability is the westward extension of the Moyar lineament (Ravindra Kumar, 1986). The rugged western margin of the plateau composed of charnockite rocks is uplifted.

Silent Valley: The Silent Valley, drained by the Kunthipuzha along the western cliff of the Nilgiri plateau is located to the north of the Palghat gap. It is one of the most well-known places in Kerala embodying a very successful environmental movement protecting a 90 km^2 tract of virgin evergreen rainforest against a proposed hydroelectric project. It has been declared as a National Park in 1985 by the Government of India. This valley surrounded by a series of peaks, namely, Vanampara (1210 m) on the south-west, Koyipara (1904 m) on the north-west, and Anginda (2382 m) in the north-east is elongated in the north-south direction and tilted

Table 3.1: Planation Surfaces in Kerala and their Occurrences

Level	*Sub-level*	*Probable* geological age*	*Altitudinal range (in metres)*	*Surface identified at*	*Remarks*
High Level Planation Surface (HLPS)	I	Early cretaceous	>2400	Anaimalai	Residual hill, Anaimudi (2695 m) the highest peak in south India
		Ecocene	1500-2000	Munnar plateau	Well developed, long slope, grass land topography
	II	Miocene	700-1000	Wayanad plateau	Easterly sloping, extension of Mysore/Karnataka plateau, most bevelled surface
				Bhavani tract	Easterly sloping narrow ridge like area, north of Palghat gap
				Nelliampathi plateau	Located south of Palghat gap, south westerly sloping, comparatively dissected
				Greater Periyar plateau	Extensive, marked in two levels, Peermade plateau included
Low Level Planation Surface (LLPS)	I	Mio-pliocene	100-300	Palghat gap	Structural-erosional surface
	II	Pleistocene	40-230	Lateritic plateau	Widely developed in north Kerala, characterised by duricrust formation
	III	Recent	0-50 m	Alluvial plain, valleys and coastal plain	Found throughout Kerala, some parts of coastal plain lateritised, hard crust formation also evident in coast (Kannur town)

Source: *Vaidyanadhan (1967), Gunnell (2001) and Soman (2002).

towards south-west drained by the southerly flowing Kunthipuzha, a tributary of the west flowing Bharatapuzha. Adjoining Attapady valley slope towards the east and is drained by the east-flowing Bhavani river following easterly tilted Bhavani shear.

Nelliampathi Plateau: This plateau extends from the southern flank of the Palghat Gap in the north to the Chalakudi basin in the south. The Peringalkuttu reserve forest marks the southern end. The northern border, along the southern flank of the Palghat gap, is a steep scarp with a near-vertical elevation of 400 m to 500 m. The rocky cliff forms the scree slope at an altitude of 700 m to 800 m. The plateau slopes towards the south-west. The general level of the plateau is in the range of 700 m to 900 m. The Minarapara Mala (1633 m) and the Padagiri (1527 m) are remnants of the upper surface. Ephemeral streams along the foothills of the northern boundary of this plateau are quite conspicuous.

Munnar Plateau: The Munnar plateau has developed around the Anamudi peak (2695 m) and the Anamudi bench (2400 m). It is located within the high altitude charnockite massif of the Anamalai-Palni-Cardamom hills. The Pambar river originating in the north-eastern part of this plateau is one of the three east flowing rivers in Kerala. Sub-humid climate due to high altitude has a definite impact on slope processes. Long smooth slopes characterize this area.

The west to east cross-section from Kurisumudi to Kilandur across the Munnar plateau presents all the planation surfaces in Kerala ranging from 300-500 m at the low level to >2200 m at the highest level (Fig. 3.3 A). The deeply incised Idamalayar gorge follows the Idamalayar lineament, which is considered to represent the shear zone (Rao, 1978). Down-thrust and up-thrust sides on both sides of Idamalayar gorge (Section A) represented by the Nalla Mala to the west and the Tera mala to the east respectively are well evident. The northern side of the Munnar Plateau associated with the Anamudi, the highest point in Peninsular India, is at a higher level compared to the southern end. The north-south cross-section from Tadaganachi Malai in the north to Vandiperiyar in the south (Fig. 3.3B) and east-west cross-section from Panel in the west to Kodaikanal in the east (Fig. 3.3C) across the Anamudi show development of all major high-level planation surfaces in Kerala.

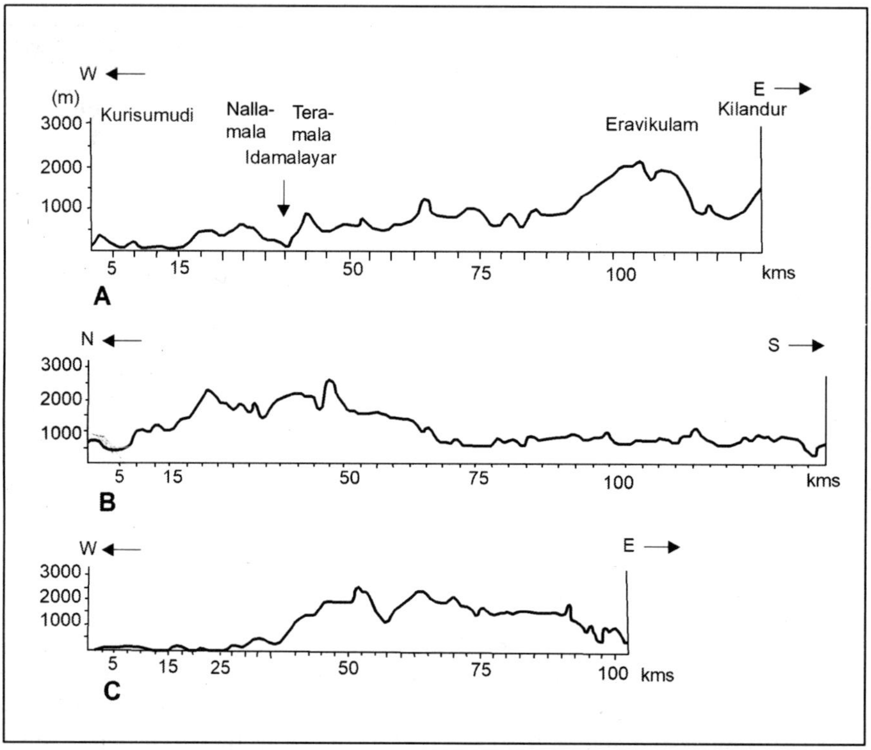

Fig. 3.3: Cross-sections across Munnar Plateau (A) and Anamudi (B and C) in Kerala.

Periyar Plateau: The Periyar plateau extends from Kakkiar Malai-Sivagiri Malai in the south to the Munnar plateau in the north. The south-western part of this plateau is also known as the Peermade plateau. The western part of this plateau slopes down gradually to the undulating landscape of the Pathanamthitta-Kottayam area. The Peermade plateau is wider towards the north and gradually tapers down towards the south. The eastern part of the Periyar plateau characterized by steep scarp slope coincides with the NW-SE trending Kambam lineament, a major shear zone (Rao, 1978; Katz, 1978). The Periyar lake is located in the southern part of the Periyar plateau, which is north-westerly tilted. The Periyar river draining this plateau flows along a major lineament, commonly referred to as the Kerala lineament. The Kottai Malai (2019 m), Pachaiar Malai (1754 m), Sivagiri Malai (1744 m), and Me-Malai (1795 m) are remnants of further higher level (>1200 m altitude) planation surfaces lying to the south of the Pariyar lake. The South western part of this plateau is drained by the Pamba river.

Ponmudi-Agasthamalai Ridge Section: This ridge section, part of the Kerala Khondalite Belt (KKB) occupies southern most segments of the Western Ghats. This area does not show any appreciable plateau formations. In all probability, they might have been eroded. There is a skeletal presence of the plateau surface around the Ponmudi hill. This hill complex is oriented in the NW-SE direction parallel to major lineaments. The Agasthamalai (1869 m) is the highest point in this part of Kerala. The ridgeline of Ponmudi-Agasthamalai is northerly tilted.

Low Level Planation Surfaces (LLPS) and Prominent Landforms

These planation surfaces are marked below an altitude of 500 m. The coastal plain is the lowest and the youngest planation surface. Most of the LLPS are covered by laterite and laterite derived materials. All low-level planation surfaces except for the Palghat gap, wherever preserved, are generally lateritic surfaces or composed of lateritic materials.

Palghat Gap: The Palghat gap with an average width of about 30 km along 76°45'E longitude and an elevation ranging from 100 m in the west to 220 m along the eastern border is a significant geomorphic feature connecting Tamil Nadu plain in the east and the Kerala plain in the west. This Gap is a well defined structural-erosional surface (Subramanian and Muraleedharan, 1985). The origin of the Palghat gap can be traced in the pre-drift Gondowana land (Thara and Soman, 1993). Fluvial erosion dominated from time to time and acted along structurally weaker planes, thereby, landform features of complex (structural-erosional) origin evolved. Bordered by the Bhavani tract in the north and the Nelliampathi plateau in the south, this pediment pass is part of the Tamil Nadu lowland surface that according to Gunnell (2001) has reached a degree of rectitude and perfection astonishing several generations of tropical geomorphologists. The pedimentation in the Palghat gap took place along the Bharatapuzha and mega fractures forming the Gap. The northern wall of the Palghat Gap shows a gradual but steep rise from 100 m at the base to the ridge crest of 1100 m, but the southern fringe rises abruptly from 300 m to 1000 m with a rocky scree face in the upper boundary. This southern face is a plateau-bounding scarp.

Laterite Surface/Mesa: Strong and deep weathering is an essential factor in initiating the planation process in humid tropics with chemical down wearing of the regolith as clays are removed from the surface by rain-wash and also sub-terraneously particularly in the low relief areas (Verstappen, 1987). The lateritic mesa surfaces are found at different altitudinal levels and on all types of rocks. Whether these surfaces are products of different erosion cycles and belong to different ages or they are part of the differentially eroded single planation surface is still being debated. Mc Farlane (1981) is of the opinion that altitudinal differences may be a part of the denudational process and need not necessarily be due to different erosional cycles. Karunakaran and Sinha-Roy (1981) attempted to link laterite profile development with polycyclic geomorphic surfaces developed at various levels.

Mesa landforms are well developed in parts of Kannur and Kasaragod districts. The satellite view of parts of the Kuppam river basin shows a straight-line course of the river indented within a mesa surface and the landform assemblages in a mesa-dominated lateritic terrain (Fig. 3.4). The river is structurally controlled. Roads and settlements developed at equidistance from the river on both sides along the central part of the mesa indicate the flat/gentle sloping nature of the top surface of the mesa. Side slopes of the mesa are steep, and therefore, they are usually avoided for construction of settlements.

Step like landforms from lower to higher elevation is well evident in the west to east transect in this area. A geomorphology map around Kundavati, Kannur district presents a classic lateritic landform (Fig. 3.5). An elongated residual hill devoid of laterite cover at an altitude of 254 m is surrounded by a lateritic mesa surface at 120-160 m. Scarp face cutting across different altitudinal range surrounding the mesa surfaces is developing through headward erosion of streams and parallel slope retreat. As a result, a single mesa surface is dissected into a number of surfaces along the weaker planes or line of denudation, manifested by valleys. The major break in slope is marked at 170 to 180 m and the minor breaks are at 140 m, 100 m, and at 90 m bordering the valleys (refer Fig. 3.5 inset).

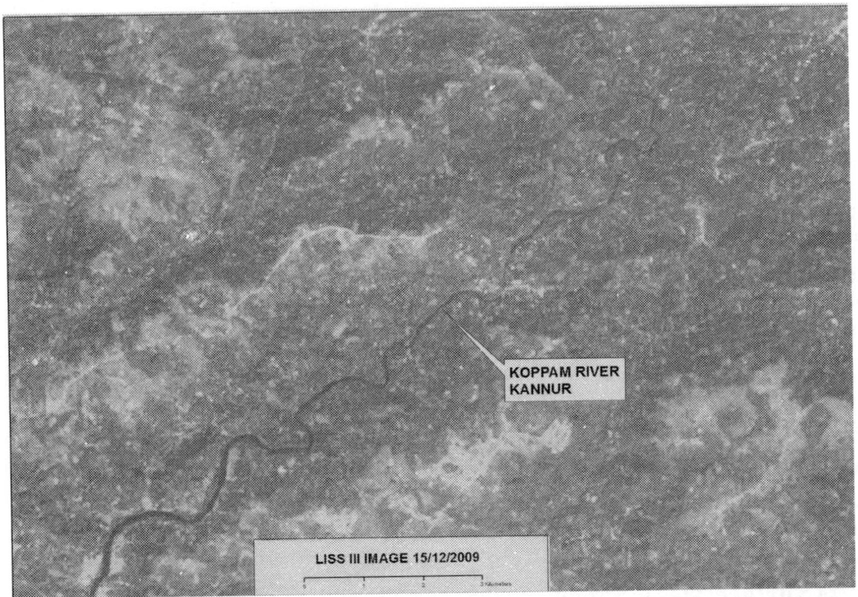

Fig. 3.4: Satellite View of a Lateritic Landscape in Kuppam River Basin of Kannur District in Kerala. Settlements are seen in Light Grey Tone.

Aggradational Landform under Fluvial Processes

Steep sloping lands, the predominance of lateritisation, narrow width, and structural control on drainage pattern and topography all together have imposed certain restrictions in the development of fluvial landscape, although Kerala receives high rainfall generating tremendous surface run-off. Valleys and floodplains in most cases are narrow with the possible exception of the Bharathapuzha and the Periyar river basins and other major river basins like that of the Pamba, Vallapattanam, etc. The Bharathapuzha having the largest catchment area within Kerala exhibits well-developed flood plain, and fluvial terraces, preserved along the tributary valleys. Paired terraces found along the tributaries indicate their recent origin.

The Periyar is the longest river in Kerala. A major part of its course lies in the highland and is structurally controlled. The lower course extending over a distance of about 30 km from the confluence to upstream exhibits the development of wide flood plains composed of various

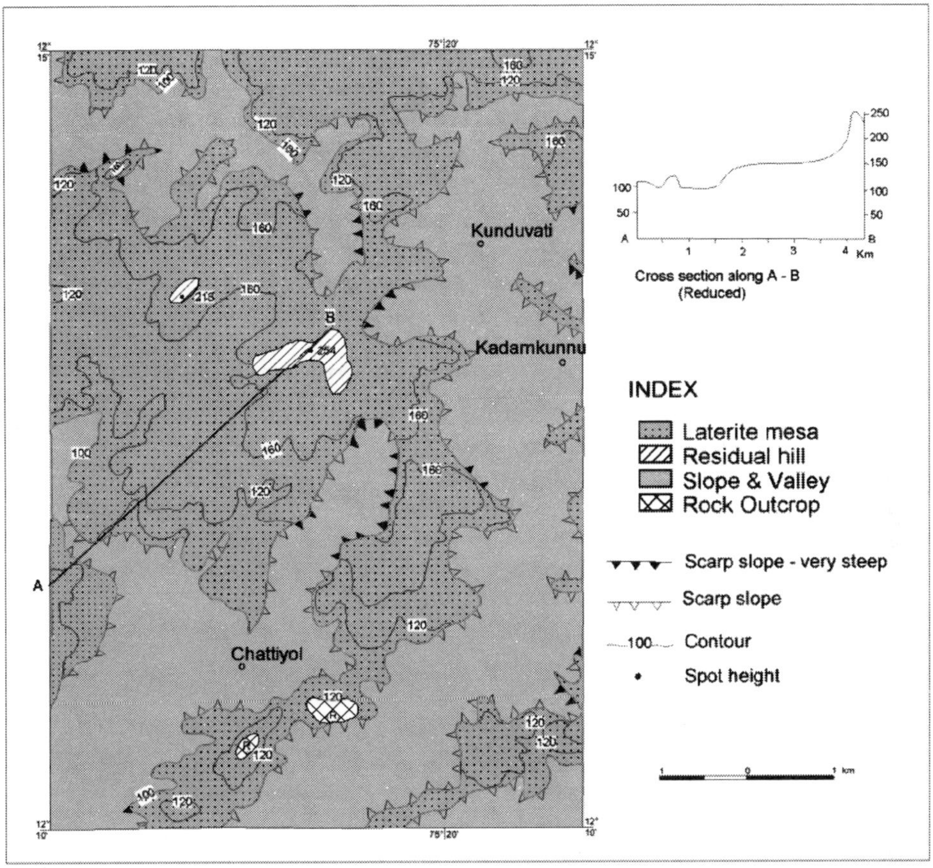

Fig 3.5: Geomorphology of a Laterite Terrain in Kannur District of Kerala.

fluvial units like a levee, back swamp, meander scar, braided bar/point bar (Fig. 3.6). The shifting course of the river can be deciphered from the meander scar or palaeo meanders. The lateritic landforms found in this area are part of the relict land. It influences river course geometry in places. The cross-section shows the relative position of different landform units in this area. Association of fluvial landform and lateritic landform can be found in many parts of Kerala particularly along the lower course of the river within lateritic terrain (Fig. 3.7). This association indicates that all laterite landforms along lower courses of rivers might not have been formed *in situ*, rather some might have been detrital in origin. Aggradational landforms are mostly indented within lateritic terrain through the headward erosion and pediplanation process.

Extensive alluvial deposits are found in the deltaic Kuttanad area mainly in the southern part of the Vembanad lake. Like other parts of the west coast, deltas are absent in Kerala, however, deltas have developed around Vembanad lake. The Pamba delta formed by the rivers of Achankovil, Pamba, and Manimala with thick alluvium deposits is noteworthy. Formation of an inland delta is possible where streams debouch into an internal basin (Vaidyanadhan, 1990). The existence of such an internal basin from the Periyar mouth in the north to the Ashtamudi estuary down south is reported from the Tertiary period. This point has been elaborated in the section on the evolution of the central Kerala coastal plain. Deltaic landscape

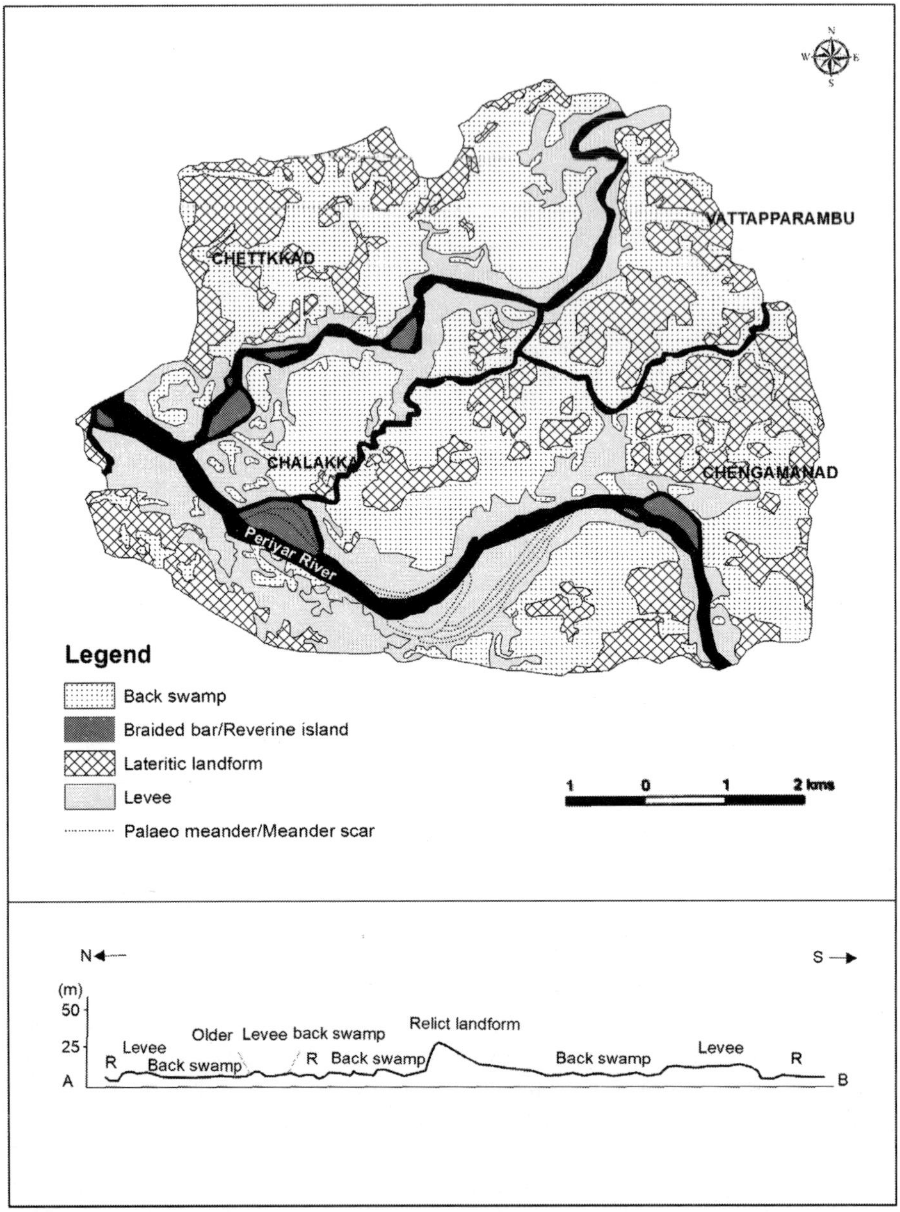

Fig 3.6: Geomorphology of Aggradational Terrain in Lower Periyar Basin of Kerala.

can also be found at the mouths of the Muvattupuzha, and the Periyar. Both these rivers are debouching into the Vembanad lake. Alluvial fans are found to be deposited along the eastern border of the Vembanad lake.

Flat bottomed valleys flanked by moderate to steep side slopes are important geomorphic features in midlands and highlands of Kerala. These valleys owe their origin partly to slope

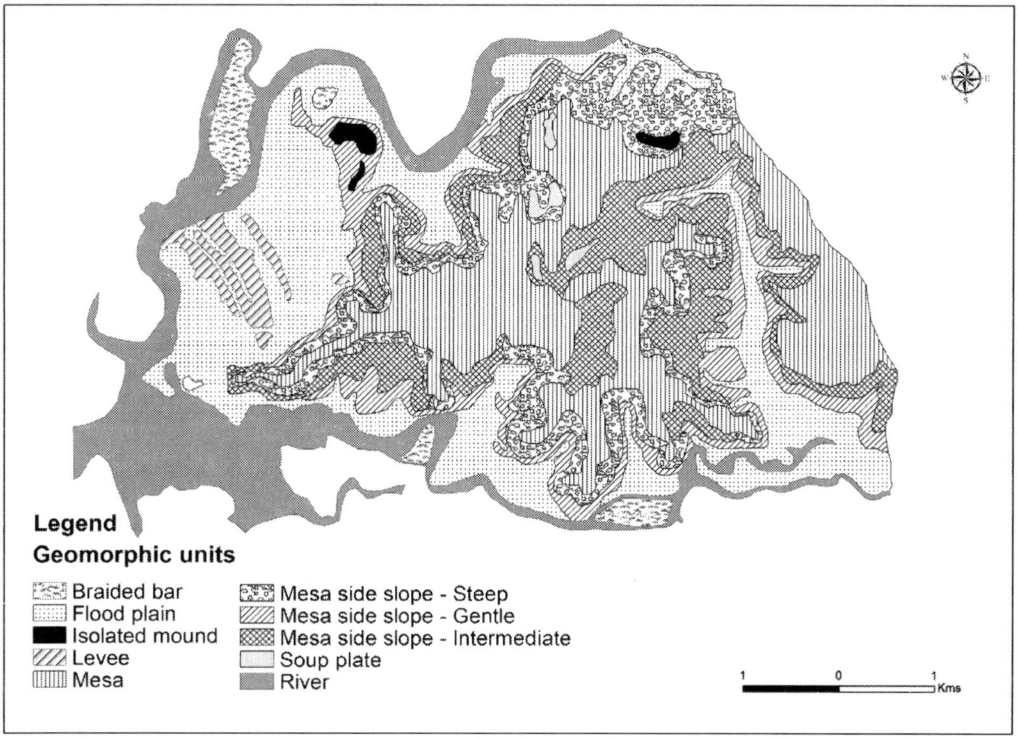

Fig 3.7: Geomorphology along Lower Course of a River in Kannur District of Kerala. Note association of fluvial features and lateritic features. Valleys are intended within lateritic landform.

wash and scarp retreat than normal fluvial action. Wayanad plateau exhibits well developed fluvial valleys. Mostly flat bottomed with shallow soil, these valleys are traditionally given for paddy cultivation. It is believed that the name "Wayanad" originated from the paddy fields. These valleys are conspicuous in Satellite images.

EVOLUTION OF WESTERN GHATS SCARP

The west-facing scarp of the Sahyadri or the Western Ghats scarp is an important geomorphic feature in peninsular India. The prominent scarp faces in Kerala are associated with two tectonic blocks of Wayanad in north Kerala and the Anaimalai to the south of the Palghat gap. Various theories have been proposed to explain mountain front or scarp land development since Davis (1899) proposed that scarps are essentially shaped by the rivers. In contrast, King (1953) maintained that scarp land evolved independently of streams. There are attempts to link tectonics with morphology and scarp development (Coque-Delhuille, 1991; Formento-Trigilio and Pazzaglia,1998).

Geological, geomorphological, and geophysical evidences suggest that the Western Ghats represents the edge of an up-raised and disrupted continental block with a date of formation during the early Miocene (Radhakrishna, 2001). The older peneplain of the Cretaceous period was uplifted during Miocene resulting in the formation of the Wayanad-Mysore plateau. As a continuation of these movements, it is possible that further block uplifts of segments of the old plateau took place resulting in further upraising of these previously uplifted plateaus to comparatively greater heights (Radhakrishna, 2001). The uplift of the Western Ghats is linked with west coast faulting, which is not a single fault; rather it is a composite unit of several sections separated from each other by horizontal shears (Balakrishnan, 2001). Geomorphologically, the Western Ghats scarp is related to faulting, erosion, and sea-level fluctuation (Dikshit, 1981).

The formation of the Ponmudi scarp in the southern most segment of the Western Ghats is related to tectonic activities (Chattopadhyay and Mahamaya, 2004). Scarp faces associated with steep slope and periodic rejuvenation due to uplift in various geological ages are subjected to continuous slope wash, which is a dominant process sculpturing the present landscape here. The position of the Ponmudi scarp in NW-SE direction and indentation in both ends associated with drainage lines perhaps indicate the role of rock resistance in scarp land formation.

While the dominance of the diffusive process can be related to King's (1953) concept of scarp retreat independent of drainage line, other field observations support the views of Davis' (1902) that scarp lands evolved due to the variable resistance of rocks to the distribution of erosion energy, concentrated mainly along the major streams. Fluvial action, slope wash, variable rock resistance, and tectonics acting in tandem contributed to the formation of scarp land of the Western Ghats. Field study in the flank of the Western Ghats revealed that waterfalls have developed in several steps, which can be attributed to successive uplift in phases, topographic rejuvenation, and headward erosion.

Scarp Retreat

Landform development in the Western Ghats fringe is associated with scarp retreat along structurally weaker planes like lineaments, fractures, foliations, etc. Analysis of valley configuration, particularly the valley wall, indicates topographic rejuvenation, valley incision, and scarp retreat along the headwater. In fact, scarp retreat is considered to be a principal mechanism leading to the development of the Western Ghats. In absence of any major fault onshore the Western Ghats scarp recession can be measured from the west coast fault (Gunnell and Radhakrishna, 2001; Balakrishnan, 2001). At 8°45' north latitude, the Western Ghats scarp (600 m altitude) is at a distance of 45-50 km from the coastline. This distance, which is an approximate measurement of scarp retreat, is about 60-65 km across the Vembanad lake segment and comes down to 35-40 km at 12°45' latitude in the Wayanad plateau area. The average scarp retreat value of about 40 to 45 km for the Kerala region is comparable to the value of 40 km worked out by Balakrishnan (2001) for the segment from Narmada to Goa.

EVOLUTION OF CENTRAL KERALA COASTAL PLAIN

The Coastal zone in Kerala is not uniform. It exhibits distinct spatial differences in material composition, geo-form, and surface features from north to south (Nair, 1987). The central Kerala coastal plain extending from north of the Asthamudi estuary in the south to the Ponnani mouth in the north is a permeable coast with sand and clay dominated recent sediments. The widest stretch of this plain is around 25 km across the Vembanad lake in the vicinity of Sherthalai from where it tapers both towards north and south.

Palaeo Coastline

Considering the eastern extension of the palaeo sand ridges, delta heads of three rivers, namely, the Pamba, the Muvattupuzha, and the Periyar river debouching into the Vembanad lake and western boundary of Trichur 'Kole' lands a line can be drawn from the Ponnani mouth in the north to the north of the Astamudi estuary in the south. This line broadly coincides with the eastern boundary of recent sediments and can be construed as palaeo coastline (Chattopadhyay, 2002). Series of shallow water bodies in the segment between the Achankovil river and the Asthamudi estuary terminates along this line. The drainage pattern to the west of this line is coast parallel, while to the east, it is perpendicular to the coast.

Vembanad Lake Basin

The Vembanad lake basin, a part of the on-land Tertiary-Quaternary basin occupied the western part of the Palaeo coastline. The present-day Vembanad lake is the remnant of a greater Vembanad lake or a mega lagoon, covering the entire stretch from Periyar mouth to Kayankulam, which got silted up progressively separating Kayamkulam lagoon and giving rise to the present landscape of Kuttanad (Nair *et al.,* 1998). It is possible that the Achankovil shear structurally determined the southern boundary of this lake basin. The northern part of the Vembanad lake is also structurally controlled. In fact, the Vembanad lake is considered as an embayment of a large fracture zone in the offshore region (Varadarajan and Balakrishnan, 1980; Mallik and Suchindan, 1984). The deepest part of this basin was in the Ambalappuzha-Alleppey stretch during Tertiary and it had shifted northwards to the vicinity of Shertallai-Pattanakkad during Quaternary (Nair, 1976; Soman, 1997). The basin, therefore, experienced northerly or north-westerly tilt, which is manifested in the flow direction of the Achankovil-Pamba drainage system.

The greater Vembanad lake used to receive sediment deposits brought down by river systems draining into it. The physiographic configuration of the lake is attained through barrier bar formation in the offshore region, subsidence, and titling. There is evidence of subsidence around the Kochi area. It is reported that relative to Kochi, Mangalore is rising by 3.22 +/–1.1mm per year (Subrahmanian, 1998). Kochi is subsiding possibly due to the accumulation of huge sediments brought down by the five rivers draining into this lake. The southern hinge of this titled basement is perhaps traceable in the vicinity of Asthamudi-Karunagapally area.

The ages of lime shell samples from two locations, Muhamma and Vechur lying along the inner boundary of the lake were estimated at 3130 ± 100 YBP and 3710 ± 90 YBP by Rajendran *et al.* (1989). Age of 7050 ± YBP is estimated for a carbonized wood from Iranimangalam, located along the eastern periphery of the lake. Barrier bar formation is the characteristic feature of the Kerala coast. From the Ponani mouth to the Asthamudi lake, there were four barrier bars/islands, which can be deciphered from the satellite images by considering the distribution of beach ridges. Historical maps also indicate the existence of four islands.

The opening at Kollam and Kochi can be approximated with the Asthamudi mouth and the Kochi mouth respectively. The outlet at Thottapally is not so active, but well identifiable due to deep clay deposits and an opening across which a spillway has been constructed to drain out floodwater and to prevent saltwater intrusion. Thick clay deposits, buried tree trunks at 2 to 3 m below ground level, and the marshy condition all indicate a shallow estuarine environment in this area. The buried tree trunks are remnants of mangrove vegetation, which flourished around 10,000 YBP. The wetter condition prevailed globally after 11,500 YBP must have started a process of forest extension that gained momentum after 10,000 BP and was virtually completed over the following 1000-2000 years (Thoen and Thonp, 1992). Well established mangrove vegetation associated with an extensive mudflat being subjected to a series of transgression and regression had given rise to the present landscape in this area. Tilting of the basin towards north-northwest as evident from the river directions has also contributed to the present configuration.

Emergence in Stages

The evolution of this landscape is not a single event rather it is through several stages following a successive transgression, regression, and neotectonic movements. Geomorphology, material, and geological structure in combination indicate the presence of three landscape blocks. The area lying to the north of Periyar mouth is one block, and the rest of the area in the south is composed of two blocks, separated by a NW-SE trend line joining Purakkad and middle course of the Achankovil river. In view of stratigraphy, Nair and Rao (1980) had divided the on-land part of the Kochi-Warkala basin into two blocks, considering a SE trending line starting near Alleppey.

Three sets of regression-transgression features (sand ridges) could be marked in the stretch from Kochi mouth to south. The set I with north-south trend extends from Kayamkulam to Mannar with an average distance of 10 km from the present-day coastline. In all probability, these palaeo beach ridges might have been extended both northward as well as southward.

The set II in NW-SE direction had developed offsetting the set I in coast parallel alignment. It might be possible to trace set I below set II and also in the offshore of Kayamkulam. Submerged terraces at depths from 10 m to 55 m belonging to the Holocene period (11,000-9,000 yrs) in the West Coast (Soman, 1997) may be the continuation of these strandlines. The concentration of offshore black sands is perhaps associated with these strands. The set-I in its southern half experienced certain deformity in alignment. If the Thenmala-Kalada lineament is extended towards the coast, it will cut across this line of deformity. However, set II does not show any irregularity. It may be inferred that set I was affected by neo-tectonic movement prior to the emergence of set II. In all probability set II emerged as a result of mid-Holocene (6000 YBP-4000 YBP) sea-level change. Although due to overlapping tectonic uplift, the net magnitude of sea-level rise cannot be ascertained, a rise of +6 m to +8 m is proposed by Raj *et al.* (1998) for Mahi valley. Raised beach ridges in the Bombay area at comparable heights of set II are not older than 5000 YBP (Agrawal *et al.*,1979; Bhattacharya *et al.*, 1987).

The set I was perhaps associated with Holocene sea-level rise following which there was a tectonic disturbance that had caused further sea level change, tilting, and also block movement in a number of places. The southern boundary of this segment is associated with horst and graben topography giving rise to the Ashtamudi estuary, which is a foundered mouth of the Kallada river. Ridge in the offshore, adjoining the Ashtamudi opening, bears testimony of this down faulting. Age data of 1330 ± 100 YBP for lime-shell from Ashtamudi estuary (Rajendran, 1987) indicates its very recent origin. Change in the orientation of sand ridges from the set-I (N-S) to set II (NNW-SSE) might indicate the tilting and directional shift of the basement. Along the northern part of this block, the clay regime is indented with a down thrust scarp line of 1.5 m to 2 m below the level of adjoining sand ridges. This indicates subsidence, most probably along a NW-SE axis.

The Achankovil river, which follows the Achankovil shear, presents the characteristics of an antecedent river in the middle stretch. Entering into the recent sediment zone, it flows in a straight course in westerly direction cutting across palaeo sand ridges. This indicates that the river is flowing in the same direction in spite of uplift. The Achankovil maintained flow direction across the indented clay belt and takes northerly turn along the periphery of sand ridges. Elbow shaped turning of the river in a pure alluvium terrain is an indication of structural control. A similar elbow shape turn can be observed in the case of the northern distributaries of the river Pamba.

The III set of sand ridges separate the Vembanad lake-Kuttanad area from the Lakshadweep Sea and is the youngest among the three sets. A well-developed spit (Palaeo) is marked to the north of Thottapally. While along the coast, the barrier bars are the youngest, some parts which emerged originally as a sand bar arc comparatively older than the surrounding landscape.

The segment lying to the north of Kochi mouth is structurally controlled along both the arms of Periyar. The seafront is a well-developed barrier bar in this block. The Kochi mouth opened sometimes in 1340-1350 AD (Menon, 1965). This opening, perhaps, coincides with a fault line. With evidence of subsidence in Kochi, it is possible that the southern arm of Periyar is more active than the northern one. This structurally controlled block is sloping towards the south. There are sand ridges identifiable to the east of the Vembanad lake. The northern distributary of the Muvattupuzha skirts a sand ridge while debouching into the Vembanad lake. These sand ridges are comparable to the set I. The evolutionary process experienced by the areas lying to the north of the Periyar mouth is somewhat different. Lowlands in this area known as Kole lands are bordered by a series of sand ridges along the west. Considering the altitude and locational aspects beach ridges in this part can be equated with the set II and set III of the southern part. The Kole land zone, the Vembanad lake, and the Kuttanad area can be considered as "Zone of unfinished" deposition. While the zone to the south of Periyar mouth is an embayment and owes its origin to block movement, barrier bar formation, and transgression and regression, the northern part had developed due to successive progradation.

The Central Kerala coastal plain has emerged in various stages associated with successive transgression, regression, and also neo-tectonic activities since the last Glacial Maximum in the Indian Ocean around 18,000 YBP. Global transgression commenced around 14,000 YBP (Millima and Emary, 1968, C.R. Bhattacharya *et al.,* 1987). Sediment supply and runoff as a consequence of climatic variabilities may not have significantly changed during the last 10,000 years except a prolonged dry spell beginning around 3500-4000 YBP and ending at about 2000 YBP (Valdiya and Rajagopalan, 2000).

Since the 11,000-9,000 YBP regression phase, there were two alternate phases of transgression and regression interspersed by neo-tectonic movement (Table 3.2). During transgression, sea level rises resulting in a change in the long profile of the river. The valleys are flooded therefore sediments are deposited in the thalwegs of the rivers. The condition reversed during regression with receding sea level and consequent lengthening of river courses. The greater Vembanad lake basin acted as a local base level for all the rivers flowing into the sea through it. This created a congenial environment for delta deposition in an inland basin. Although features evolved in the course of one event were overlapped by the next event, some of the indications are discernible with definite geomorphic signatures.

Table 3.2: Probable Sequence of Events and Emerging Landscapes in Central Kerala Coastal Plain

Events	*Approximate Time Scale (YBP)*	*Evidences*	*Geomorphic Signatures*	*Remarks*
Regression 1	11,000-9,000 1	Submerged beach ridges, marine terraces, exposed sea bed including greater Vembanad basin, lengthening of river course	Set I sand ridges, chain of Wetlands	Flourishment of mangrove vegetation (10,000 YBP) 2
Neotectonics transgression 3	8,000-6,000 3	Peat bed formation, antecedent river, buried mangrove, submerged beach ridges	Deformity in Set–I sand ridge alignment, drowned valleys, imposed maturity, barrier bar formation, delta deposits	Subsidence tilting and horizontal displacement
Regression 4 Spell of dryness 5 Neo Tectonics activity	5,000-3,000 4 4,000/3,500 to 2,000 5 <2000 6	Raised beaches Lengthening of river courses Drying up of lakes Block faulting–Buried river mouth; Foundering of Asthamudi mouth, river course change and flooding; Formation of present coastline Eustatic changes	Set II sand ridges formation of coastal plain unit, Delta deposits. Submerged ridges in front of Asthamudi lake, imposed maturity Emergence of coastal plain; Loss of river gradient in lower course, Sand Bar formation	Exposure of Vembanad lake basin; Incised valleys; Barrier bar formation

Source: 1. Nair and Hashimi (1980), 2. Van Campo (1986), 3. Agrawal *et al.* (1970), Rajendran et al (1989) and Raj et al (1998), 4. Agrawal et al (1979) and Rajendran *et al.* (1989), 5. Valdiya and Rajagopalan (2000), 6. Soman (2002).

CHAPTER

4

Water Resources, Utilization and Management Challenges

SURFACE WATER YIELD

The availability of clean and quality water is a major component of Kerala's natural resource endowment. The average rainfall in the state is 305 cm, which is two times more than that of the national average (Government of Kerala, 2017). The spatial distribution of rainfall in Kerala varies from coast to foothills and also from south to north. Most of the rivers in Kerala are perennial and have base flows throughout the year. The total annual yield of all rivers in Kerala is estimated to be 70,323 Million Cubic Meter (MCM), in which the west-flowing rivers contribute 64,263 MCM and the rest comes from three east flowing rivers (Table 4.1) (Government of Kerala, 1974).

The total annual yield of water in Kerala is about 30 per cent less than that of a single river like the Godavari. On average, one square kilometer of catchment area yields 1.81 MCM of water of which 60 per cent is usable. A recent estimate indicates that total surface water availability in Kerala after accounting for losses due to evapo-transpiration and interception is about 54,410 MCM. Of which 41,000 MCM flows as surface run off, 7900 MCM infiltrates to recharge groundwater and the rest is stored in reservoirs (Planning Commission of India, 2008a).

It is not appropriate to plan for using the entire quantity of 41,000 MCM of water. Some amount of freshwater is required as environmental flow to maintain the functioning of the hydro ecosystem of all water bodies, like rivers, lakes, and backwaters. A certain amount is required to prevent saltwater ingress from the sea and some flow should be maintained to flush out wastes, regularly discharged into the rivers from human settlements. The water requirement for flushing wastes is increasing with incremental waste discharge as a consequence of growing consumption. It is generally estimated that 40 per cent of flow should be allowed for these functions and the rest 60 per cent can be considered as annual utilisable water. However, this may not be feasible because rivers are rain-fed in nature and hence experience wide variations in flow between lean season and monsoon. Drawing 60 per cent water during non-monsoon months may not be practical as rainfall is the main source of all available water and it is mostly concentrated during monsoon months.

RESERVOIRS

Many of the rivers have been dammed in their upstream sections for irrigation and hydroelectric projects. There are 54 reservoirs of all sizes covering an area of 429 sq. km. The total stored

Table 4.1: Surface Water Resource Potentials of the River Basins in Kerala

Name of Basin	*Annual yield (MCM)*			*Annual utilisable yield (MCM)*				
	Total	*In Kerala*	*Outside*	*Total*	*Utilised per cent*	*In Kerala*	*Utilised per cent*	*Outside*
Manjeswar-Uppala	698	309	389	379	54.2	106	34.3	273
Shiriya	1337	620	717	973	72.7	358	57.7	615
Chandragiri-Morgal	3964	1718	2246	3129	78.9	1218	70.8	1911
Chittari	254	254	Nil	100		100		
Nileswar-Karingode	1710	1356	354	1238	72.3	937	54.8	301
Kavvayl-Peruvamba Ramapuram	1143	1143	Nil	603	52.7	603	52.8	Nil
Kuppam	1516	1236	280	1024	67.5	786	63.6	238
Valapattanam	4092	2784	1308	2938	71.7	1823	65.5	1115
Anjarakandy	986	986	Nil	503	51.0	503	51.0	Nil
Tellicherry	251	251	Nii	122	48.6	122	48.6	Nil
Mahe	803	803	Nil	445	55.4	445	55.4	Nil
Kuttiyadi	1626	1626	Nil	1015	62.4	1015	62.4	Nil
Chaliyar-Korapuzha-Kallayi-Kadalundi	7775	7135	640	3160	40.6	2616	36.6	544
Tirur	165	165	Nil	60	36.3	60	36.3	Nil
Bharathapuzha	7478	6540	938	4146	55.4	3349	51.2	79,
Keechieri-Puzhakkal	1024	1024	Nil	345	33.6	345	33.6	Nil
Karuvannur	1887	1887	Nil	963	51.0	963	51.0	Nil
Chalakudy	3121	2541	580	2033	65.1	1539	60.6	494
Periyar	11607	11341	266	8232	70.9	8004	70.6	226
Muvattupuzha	3814	3814	Nil	1812	47.5	1812	47.5	Nil
Meenachil	2349	2349	Nil	1110	47.2	1110	47.2	Nil
Manimala	1829	1829	Nil	1108	60.5	1108	60.6	Nil
Pamba	4641	4641	Nil	3164	68.1	3164	68.1	Nil
Achancovil	2287	2278	Nil	1249	52.4	1249	52.4	Nil
Kallada-Pallickal	2770	2770	Nil	1368	60.2	1368	60.2	Nil
Ithikkara	761	761	Nil	429	56.3	429	56.3	Nil
Vamanapuram-Ayroor-Mamom	1324	1324	Nil	889	67.1	889	67.1	Nil
Karamana	836	836	Nil	462	55.2	462	55.2	Nil
Neyyar	433	433	Nil	229	52.8	229	52.8	Nil
Kabbini*	4333	4333	Nil	4333	100	4333	100	Nil
Bhavani*	1019	1019	Nil	1019	100	1019	100	Nil
Pambar*	708	708	Nil	708	100	708	100	Nil
Total	78041	70323	7718	49286		42772		6514

Source: Water Resources of Kerala, PWD, 1974.
* East flowing rivers.

surface water in Kerala is estimated at 5,500 MCM. Water is diverted through canals for irrigation during the dry season. It also helps recharging wells in certain areas. But the reservoirs are of mixed blessings. On the one hand, they facilitate the generation of hydroelectricity, provide irrigation water for agriculture, facilitate pisciculture, and therefore contribute to development. On the other hand, they restrict silt movement and the coastal area is deprived of sediment supply from the upland through these rivers. The distribution of sediment is affected in the process, which in many cases causes an imbalance in coastal sediment distribution. Most of the reservoirs in Kerala are silted up. Impounding of new reservoirs is not feasible due to various reasons. Furthermore, impounded water may lead to Reservoir Induced Seismicity (RIS) causing minor earthquakes. There are debates throughout the world about the viability of the large reservoirs, particularly in the areas with high seismic potential.

GROUNDWATER

Six hydrogeological formations, namely sandy alluvium, valley fills, silt/clayey alluvium, granites, laterites, and weathered crystallines and associated rocks characterise Kerala (GWD and CGWB, 2014). The unconsolidated or poorly consolidated sediments characterise the coastal zone, which is relatively wide in the central part of the state. The deep coastal aquifer system is characterized by four beds in the Tertiary group of formations, viz., Alleppey, Quilon, Vaikom, and Warkallai beds. Open wells, filter point wells and shallow tube wells are the major sources of groundwater extraction structures in the top coastal unconfined aquifer.

Groundwater occurs in the hard crystalline rocks in the weathered, partly weathered, jointed and fractured rocks. Some of the bore wells drilled in the hard rock terrains also show deep-seated fractured zones under semi-confined and confined conditions. These weathered, jointed and partly weathered rock aquifers are tapped through dug wells. Borewells are feasible along fracture zones and are site specific. Along the hill ranges, the hard rocks are covered by thin weathered zone, not suitable for deep dug wells. But thick weathered zones found along the valley fills in the midlands and highlands can be developed for medium capacity dug wells for drinking and irrigation.The thick laterites in midland stretching till foothill zone are potential sites for storage by virtue of their high porosity. However, hard crust formation as is evident in Kasaragod and Kannur limits infiltration and there is little percolation. The little water that is stored is readily released by seepage and through springs in the valley heads due to a high degree of topographic dissection. The thickness of the lateritic aquifer is found to be the highest in Kozhikode, Malappuram, and Palakkad districts. The recharging potential of groundwater depends upon biophysical characteristics like geology, structure, topography, and slope, infiltration capacity of the soil, transmissivity, and rainfall.

Groundwater Potential and Use

Kerala has good groundwater potential with a total annual groundwater recharge of 6251.31 MCM and net availability of 5651.53 MCM (GWD and CGWB, 2014). There is a provision for 599.78 MCM for natural discharge. The average net availability is in the order of 0.145 MCM/Km^2. Gross groundwater draft for all uses is 2635 MCM of which 45 per cent are used for irrigation and the rest 55 per cent are used for domestic and industrial supply (*ibid*). The use of groundwater is more for irrigation in the districts of Palakakd, Thrissur and Kasaragod (Table 4.2). These three districts together account for 50 per cent of all groundwater use in the irrigation sector in Kerala. Palakkad, and Thrissur are well known for irrigated agriculture. Although these two districts are served by canal irrigation, there is increasing dependence on groundwater as the canal water supply is not sufficient. In the case of Kasaragod district, which uses more than 70 per cent of its groundwater draft for irrigation, canal irrigation has not been developed. Kozhikode, Wayanad, Alappuzha, and Thiruvananthapuram districts use more than 70 per cent of their groundwater draft for domestic and industrial purposes. Traditionally, people in Kerala depend on open wells for meeting their domestic needs and also for irrigating homestead gardens. There are about 4.5 million open wells in Kerala. The density of open wells is very high (around 150/km^2), perhaps the highest in the country. On average, more than 90 per cent of wells are used for domestic purposes. The stage of development of groundwater is the highest in Kasaragod district (70 per cent) followed by Tiruvananthapuram (60 per cent) and Malappuram (58 per cent). The lowest development (19 per cent) is in Wayanad district (Table 4.2).

Table 4.2: Groundwater Resources and Use by Districts in Kerala, 2013

District	Net annual groundwater availability (MCM)	Existing gross ground-water draft for all uses (MCM)	Gross ground-water draft for irrigation (per cent)	Gross ground-water draft for domestic and industrial use (per cent)	Stages of groundwater development (per cent)
Thiruvananthapuram	296.24	178.54	29.6	70.4	60.27
Kollam	364.55	162.77	30.7	69.3	44.65
Pathanamthitta	254.11	92.92	39.2	60.8	36.57
Alappuzha	431.61	137.48	28.3	71.7	31.85
Kottayam	407.74	131.32	36.9	63.1	32.21
Idukki	200.43	101.53	54.3	45.7	50.65
Ernakulam	538.72	198.78	38.7	61.3	36.9
Thrissur	581.75	324.08	62.4	37.6	56.4
Palakkad	648.93	352.85	61.9	38.1	54.37
Malappuram	516.87	302.14	31.6	68.4	58.46
Kozhikode	331.72	178.39	24.7	75.3	53.78
Wayanad	280.96	54.74	28.8	71.2	19.48
Kannur	469.01	185.79	43.1	65.9	39.61
Kasaragod	328.89	229.58	71.4	28.6	69.81
Kerala	5651.53	2634.91	44.9	55.1	46.68

Source: GWD and CGWB, 2014: Dynamic Ground Water Resources of Kerala (2013).

Based on stages of groundwater development (100%, 90% to 100%, 70% to 90% and <70%) and declining trend of groundwater level during pre-monsoon and post-monsoon over a period of 10 years (2003-2012) Community Development (CD) Blocks are grouped under four categories, namely, overexploited, critical, semi-critical and safe (GWD and CGWB, 2014). It emerges that Chittur Block in Palakkad district is overexploited; Kasaragod block in Kasaragod district and Malampuzha block in Palakkad district are in critical condition. There were 18 blocks distributed in nine districts in the semi-critical category and the rest 131 blocks were safe by 2013. The number of semi-critical blocks has increased to 30 and the number of safe blocks has come down to 119 by 2019. Groundwater availability has declined from 5651 MCM in 2013 to 5212 MCM by 2017.

GROUNDWATER RESOURCES ACROSS PHYSIOGRAPHIC ZONES

Availability of groundwater resources record variations across the physiographic zones following hydro-geological conditions and the amount of rainfall received. Depth to water level also varies seasonally and across physiographic zones as elucidated from observations of well data monitored for groundwater measurement (GWD and CGWB, 2014). In the coastal region, groundwater is estimated at 0.3 MCM/Km^2 and utilization is 0.12 MCM/km^2 (Soil Survey Organisation, 2007). Although utilization is around 40 per cent, there are pockets of over-utilization resulting in various problems. Saline water intrusion along the rivers during summer months is a major issue in the coastal tract. There are reports of seawater reaching up to 30 km upstream from the river mouths. Temporary check dams are being constructed to prevent salt-water intrusion in such cases. Many households use electric pumps for lifting water from open wells. However, there are also reports of saline water ingression in wells in some parts of the coastal belt due to overpumping.

In the case of midland and highlands, the levels of utilization are in the order of 0.05 MCM/km^2 and 0.01 MCM/km^2 against the availability of 0.4 MCM/km^2 and 0.2 MCM/km^2 respectively (Soil Survey Organisation, 2007). During pre-monsoon season, the overall range of water level is between 0.30 m to 25.75 m bgl and most of the area experienced water level falling in the range of 5 m to 15 mbgl. In the coastal tracts of Alappuzha, Ernakulam, and Thrissur, the water level is noted in the rages of 0.3 m to 2.0 m bgl. Valleys and valley fills in the highlands like that in the Idukki district also show a similar shallow water level. The midland section shows water level varying between 3 m to 12 mbgl. In the highland section with thick laterite columns, the water level falls well below 20 m from ground level. The scenario changes during monsoon with the onset of rainfall. The overall water level during monsoon varies from 0.10 m bgl to 23.75 m bgl and most of the area experiences water level in the range from 3 m to 10 mbgl. The shallow water zones in the coastal tract and highlands record values ranging from 0.20 m to 2.0 m bgl. The midland section shows the water level in the range of 2-10 m bgl and in the highland areas of thick lateritic overburden like that in Thiruvananthapuram, Kannur, Kasaragod etc., the water level is noted at 20 mbgl. During the post-monsoon season in January, the overall water level ranges from 0.30 m to 25.0 m bgl, and the majority of the area shows the water level in the range of 3 to 10 m bgl. The zones of shallow water level experience water depth <2 m bgl. There are zones of 10-20 m bgl and >20 m bgl noted in different parts of the state in the midlands and highlands.

Depth of water table depends more on the site than the physiographic zone as a whole, therefore site characteristics are important considerations for sinking wells. Due to the sloping nature of terrain, steep slopes, and thin soils, there is limited scope for storing water in most parts of Kerala. Run-off is faster than infiltration. Open wells are not common in the highland section as digging wells are not economic and the groundwater table is deep (>15 m) in most localities. Springs are widely used in highland areas but they dry up during summer months.

Doubts are raised about the wisdom to assess groundwater potential of Kerala following conventional norms developed at all-India level as 90 per cent of Kerala's landmass is characterised by hard rocks. It is felt necessary to evolve/adopt a suitable method for assessing Kerala's groundwater potential.

WETLANDS AND COASTAL ZONE

Wetland is a dynamic eco-system having complex interrelationship of soil, hydrology and vegetation. They occupy the transitional zone between the aquatic and the terrestrial environments sharing characteristics of both. They are constituted by small and large lakes, rivers, streams, seasonally flooded plains, estuaries, deltas, mangrove areas, coral reefs, swamps, low-lying paddy fields, and seasonally flooded forests. Their common denominator is that they have a soil or substrate that is at least periodically saturated with or covered with water. The wetlands in the state have been broadly classified into 'coastal wetlands' and 'inland wetlands', which are either natural or man-made.

The coastal wetlands, locally known as backwaters distributed from the north to south of Kerala and covering an area of 2426 km^2 perform a variety of functions that directly or indirectly benefit human settlement. They provide fish; hold excess monsoon run-off; facilitate water percolation; purify water; maintain the nutrient cycle; provide sites for biodiversity;

easy transport; retting of coconut husk; and absorb pollutants. As most of Kerala's rivers pass through the backwater system before emptying into the Lakshadweep Sea, the backwaters tend to be enriched with nutrients brought down by the rivers. The sediment and nutrient loads flowing down with the river water first settles (or is naturally panned) in the backwaters before they move to the sea. As a result, the productivity of the coastal waters is linked to the nature of the settling process in the backwaters. The presence of the backwater system in the coastal plain has added to the diversity of the resource base of Kerala. It created an additional opportunity for the fishermen to have a choice other than the sea for fishing activity. The Vembanad lake with an area of 205 km^2 and spreading over Ernakulam, Alappuzha, and Kottayam districts is the largest coastal wetland not only in Kerala but also on the west coast of India. The Asthamudi lake (64 km^2) located in Kollam district is the second-largest coastal wet land in Kerala. Both the Vembanad and Asthamudi lakes are Ramsar sites. All coastal wetlands are shrinking due to reclamation all-around the periphery and are losing their depth due to siltation.

FRESH WATER LAKES

While most of the natural wetlands are brackish, there are few freshwater bodies too. The important ones are, Vellayani, Sasthamcotta, Pukot, Enamakkal, Manakkodi, Muriyad, Eravikulam, Devikulam, and Elephant Lake. The total area of these nine lakes is around 16 km^2. Among these nine freshwater lakes, three namely, Sasthamcotta Lake in Kollam district, Vellayani Lake in Thiruvananthapuram district, and Pukkot Lake in Wayanad district are significant. While Pukkot Lake is used for pisciculture and irrigation, Vellayani Lake and Sasthamcotta Lake supply drinking water for urban areas. Kollam municipality depends upon the Sasthamcotta Lake for its drinking water need. Water budgeting for Sasthamcotta Lake has indicated that the annual yield of this lake is to the tune of 4.7 million cubic meters, contributed by direct rainfall, run-off, and groundwater recharge (CESS, 1998). A large sub-surface catchment feeds this lake.

The wetlands are under extreme threat. There are various reasons contributing to this scenario. Few instances like (i) the steady encroachment going on in all the lakes, including the fresh water lake of Vellayani, (ii) the near leveling of Kayakulam Lake, (iii) siltation and pollution of the freshwater lake like Sasthamcotta, and (iv) choking of most canals with water weeds, are noteworthy. Besides, there are numerous ongoing and proposed reclamation projects in and around Kochi as the Goshree project, southward extension of Wellington Island, foreshore reclamation programme, extensive reclamation of wetlands situated to the east and north of Ernakulam towards Edappally and Thrippunithra, and so on.

PONDS AND TANKS

Ponds and tanks are traditional rainwater harvesting structures, used widely in tropical countries. The ponds in Kerala may be grouped under three types: (a) natural depressions (locally known as *chiras* in some parts of the state), (b) excavated and man-made ponds or tanks, and (c) valley-head ponds (or *Talakkulam*). While the first two types are located in the low-lying areas, the third type is located at the valley head. Water is impounded by constructing a barrier or dam across the narrow valley in relatively low sloping areas. The main valleys are converted to agricultural fields pushing the stream channel from the centre of the valley to the side along the break-in-slope that demarcates the boundary between the valley and the side slopes. This is a common sight throughout Kerala. Valley-head ponds are used mainly for irrigation purposes supplementing rainwater to grow rice or other crops after the monsoon. Many of the valley-head ponds in Kerala are associated with springs. Spring-fed ponds are perennial; however, all springs do not yield large quantities of water. Kerala has 995 large ponds with a minimum water storage capacity of 1500 m^3 each (CWRDM, 1989). Of these, 287 ponds are public and 137 belong to temples. These ponds are distributed throughout the state, but a sizable number are concentrated in Palakkad district, which is drought-prone. Rainfall in this area is lower than in other parts of Kerala. Even the Thiruvananthapuram Corporation area can boast of as many as 80 ponds, some of which are still in use.

Most ponds and tanks in Kerala are small structures conforming to the region's more variegated topography. Nevertheless, they have helped the state to augment its water resources from the early days.

USE OF WATER RESOURCES

According to the PWD estimate, Kerala has a utilizable yield of 42,672 MCM. The major uses are for drinking and domestic purposes, irrigation, industry, and generation of hydel power. Water recreation is another important emerging sector associated mainly with tourism.

Demand and Supply of Water

Prior to the development of modern water infrastructure, people used to draw water from traditional sources like streams, springs, open wells, ponds, and tanks for daily use. There were designated water bodies to be used exclusively for drinking purposes. However, conditions have changed significantly over the years with population growth and socio-economic development. There is tremendous pressure on water resources. With the development of science and technology, society mastered the art of extracting/harvesting surface and groundwater through various means and distribute water through pipes. Due to prolonged use of water and indiscriminate discharge of wastes, the water quality of both surface and groundwater has deteriorated and traditional sources have been mostly destroyed. People in Kerala now turn to pipe water supply even in rural areas.

Based on the Government of India directives Kerala Water Authority (KWA) has developed certain norms to assess water requirements (Planning Commission of India, 2008a). For rural households without house connection, water requirement is estimated at 40 litres per capita per day (lpcd), and with house connection, it will be 70 lpcd. For an urban area, the water requirement has been pegged at 125 to 200 lpcd for communities with a population of more than 50,000. By 2017, KWA operated 1073 water supply schemes and had a total installed capacity of 3367.13 MLD (Government of Kerala, 2018). Per capita availability of piped water in Kerala is 104.2 litres/day. The bulk of KWA supply (77%) goes for domestic purposes followed by 15 per cent for the local body, 7 per cent for non-domestic use, and 1 per cent for industrial use (Government of Kerala, 2005). Meeting pipe water demand covering the entire population in Kerala will be a herculean task. An integrated strategy is required by combining all sources of water, local traditional sources, and modern technology. The wisdom of supplying pipe water to all households in Kerala may also be revisited.

Drinking Water

Traditionally, people used to depend on dug wells for meeting their needs for drinking and domestic water. In general, every household in the rural area has at least one well. Some of the urban households still maintain wells in their house compounds, in spite of getting tap water service. Kerala Water Authority has a total number of 18,12,528 water supply connections in the state, of which 93.61 per cent are domestic (Government of Kerala, 2018) The existing water supply schemes cater to the need of 181 lakh persons or 54.27 per cent of the total population in the state (Table 4.3).

Distribution of households having water supply schemes in per cent to total households in the districts indicates that coverage varies from 20.93 per cent in Kasaragod district to 82.63 per cent in Ernakulam district and there are north Kerala and south Kerala differences in the matter of coverage. All districts in north Kerala record less than 50 per cent coverage except Wayanad, which has coverage of 56 per cent. In south Kerala, all districts have attended >50 per cent coverage barring Idukki, which records only 30 per cent coverage. One of the striking features of water supply schemes in Kerala is that both rural and urban areas are equally served at the state level with coverage of 54.19 per cent and 54.36 per cent respectively. However, there are wide variations at the district level. The coverage of the urban population varies from 10 per cent in Kasaragod district to 86 per cent in Ernakulam district. There are eight districts, each of which has more than 50 per cent of urban households catered by water supply schemes. In the case of the rural population, coverage varies from 28 per cent in Kasaragod district to 75 per cent in Kollam district. There are nine districts showing coverage of the rural population by >50 per cent. In some districts, coverage of the urban population is less than that in rural areas (Table 4.3). The rural-urban difference in coverage ranges from 0.13 per cent point in Alappuzha to 36.75 per cent point in Kollam district. Kerala is trying to provide basic services to all sections of society, be it rural or urban. This is one of the notable characteristics of Kerala's development initiatives.

Around 46 per cent of people in Kerala depend on mainly wells and other sources including tube wells, bore wells, hand pumps, rivers, canals, and springs. The majority of wells remain uncovered. The use of spring water is common among people in Wayanad and Idukki districts. The Census data brought out that among the covered households, 77.7 per cent of households have drinking water sources within their premises, 14.1 per cent have near the premises and the rest 8.2 per cent have to travel some distance to fetch water. In the coastal area, around 70 per cent of dug wells are used for household purposes. The estimates show that 17 per cent of the villages in Kerala do not get any benefit of protected water supply and 69 per cent are partially covered. In partially covered areas, 14 per cent get less than 10 litres/day and 55 per cent get less than 20 litres/day. Grassroot level analysis in some places show even a more dismal picture than this. Some Panchayats in Palakkad district reported that 70 per cent of households do not have their own source of safe drinking water and 41 per cent of households have to walk more than 50 meters to collect water.

Kerala Rural Water Supply and Sanitation Agency (KRWSA)/Jalanidhi Scheme

The Kerala Rural Water Supply and Sanitation Agency (KRWSA), popularly known as *Jalanidhi* was launched and entrusted with the responsibility to facilitate the implementation of rural water supply systems as part of the World Bank-supported project. The first phase of this

Table 4.3: Population Covered by Water Supply Schemes by Districts in Kerala, 2017

District	Total population covered	Coverage as per cent of total population	Per cent of rural population covered	Per cent of urban population covered	Rural-Urban difference (per cent)
Thiruvananthapuram	2377000	71.90	66.60	76.40	9.80
Kollam	1548500	58.90	75.50	38.70	36.80
Pathanamthitta	644277	53.90	53.60	56.60	3.00
Alappuzha	1445070	68.10	68.20	68.00	0.20
Kottayam	1186304	59.90	64.20	49.30	14.90
Idukki	337663	30.50	28.90	63.50	34.60
Ernakulam	2710244	82.60	74.30	86.50	12.20
Thrissur	2215297	70.80	67.90	72.20	4.30
Palakkad	1391252	49.50	42.30	72.10	29.80
Malappuram	1467996	35.70	50.30	17.20	33.10
Kozhikode	1012273	31.90	36.20	29.70	6.50
Wayanad	457401	56.00	56.20	52.50	3.70
Kannur	1108474	43.90	40.60	45.70	5.10
Kasaragod	272643	20.90	27.80	10.10	17.70
Kerala	18174394	54.30	54.20	54.40	0.20

Source: Government of Kerala, 2018.

project also known as '*Jalanidhi*' was launched during the period of 2000-2008. During this phase, 3663 villages and 1.13 million people were covered. This was well above the target. By March 2017, 5072 Jalanidhi schemes were in operation under Phases I and II covering 1.49 million people. The highest number of schemes (929) was in the Malappuram district covering 3.16 lakh people. The assessment report of project performance prepared at the behest of the World Bank (IEG Public Sector Evaluation, 2013) considered Kerala's performance satisfactory, however, highlighted a couple of points which merited close scrutiny. These were: (i) quality of water—70 per cent of beneficiaries boil their drinking water as chlorination is not preferred due to bad taste, (ii) lack of technical and executive hands for maintenance, (iii) dearth of experienced NGOs, (iv) dependence on donor agency for maintenance cost, and (v) institutional problems related to repair of major damage to community schemes. The last one is noted as having a significant risk to development.

Efforts of 50 years of public intervention in creating drinking water facilities have not brought the desired results in many places (Nair and Chattopadhyay, 2005). In some Panchayats, drying up of wells during summer months is a common phenomenon that affects sizable households depending on well water for drinking purposes. The worst sufferers are poor and landless households, small landholders, Scheduled Castes, and Scheduled Tribes. It is the women who bear the brunt of the ordeal of fetching water from faraway places. Caste and gender-based inequities and economic inequalities are dominant in several Panchayats.

Irrigation Projects

Irrigation was a priority sector of intervention in the agricultural development of Kerala since independence like the rest of the country. The thrust on irrigation was aimed to stabilise agricultural production, particularly the 2nd crop. Owing to topographic conditions, a major network of canal irrigation was built up in Palakkad district and parts of Thrissur districts. Subsequently, other districts were also covered. Irrigation infrastructure is controlled through

18 dams, of which 13 have storage capacity and the rest five are barrages (Government of Kerala, 2010). For the comprehensive and systematic development of irrigation, an autonomous body—Command Area Development Authority (CADA) was formed in 1985 and it started operation in 1986.

Performance of Irrigation Projects

The viability of major irrigation projects in Kerala is being debated in recent years. It is well documented that the canal irrigation projects could not perform to the expected levels and the envisaged benefits were not realised. The undulating nature of the topography, alternate ridge-valley arrangement, and current change in cropping pattern have raised number of issues in the future expansion of canal irrigation projects. As the irrigation projects in Kerala are conceived and technically designed to irrigate mainly rice fields, the substantial reduction in area under rice and non-cultivation of rice have contributed to the underperformance of the irrigation projects. Field investigation in the Pamba Irrigation Project area indicated that irrigation canals created a waterlogging situation, wherever, it (canal) cut across the valleys. There are areas that have never received irrigation water although distributary canals were laid (Chattopadhyay, 2002). The performance of the irrigation project is related to terrain type, cropping pattern, organizational aspects, operational constraints, and finally policy issues.

Beneficiary participation in irrigation water management form an integral part of CADA projects, which are also considered second-generation projects in contrast to first-generation projects which had been concerned mainly with the construction of reservoirs and canal network. The formation of the beneficiary farmers' association (BFA) is skewed over time. Around 86 per cent of the BFAs were formed during 1986-87, possibly to utilize the allocated funds. The performance of the BFA is not up to the desired level. There are various underlying factors. The higher level of interest in augmenting irrigation facilities was evinced by the cultivators with large holdings. Initiatives to form BFA came from these large cultivators, CADA officials, and prospective office bearers. Proper management of irrigation water requires the active participation of the BFAs and the introduction of decentralized administration and functions of the irrigation officials.

Irrigation Status

Irrigation sources in Kerala comprised of government canals, private canals, tanks, wells, and other sources like rivers, springs, etc. They have been classified as major and minor irrigation projects. Traditionally, the classification of major and minor irrigation was based on the amount of investment. With the introduction of decentralized planning since the 9th Plan period all minor irrigation projects having a cultivable command area of more than 2000 ha were vested with the Panchayati Raj Institutions (PRIs). In recent years with the enactment of the Kerala Irrigation and Water Management Act, 2003, there was further reclassification. Irrigation works benefiting an area of less than 15 ha only are now vested with the PRI. All other works having a cultivable command area of more than 15 ha are brought under the Water Resources Department as medium irrigation projects.

Irrigation water management is a big challenge in Kerala. There are limitations to expand canal irrigation due to undulated topography and soil character. Dependency on groundwater is increasing thereby questions are often raised about the wisdom of investment in canal irrigation. Ponds, tanks, springs, and rivers provide significant prospects. Minor irrigation deserves more attention. In this context, watershed management and watershed-based irrigation planning are necessary. Discussion on the source and crop-wise irrigation has been provided in Chapter 6 on Primary Sector-I: Agriculture and Performance of Crops.

Industry and Recreation

Apart from domestic use, there are water requirements for non-domestic use and industries. Water connection for industrial use accounts for only 0.08 per cent of all connections installed by KWA (Government of Kerala, 2018). One of the estimates indicated that domestic/industrial use of water would increase from 200.35 MLD in 2001 to 300 MLD in 2051 (Government of India, 2008). Many of the industrial estates try to make their own provision for water supply. It is highly questionable, whether the water demand can actually be met by the government at all. The possible long-term solution for meeting most of the domestic and industrial water needs in the future is to draw water from all existing and future reservoirs.

The sprawling backwaters of Kerala invariably find an important place in any discourse on the tourism potential. The development activities in the field of tourism, especially those initiated by the government, such as the setting up of modern mooring facilities and the introduction of houseboats with modern amenities have only touched a part of the potential of backwaters. However, there are several guiding factors before actually implementing the recreational facilities in the backwaters or the seashore. Views of local inhabitants warrant due consideration while planning any project that intends to use water resources in which local people have a stake. Any new idea on new accommodation, transport, or entertainment should first necessarily consider the present status of these facilities as enjoyed by the residents of the area.Tourism is an emerging important industry in Kerala, and water sport or any other water adventure facilities would add to the state revenue and increase per capita income. Water-based tourism has also been developed in some rivers.

Hydel Power Projects

Kerala has a long tradition of using hydel power since 1940 when the first hydel power project was installed in Munnar. By 2017, hydropower accounted for 71 per cent of the total installed capacity in Kerala. This is in sharp contrast to India's dependence on hydel power for 14 per cent of the demand. High rainfall, an abrupt rise in slope along the Western Ghats scarp, and high altitude plateau formation have provided Kerala certain unique opportunities to impound reservoirs and set up hydel power projects taking advantage of natural head due to high altitudinal difference in short distances. However, on account of overdependence on hydel power, the power sector in Kerala is vulnerable to rainfall variations. Any fluctuation in rainfall directly impacts power generation. The drought in the year 1982-83 witnessed an unprecedented power shortage in Kerala. Till that year, Kerala was considered as a power surplus state. Kerala faces a power shortage due to frequent drought years coupled with inadequate installed capacity. Dependence on hydel power makes Kerala's energy scenario green, however, there are geographic limitations to continue the installation of hydel power projects, particularly large projects. Detailed discussion on hydel power projects in Kerala is provided in Section on Energy Resources under Chapter 8 covering 'Industrial Development'.

CHALLENGES OF WATER MANAGEMENT

The Global Water System (GWS) Project raised three thematic questions: (i) the magnitude of anthropogenic and environmental changes in the global water system and the key mechanism by which they were induced, (ii) the main linkages and feedbacks within the earth system arising out of changing GWS, and (iii) resilience and adaptability of GWS to the change and strategy for sustainable development (Vorosmarty *et al.*, 2004). The challenges for water management are to understand the nature of human activities and the natural forces that generate pressures influencing the water system. Many of these are highly dynamic; they change at a faster pace influencing water management strategies and policies (UNEP, undated).

Increasing population pressure and higher consumption is a major challenge of water management in Kerala. Due to the high population density per capita availability of water in Kerala was 12,500 lpcd (litre/capita/day) by 2011. This is decreasing with the growth of the population. Besides, some parts of the state already face scarcity of drinking water in certain periods of the year.

The steep slope and undulated topography characterise the provenance of all rivers in Kerala. The rivers are swift-flowing and the residence time of water is short. There is not much scope to construct any more reservoirs. Existing reservoirs are silted up to various measures. Deforestation, diversion of land from agricultural to non-agricultural use, shrinkage of wetlands and flood cushioning areas, increased surfaced/paved area due to expansion of the built-up area, and growing urbanization all in combination contributes to stronger surface run-off and lower infiltration and thereby constraining groundwater recharge. The depths of rivers have increased due to indiscriminate sand mining from the river bed. This has lowered the water table in the river. The river water level controls the water table in the adjoining dug wells, which are the main sources of domestic water in rural areas. Along with the fluctuation of the water level in the rivers, the level of the groundwater table also fluctuates. Maintenance of the base flow in the rivers during non-monsoon months is, therefore, very significant for water security. Rainfall is the major recharging source of groundwater, and the rise in water level in the phreatic aquifer is directly related to precipitation. A falling trend of the water level in the post-monsoon months indicates that rainfall is not sufficient to recharge the aquifer to required level. Climate change can complicate this issue further.

Another important matter to consider is deteriorating water quality and pollution of water sources due to anthropogenic interventions. Incoming water is more or less constant therefore quality degradation will reduce availability and impose further pressure on the already constrained resource base. It may be noted that good quality water and its availability have contributed to improved human health conditions and high quality of life in Kerala. If the water resource deteriorates and quality problem restricts easy availability, the very foundation of the Kerala mode of development will be in jeopardy. Besides, there is impending challenge on account of climate change. Therefore, the management of water resources assumes significance for sustaining and strengthening Kerala's development.

Sustainability of Water Sources

Demand for water resources is increasing and at the same time, water availability is declining due to deteriorated water quality of surface and sub-surface sources and also drying up of traditional water structures like ponds and springs. Availability and demand calculations of CGWB indicated that Kerala faces a shortage of 7.10 bcm during the summer season. As the water retention capacity of the river catchments is coming down due to change in natural vegetation cover, removal of soil, excavation of hills, and spread of surfaced area, the summer water availability is decreasing. In-stream interventions and removal of sands and clay from flood plains also contribute to the declining water holding capacity of the river and flood plain systems. Springs, catering to the drinking water need of more than one lakh families in Kerala are facing problems of drying up. It is reported from Wayanad and Idukki districts that springs are drying up in many places. The drinking water situation is grim in both these districts, particularly in the rugged hilly areas. One of the major challenges of water management in Kerala in the coming years will be the sustainability of water sources.

Improvement and Maintenance of Water Quality

Deterioration of water quality, both of surface water and groundwater is a major issue. The Kerala State Committee on Science Technology and Environment (KSCSTE) report indicated that water of almost all of 44 rivers in Kerala is being increasingly polluted by industrial and domestic waste, pesticides, and fertilisers used in agriculture. Kerala State Pollution Control Board (KSPCB) classified all rivers according to best use. It is found that hardly there is any stretch of river in Kerala from where water can be directly used. Improving the water quality of both surface and groundwater is a huge task as it is also linked with solid and liquid waste management. Change in water quality has been dealt with in Chapter 15 on 'Changing Geography of Kerala'.

Operation and Maintenance

The operation and maintenance of the drinking water supply structure is another serious challenge. Treatment of drinking water supplies to maintain minimum residual chlorine of 0.5 mg/l is necessary. Experimental studies indicated that the quality of drinking water supplies in Kerala clearly indicates a high level of bacterial contamination. This poses a serious risk to public health which is evident from the high incidence of acute diarrheal diseases and other water-borne infections among the people especially the poor sections of the community. Continuous disinfection of drinking water and maintenance of the potable quality are necessary prerequisites. It is also reported that there is a high transmission loss of drinking water due to leaking pipe-lines. As sewerage lines run underground more or less parallel to the drinking water supply lines, particularly in urban areas, there are possibilities for contamination. Both source and supply lines require proper maintenance and operation.

Water Governance

The Public Health Engineering Department (PHED) of the Government of Kerala was the controlling authority of public water supply in the Urban sectors untill 1984. Kerala Water Authority (KWA) was formed in April 1984 as an autonomous body entrusted with the job of planning, implementation and management, revenue collection, and maintenance of the water supply systems. Over the years public water supply was introduced in the rural areas also. The KWA emerged as the largest institution in the field of water supply and implemented thousands of small, medium, and large urban and rural piped water supply schemes. However, as the demand for piped water supply in the rural areas grew, KWA was found wanting to manage the water supply schemes to the satisfaction of users. There was increasing demand for a decentralised water governance system. This demand was further strengthened following the People's Plan Campaign introduced in Kerala in 1996. At the same time, the new policy initiatives of the Government of India following globalization tried to promote a demand based, community managed decentralised water supply run jointly by the government, civil society, and people/beneficiary. The *Jalanidhi* programme is part of this initiative. In spite of the successful implementation of local self-governments in Kerala, water management is still centralized and department centric.

NEED FOR CHANGE IN WATER MANAGEMENT

Water is generally considered merely as a resource to be withdrawn and utilized as desired. The need for changing this perception is globally recognized; however, the alternative approach is yet to emerge to the satisfaction of all stakeholders. Water is a decentralized resource and its management also requires a decentralised approach. Adjustment and reorientation of existing practices, as advocated by many quarters, cannot possibly yield the desired results as the commodification of water, management decisions on that premise, and mechanical interventions, so far followed, have taken the society to a crisis-like situation. With present-day technology, it is hardly possible to intervene in the domain of incoming rain; therefore, the entire emphasis will have to be on the management of water. To meet this emerging challenge some fundamental changes are necessary for our approach to water resource management. Water is an essential component of the environmental resources and its management cannot be divorced from total management of land and biomass. Micro watersheds based approach within the framework of river basins is important. In fact, drainage basin security is now considered as the foundation for planning water security. Demand, supply, availability, and distribution are all part of this integrated management paradigm. Understanding this reality forms the base for future action.

CHAPTER

5

Land Resources, Land Use and Landholdings

DEFINITION OF LAND

The land is the uppermost part of the lithosphere that performs several ecologic, economic, and social functions to sustain life on earth. All human activities take place on land, which is considered as the prime resource. Access to land is a basic means of livelihood. By definition, the land is a delineable area of earth's terrestrial surface and refers to a complex of surface and near surface attributes significant to human (FAO, 1995). The formation of land depends upon the interplay of endogenic and exogenic processes. Climate, geology, geomorphology, surface and subsurface hydrology, topography, vegetation, and soil characterise the land. Demand for land is escalating with a growing population, development initiatives, and increasing human activities. A land unit is integrated in nature and distinctly differs from its surroundings. Each land unit has a production potential and multiple uses. Often, these uses are determined through economic and political considerations with little attention to ecosystem services and sustainability issues resulting in land degradation and loss of this capital resource.

LAND RESOURCES

Kerala is rich in land resources, which supports a wide range of crops from subsistence crops like paddy, tapioca, oilseed like coconut to plantations, and high-value spices crop of cardamom. Kerala is also one of the states in India where access to land resources has been ensured for a large section of people through land reforms. Arguably, land resource base and access to land resources have contributed to Kerala's high human development and poverty alleviation (Chattopadhyay and Mahamaya, 2020). The land is intensely used with multi-tiered crops, and it is well suitable for high-value cash crops. The average agricultural income from one hectare of land in Kerala is around two times higher than that of the all-India average.

Kerala falls in a single agro-climatic region of the West Coast Plain and Ghat Region in all-India classification. The committee on Agroclimatic zones and cropping pattern, constituted by the Government of Kerala at the behest of Planning Commission, Government of India, identified 13 agro-climatic zones based on four principal parameters, namely, altitude, rainfall, soil type, and topography (Table 5.1). Land utilization has been worked out according to these zones.

Table 5.1: Agro-Climatic Zones in Kerala

Name of Zone	Area in per cent	Soil Type	Principle crops
Highland	27.4	Laterite without B-horizon	Rubber, coconut, cashew-nut, arecanut
High ranges	15.9	Red loam	Tea, coffee, cardamom
North midland (Malappuram type)	13.1	Laterite	Coconut, rice, cashew, arecanut, rubber
North midland	11.6	Laterite	Coconut, rice, cashew-nut, rubber, arecanut
Southern midland	10.0	Laterite with B-horizon	Rubber, coconut, rice, tapioca, arecanut
Central midland	8.2	Laterite	Rubber, rice, coconut, tapioca, arecanut, banana.
Coastal Sandy	4.8	Sandy loam	Coconut, rice,
Palakkad	3.9	Red loam	Rice, cotton, groundnut, millet
Chittoor Block	1.6	Black soil	Rice, sugarcane, cotton, groundnut, millet
Onattukara	1.6	Sandy loam	Rice, coconut, tapioca, arecanut
Red loam	1.0	Red loam	Coconut, tapioca, rubber, rice
Kuttanad	0.9	Peat	Rice, coconut
Riverbank alluvium	-	Alluvium	Rice, coconut, sugarcane, arecanut

Source: Report of the Committee on Agro-climatic Zones and Cropping Patterns, Department of Agriculture, Government of Kerala (1974).

Note: Total area covers 32,382 sq. km. The rest is under forest and not considered for this zoning.

The Indian Council for Agricultural Research (ICAR) has divided Kerala into five agro-climatic zones like northern zone, central zone, high altitude zone, problem area zone, and southern zone for conducting agro-climatic zone-specific research (ICAR, 1991). Chattopadhyay (1985) identified 24 landscape ecological zones under three primary zones of *Coastarea*, *Cultivarea*, and *Naturarea*. Following the USDA scheme of three-tier land resource categories, like, Land Resource Regions (LRR), Major Land Resource Areas (MLRAs) and Land Resource Units (LRUs), the Soil Survey Department, Government of Kerala has identified 28 major land resource areas under 10 geographic units (Government of Kerala, 2007). MLRAs are identified based on physiography, geology, climate, water, soil, biological resources, and land use practices. Major Land Resource Areas are geographically associated land resource units. Identification of MLRAs is important for state-level agricultural planning and also for regional and national planning.

TERRAIN UNITS/LAND RESOURCE UNITS

The importance of incorporating process data in land resource management has been increasingly stressed during the last couple of decades (Chattopadhyay and Mahamaya, 1995). The topographic grain of Kerala is longitudinal. Three major longitudinal terrain provinces in Kerala are: (i) Coastal areas and associated landscape mostly confined to <20 m dominated by marine/ coastal and estuarine processes, (ii) the dissected pediplain mostly lateritic including foot slopes formed by interacting fluvial, denudational, and chemical weathering processes, and (iii) the Western Ghats characterized by scarp slopes, plateaus, and mountains sculptured by the combined action of scarp retreat, slope wash, mass wasting, and other denudational processes. Considering process-response functions 22 terrain units have been identified under five main morphogenetic processes namely marine/coastal, estuarine, fluvial, denudational cum accumulational, and denudational (*ibid*). There are specific agents like erosion and accretion operating under each of these main processes. A land unit under marine processes can be either erosional or accretional or a combination of both. A hierarchical classification system was followed to identify these 22 terrain units and characteristic features have been worked out. It is found that around 85 per cent of the total area of Kerala falls under the denudational category. The rest 15 per cent of the land has developed under marine, coastal, estuarine and fluvial processes. These terrain units manifest macro-scale variability of land resources.

MICRO LEVEL DIVERSITY OF LAND TYPES

The diversity of the land is also evident even at the micro-level. In coastal area, ridge runnel topography runs parallel to the coastline. Sandy ridges accommodate settlements, roads, and tree crops, and the runnels are given for seasonal agriculture—mostly rice and vegetables. The riverine alluvial tract adjoining the sandy coastal plain to the east shows large paddy fields. They are quite extensive in the Kuttanad area of Alappuzha and Kottayam, Kole lands of Ernakulam and Thrissur, and Bharatapuzha valley of Thrissur, Malappuram, and Palakkad.

Towards the east, the main valleys are in east-west direction indented within lateritic terrain and extending up to foothills. Alluvial valleys/flood plains gradually narrow down towards upstream. All these arrangements are like ribbon patterns and recurring (Fig. 5.1). Valleys are continuously enriched with sediment, other mineral nutrients, water, and organic matter moving down from surrounding elevated areas through colluviation and alluviation. This diversity of land even at the micro-level is significant. It has facilitated diverse crop practices, enhanced productivity, and contributed to livelihood security. Interdependence between valleys and ridges are well manifested in these spatial arrangements.

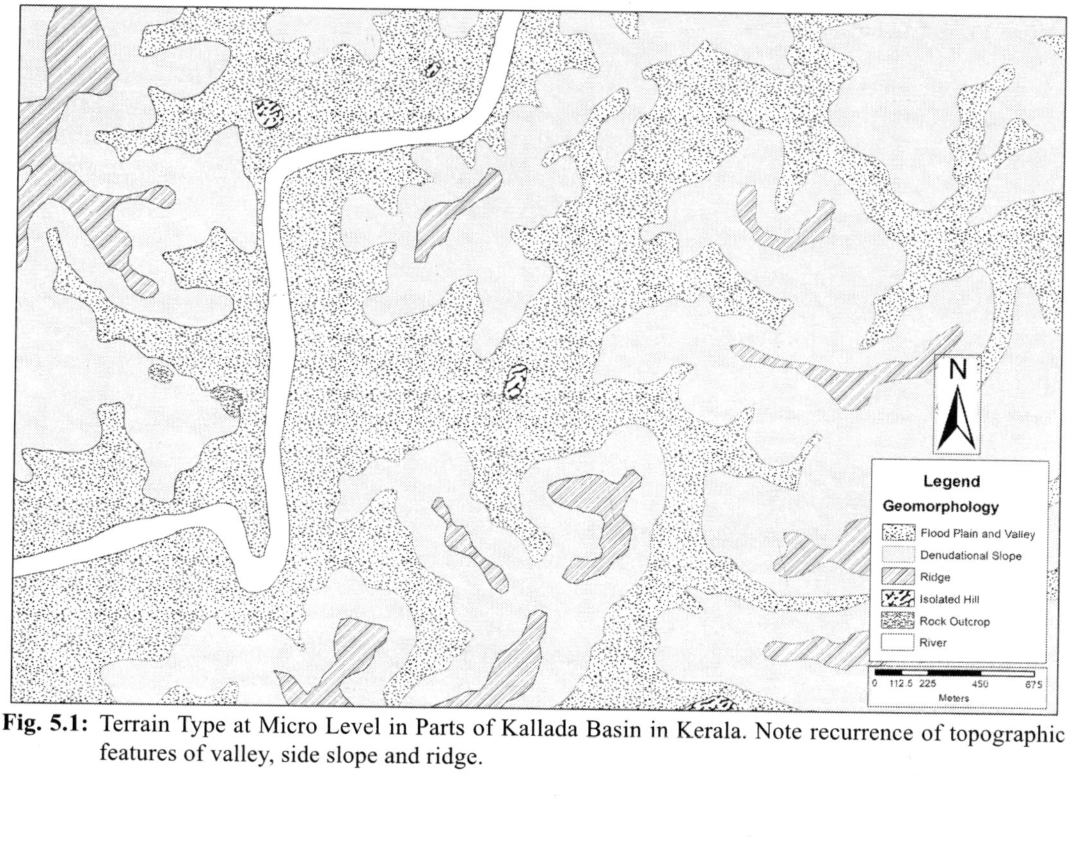

Fig. 5.1: Terrain Type at Micro Level in Parts of Kallada Basin in Kerala. Note recurrence of topographic features of valley, side slope and ridge.

LAND USE

Land in Kerala is intensely used. Settlements are distributed in all types of land use categories barring deep forests. Hardly, there is any vacant or unused land in the state. The Land use pattern of Kerala is quite different from the rest of India. Land use statistics have been published for the state from the period beginning 1908-09. At that time, different parts of present-day Kerala had different administrative arrangements with some parts under Princely Rulers and some other parts under the Madras Presidency of the British rule. Therefore, there are certain problems with the compatibility of data. To overcome this problem only three broad categories—forest, non-forest, and agriculture have been identified during that period. In those early years, the land survey had not covered the entire state of Kerala. Available data for 1908-09 showed that the total area reported was 38,580 km^2, of which agriculture accounted for 53.5 per cent. The rest 46.5 per cent comprised of un-surveyed area (mostly forest) (25.6%), forest area (6.5%), and non-agricultural area (14.4%). In 1956-57, when modern Kerala formed, the area under agriculture covered 61.2 per cent of the total geographical area of the state. Most parts of Kerala were surveyed by that time and the area under agriculture increased by 7.7 per cent. By 2016-17, the area under agriculture including fallow lands accounted for little over 55 per cent of total area of the state. Agriculture occupying more than 50 per cent of the total area since the first decade of the 20th century indicates the extent of pressure on land experienced in Kerala even more than 100 years back. High pressure on land is further evident from the fact that the population density was 165 persons/km^2 in 1901 which increased to 184 persons/km^2 in 1911. The corresponding figures for India were 77 and 82 respectively.

Nine Fold Classifications

A nine-fold land use classification for Kerala is available since 1957. The Department of Economics and Statistics, Government of Kerala, continues to publish land use data following this standard classification scheme till 2004-05. Since 2005-06, four more categories, namely, marshy land, still water, waterlogged area, and social forestry were added to the existing nine-fold classification, thereby increasing the number of classes to 13 (Table 5.2). A brief description of all these land use classes is provided here.

Forest: Before the reorganization of the state in 1956, the forests in Kerala were owned both by the government and individuals. Once the state was formed, all forest lands were vested with the government and were brought under the administrative control of the Department of Forests. In 1956-57, the forest area was 10,180 km^2 and constituted 26.39 per cent of the total geographical area of Kerala. There were some readjustments in the surveyed area over the years. Forest area was 10,640 km^2 in 1960-61 and since 1980-81 to date (2016-17) the reported area under forest was 10,815 km^2 or 27.83 per cent of the total geographical area of Kerala. It may be noted here that the area under forest is actually the area under the administrative control of the forest department and not the area under natural vegetation cover. Besides natural vegetation, there are forest plantations, agriculture, and other land uses practiced within the forest administrative area. Natural vegetation is mostly concentrated in the rugged Western Ghats slopes.

Littoral forests are limited to few patches along the coastline and estuaries. There are six districts in Kerala with more than 30 per cent of their geographical area under the control of the forest department. One district among those six districts, namely, Pathanamthitta has as much as 59 per cent of its total geographical area under the control of forest department. Alappuzha district confined to coastal plain and low lands does not have any appreciable natural forests.

Table 5.2: Distribution of Area under Different Land Use Classes in Kerala (1960-61, 2016-17)

Land use Class	*1960-61*		*2016-17*		*Change*
	Area (in hectares)	*per cent*	*Area (in hectares)*	*per cent*	*per cent*
Forest	1,064,000*	27.4	1,081,509*	27.83	+ 1.64
Land put to non-agricultural use	206,000	5.3	441,934	11.37	+114.50
Barren and uncultivable land	155,000	4.0	11,780	0.30	– 92.40
Permanent pastures and grazing land	47,000	1.2	0	-	–100.00
Land under miscellaneous tree crops	206,000	5.3	2,450	0.06	–98.8
Cultivable waste	144,000	3.7	1,01,379	2.61	–29.60
Current fallow	67,000	1.7	72,008	1.85	+13.80
Fallow other than current fallow	62,000	1.6	55,530	1.43	–10.40
Marshy land$			106		
Stillwater$			98,343	2.53	
Waterlogged area$			3,210	0.08	
Social forestry$			2,556	0.07	
Net area sown	1,934,000	49.8	20,15,482	51.86	+4.21
State Total	3,885,000	100.00	3,886,287#	100.00	

* There was some area readjustment in the forest over the years.

Area reported as per village records.

$ Area under these categories enumerated since 2005-06.

Note: 100 hectares = 1 square kilometer. Figures in various tables are in one or the other of these measurement units.

Source: Government of Kerala, 2017a.

Land put to non-agricultural use: This category includes buildings and courtyards, parks, play grounds, industry, roads, railways, and other non-agricultural uses like brick kilns, sites of clay, and sand mining in the flood plains and water bodies used for pisciculture. Agricultural statistics of 2016-17 recorded 4420 km^2 or 11.5 per cent of total geographical area of Kerala under this category. There is a steady increase in area over the years. A positive correlation exists between population density and this category of land put to non-agricultural uses. Analysis of time series data indicates that the area under this class increased by 34 per cent between 1960-61 and 1970-71. However, from 1970-71 to 1991-92, the rate of increase fell to 9.5 per cent, and again, it rose to 42.9 per cent between 1991-92 and 2004-05. The later period (1991-92 to 2004-05) witnessed a rapid expansion of construction activities, specially buildings and roads, and clay mining from flood plains. In addition to these, aquaculture is also spreading over. Since the year 2005-06, when water-related uses were separately accounted for, and this land use category of land put to non-agricultural uses represented more or less only the built-up area, it was found that urbanized and industrial tracts like Ernakulam-Thrissur-Palakkad and Thiruvananthapuram and surroundings were experiencing higher growth rates under this category. This trend continues.

Barren and uncultivable land: This category of land use includes rocky and sandy wastelands, marginal lands, and degraded areas. Kerala had 118 km^2 or 0.30 per cent of the total area under this category by 2016-17. Around 40 per cent of total barren and uncultivable land in Kerala is confined to the Kasaragod district. These barren lands are mostly hard crust laterite exposures. The area coverage of this category declined from 4.02 per cent in 1960-61 to 1.42 per cent in 1991-92 and 0.30 per cent in 2016-17. Idukki experienced a maximum reduction. Most of the usable and semi-usable lands have been brought under some uses. Hard rocks are used as a foundation for building construction in Kerala, therefore, there are rock quarries distributed throughout Kerala. The environmental impact of rock quarrying is a serious issue drawing the attention of the concerned people from time to time. The barren hard crust lateritic wastelands in north Kerala are utilized for building purposes. Sandy patches are used for settlement and industry. Some of the barren areas are given for social forestry. The reduction of area in this category indicates increasing utilization of marginal lands.

Permanent pastures and grazing land: There was no land reported as permanent pasture and grazing land in Kerala during 2016-17. However, in 1960-61, Kerala had 470 km^2 or 1.17 per cent of the total geographical area under this category. The land is so valuable and productive that all pasture lands have been brought under some productive use. The reduction of area under this category indicates the loss of village commons, which used to support livestocks and supplied fuel-wood to the marginal sections of society.

Land under miscellaneous tree crops: This land use class covered only 24.5 km^2 of the area in Kerala by 2016-17. Palakkad is the only district that has a 10 km^2 area under this category. State-level area coverage of this category has come down from 2055 km^2 in 1960-61 to the present level. The average annual reduction is around 36.25 km^2. Between 1960-61 and 1991-92, the annual rate of decrease was around 56 km^2, and the period from 1991-92 to 2016-17 witnessed a reduction rate of 19 km^2/year. This category falling under drylands mostly covers garden lands. Varieties of tree crops were formerly grown in dryland areas. The scenario has changed with the expansion of agriculture and most of the lands under this category were transformed mainly for raising cash crops. The decline in this category of land indirectly indicates that village-level biodiversity is being lost. The range of crops raised in a village is now reduced. Loss of this type of land also reduces the chance of free firewood availability to the poorer sections of society. Idukki district, which once had around 113 km^2 area under this category, is now left with only 2.3 km^2 area.

Cultivable waste: This category includes the type of land that could have been brought to cultivation easily; however, it has now been left as wasteland. It accounted for 2.6 per cent of the total geographical area of Kerala in 2016-17. The share of this category of land was 3.73 per cent in 1960-61. There was an overall reduction of 29.6 per cent by 2016-17. However, time-series data indicated that the trend was fluctuating. The declining trend continued till 2005-06 and subsequently, it started increasing. There was a 53 per cent increase in cultivable wasteland from 2005-06 to 2016-17. As many as nine districts experienced an increase in area under cultivable waste during this period. Substantial increase in the cultivable wasteland in the districts of Alappuzha, Pathanamthitta, Kottayam, Ernakulam, and Thrissur is a matter of serious concern as most of the uncultivated paddy fields in these districts constitute the bulk of this class of land. It would be expected that with an increase in population pressure and demand for land, the areas of cultivable waste would be substantially reduced. This was not the case, however. One reason might be that the land is only cultivable for a certain type of crop that the cultivators do not favour. It is also possible that the size of the holding is too small for profitable operation. In the case of highly urbanized districts like Ernakulam, Kannur, etc., cultivable land might have been acquired/proposed to be acquired for some other use, and for the intermittent period, it was left as wasteland.

Current fallow: This category includes seedbeds or fields which could not be put to cultivation for reasons related to agriculture. Over the seasons and years, the spatial distribution of fallow land varies. The same lands are not kept as fallows year after year. This land use class accounted for 1.85 per cent of the total area of Kerala in 2016-17. Ernakulam, Thrissur, and Palakkad districts together recorded >50 per cent of all current fallows in the state. There was a substantial reduction in area under this category from 670 km^2 in 1960-61 to 230 km^2 in 1970-71. This was a period of agricultural expansion, the introduction of high yielding varieties, and modernization of agricultural practices. The maximum area under paddy cultivation was reported in 1976-77 when the Government of Kerala considered a proposal to make Kerala self-sufficient in food grain production. Subsequent years witnessed a gradual increase in area under current fallow. By 1991-92 area under current fallow rose to 439 km^2 and in 2010-11 it further went up to 760 km^2 and in 2016-17 it stood at 720 km^2. However compared to 2006-07, when the current fallow was 816 km^2, there was a reduction of 96 km^2 by 2016-17. While the area under current fallow increased in every district except Wayanad and Kannur, the rate of increase was different. An increase in area under current fallow indicates that cultivable lands are not fully utilized. Most of the current fallow lands are uncultivated paddy fields. The recent reduction in current fellow is due to strong initiatives undertaken by the Government to promote paddy cultivation.

Fallow other than current fallow: This category includes lands that are kept fallow for a longer period. These lands are suitable for agriculture but are not put to use for various reasons that may or may not be directly related to agriculture. A little over 555 km^2 of the area has been estimated under this category in 2016-17. The concentration of this category of land is mostly in the rice-growing areas spread over Alappuzha, Ernakulam, Thrissur, Palakkad, and Malappuram districts. Both categories of fallow lands are common features in rice growing districts. There are various socio-economic reasons underpinning this scenario. The time series trend indicates a reduction of area under this land use class from 1.65 per cent in 1960-61 to 0.6 per cent in 1970-71. After 1970-71, the rate levelled off at 0.7 per cent till the year 1992-93. Subsequently, the area under this category started rising, and by 2016-17, it accounted for 1.43 per cent of the total area of Kerala.

Net sown area: Kerala had 20,154 km^2 area or 52 per cent of the total geographical area estimated as net sown area (NSA) in 2016-17 against all-India average of 41 per cent. NSA includes wetland agriculture, dryland agriculture, tree crops including coconut, and agricultural plantations. There is very little scope for further expansion of NSA in most of the districts. Water bodies restrict the expansion of the net sown area in the coastal plain and there are forests and rugged terrains of the Western Ghats to limit the expansion eastward. In 1960-61, the net sown area in Kerala was 50 per cent, which rose to 58 per cent in 1991-92 and subsequently declined to 52 per cent in 2016-17. It varies from 30 per cent in Pathanamthitta district to 73 per cent in Kottayam district (Fig. 5.2).

Initial growth in NSA had been at the expense of almost every other category of land. Expansion of rubber cultivation has contributed significantly to the expansion of NSA and the decline of land under other categories. The subsequent reduction since 1991-92 is a post-globalisation/economic liberalisation phenomenon associated with the growth of the real estate business. An increase in area under non-agricultural use and leaving cultivable land fallow for a longer period are two contributing factors of decreasing NSA. Compared to 1991-92, all districts, except Idukki district, recorded a decline in NSA by 2016-17. However, the rate of reduction varies spatially. Idukki shows a more than 9 per cent increase under this category, which is at the expense of all other categories, particularly barren and uncultivable land, cultivable waste, and land under miscellaneous tree crops. Bringing land from cultivable waste to NSA is a positive trend. However, the use of barren and uncultivable land is tantamount to expand agriculture to marginal lands, which may not be productive in the future. Similarly, reduction in area under miscellaneous tree crops also indicates loss of biodiversity in non-forest areas.

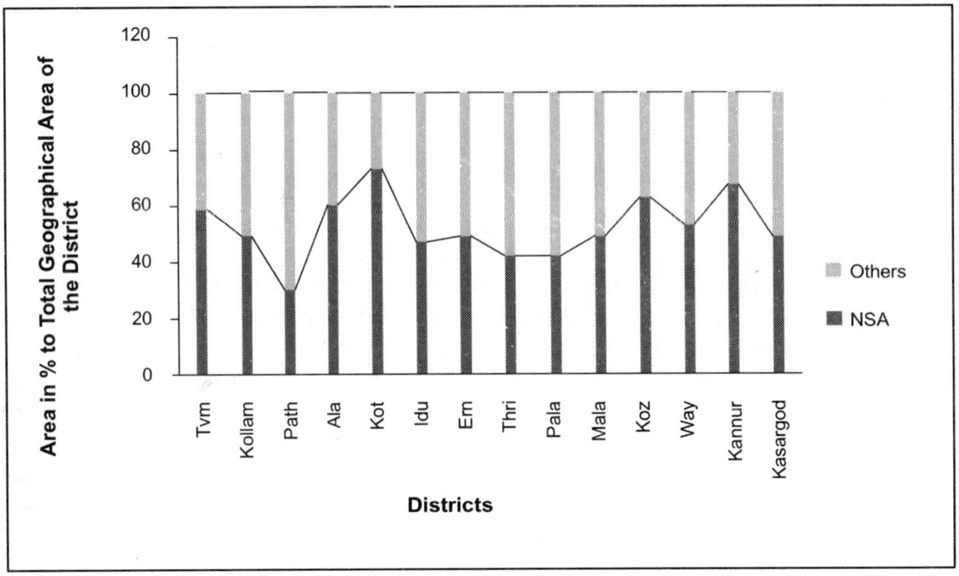

Fig. 5.2: Distribution of Net Sown Area by Districts in Kerala (2016-17).

Data Source: Government of Kerala, 2017a.

Additional Land Use Categories

Since 2005-06, four additional categories have been enumerated (Government of Kerala, 2006a). These are marshy land, still water, waterlogged area, and social forestry. All these categories were earlier counted under the class 'Land put to Non-Agricultural use'. Among these four categories only Still Water covered 983 km^2 by 2016-17. The category of still water includes lagoons, lakes, reservoirs, and similar other large water bodies. The Vembanad lake is the main still water body in Alappuzha, Kottayam, and Ernakulam districts, and in Idukki district, there are a series of impounded reservoirs like Idukki, Pallivasal, Sengulam, and Periyar lake forming still water bodies. The waterlogged area accounted for 32 km^2 followed by Social forestry (25.5 km^2) and Marshy land (1.0 km^2). The waterlogged area covers permanently waterlogged areas, both natural water bodies like ox-bow lakes, left out channels, and shallow water bodies, and also man-made waterlogged areas mainly referring to the abandoned quarries caused by sand, clay, and rock mining. These areas may be productively used through better management. Social forestry is a major organized social intervention to increase natural vegetation cover as part of environmental management. Waste lands, vacant lands along roads and railways, and degraded forest areas are usually given for social forestry. It is expected that once these vegetations are well established they will be included under forest cover.

LANDUSE AND LANDFORM

The relation between land use and landform is quite explicit in Kerala. Nature of land use varies with altitude. The land arrangement, often noticed in lower elevations of Kerala consists of an alluvial plains/valleys mostly given for rice cultivation bordered by banana and areca nut, surrounded by lateritic slopes with tapioca, coconut, mixed tree crops, and settlements which finally merge with lateritic ridge/mound with plantations of rubber, tree crops, settlements, and roads. The high hills are placed further up and dominated by forests and plantations (Fig. 5.3). The coastal plain and the adjoining low lands are dominated by coconut and rice. This segment of land is densely populated. All major urban centres are located in the coastal plain. The main transport lines, railways, and roadways have developed in the coastal plain and adjoining lowlands. The undulated lateritic terrain hosts crops like coconut, varieties of fruit trees, rubber, cashew, tapioca, pepper, etc. Coconut dominates the land up to 100 m altitudes. Rubber plantation is an important component of land use up to 500 m altitudes. Settlements are scattered amidst mixed tree crops. Further up, between 600 to 900 m altitudes, there are tea and coffee plantations. Cardamom is raised above 1000 m altitude. The rugged scarp slopes and high hills are forest-clad. The boundaries between land use categories are not razor sharp. There are overlaps; nevertheless, a broad zonation is well evident. Due to pressure on land and multiple usability of land, crop husbandry continues with little respect to landscape ecology or land suitability. This impedes the proper use of land resources.

Fig. 5.3: A Panoramic View of Land Use across Different Landforms in Central Kerala. Photo by Mahamaya Chattopadhyay.

ARABLE LAND: AVAILABILITY AND DEMAND

Kerala is a land-scarce state with per capita availability of arable land standing at 0.10 ha against all-India average of 0.11 ha by 2017. The average per capita requirement of arable land is 0.4 ha for sustaining a family as suggested by Mukherjee (1938) and recently adopted by Radha Devi and Ajit Kumar (2011). The field survey also corroborates this assumption. With technological improvement, use of high yielding varieties of seeds, fertilizer, and water management, and by having the provision of multi-tier cropping, there may be contest about the requirement of the amount of per capita arable land for family sustenance. However, given the improved standard of living conditions and higher demand for non-food items, the area requirement of 0.4 ha is reasonable. Even in the case of North America and Western Europe, where food basket contains a large proportion of meat an area of 0.5 ha of arable land is proposed as the minimum requirement (Myers, 1999). "It is realistic to suppose that the absolute minimum of arable land to support one person is mere 0.07 ha, and this assumes a largely vegetarian diet, no land degradation or water shortage, virtually no post harvest waste and farmers who know precisely when and how to plant, fertilise, irrigate etc." (Myers, 1999; FAO, 1995). However, this refers to practicing precision land management, which is an ideal condition. A society may strive to attain this level of land and water management in due course. Considering per capita land requirement (0.40 ha) and availability (0.10 ha), it may be surmised that Kerala experiences four times pressure on arable land, and therefore, she has to depend on external sources for sustenance of her inhabitants.

ACCESS TO LAND

Access to land is an important factor for sustainable land management. Like the rest of the country, Kerala inherited a feudal system of land tenure. The Government of Kerala initiated the process of land reforms immediately after the formation of the state in 1956, and finally in the 1970s, land reform was introduced. It provided access to land resources to a large section of people, who were involved in land management but without any ownership rights. Through a ceiling provision and a transfer of ownership right, 1.29 million tenants got title deeds to 1.97 million acres of rice land, about 1.5 acres, or 0.6 ha each. Another 340,000 households, each received house compound land of about 0.08 acres or 0.03 ha (Franke, 1996). Land reforms have accorded ownership rights to people, who were not traditionally landowners. This has altered the land-based power structure, hitherto prevailing in Kerala. However, land reform is yet to be fully completed. There are issues like exclusion of plantations from the ambit of land reforms and also non-implementation of the land ceiling Act. Besides, the task to accord entitlement to the underprivileged and tribal groups remained incomplete, which is causing social tension as evident from tribal movements from time to time (Scaria, 2010). The Government of Kerala launched Zero Landless Kerala programme. Kannur district was declared as the first district in India with Zero Landless Citizen in 2013 (Government of Kerala, 2013a). Providing access to land for landless people and marginal groups is a challenge.

OPERATIONAL HOLDINGS

The size of operational holdings plays an important role in agricultural productivity. Land management turns difficult when the size of the operational holding is too small. Small size farms are increasing in Kerala due to a variety of reasons like increasing population density and concomitant increase in number of households, breaking down of joint families, implementation of land reforms and distribution of excess lands. Growing urbanization and increased investment in real estate also play their role in the fragmentation of landholdings. This may lead to the under-utilization of land resources, which is quite limited in Kerala given the high population density and intense use of land. Distribution of operational holdings bears close relation with physiography. The size of holdings increases towards highlands.

Growing Fragmentation of Holdings

The average size of landholdings in Kerala is small and it is getting further reduced over the years. Average size of individual and joint holdings decreased from 0.37 ha in 1990-91 to 0.22 ha in 2010-11 (Government of Kerala, 2013b). The fragmented nature of landholdings in Kerala is well manifested in size class distribution. A number of operational holdings under different size class and their area coverage decrease with an increase in the size of holdings. Among the five size classes that the Department of Economics and Statistics recognised for presenting data on operational holdings, the marginal group (up to 1.0 ha) accounted for 96.33 per cent of all operational holdings and 58.62 per cent of total area coverage in 2010-11. If the small group (1 to 2 ha) is added to this class, the number of operational holdings in Kerala will rise to 98.9 per cent and the area coverage will be little over 77 per cent (Table 5.3).

Table 5.3: Operational Holdings in Kerala (1966-67 to 2010-11)

Size class	*Number of holdings (per cent)*				*Area covered (per cent)*			
	1966-67	*1985-86*	*2000-01*	*2010-11*	*1966-67*	*1985-86*	*2000-01*	*2010-11*
Marginal (<1 ha)	81.79	91.53	95.17	96.32	31.07	46.02	56.23	58.64
Small (1-2 ha)	10.09	5.76	3.41	2.64	19.63	21.55	19.10	18.66
Rest (> 2 ha)	8.12	2.77	1.42	1.04	49.30	32.43	24.67	22.70

Source: Government of Kerala, 1976, 2013b.

The number of operational holdings in the size class of = ≤1 ha increased by 237 per cent in a span of 30 years from 1970-71 to 2000-01 against the overall growth of 189 per cent for all size classes together during the same period. All other categories showed a decrease in the number of operational holdings. The maximum percentage decline was in the size class of >10 ha and the decline in number was in the size class of 2-4 ha. At the all-India level, the number of operational holdings and areas covered under the size class up to 1 ha was 46.02 per cent and 13.20 per cent, respectively, in 1985-86. Data available for 1995-96 showed that, in Kerala, the number of holdings below 0.02 ha was 7.48 lakh, which covered only 9741 ha of area. This indicates an average holding size of 0.013 ha or 3 cents for nearly 7.5 lakh holdings.

Fragmentation of Holdings: A Topographic and Social Reality

Fragmentation of holdings is another major issue in planning for sustainable land management. High fragmentation of landholdings in Kerala threatens to undermine the beneficial effects of the land reform in terms of the distribution of wealth-producing assets. Topographic control on the fragmentation of land is well evident. The size of plots is small in the valleys and narrow flood plains, usually known as wetlands, and mostly given for paddy cultivation. Plot size increases along the side slopes and ridges, which are considered as garden lands or drylands, largely given for tree crops, settlements, and plantations. Before land reform, almost all the land-owners used to possess some portions of wetlands and some portions of garden lands. With an increase in population and the consequent division of property, both types of lands were divided among the heirs. Apart from population growth, the disintegration of joint family systems among upper castes contributed to fragmentation. Growth of number of operational holdings during the 1990s and subsequent years can be attributed to liberalisation, increased money flow, international economic melting down, problems in Gulf countries, and a growing tendency of people to invest in real estate. The age structure of population also plays an important role in fluctuating growth of operational holdings. As the valleys accommodating paddy fields are limited in Kerala, the intensity of fragmentation is relatively high in this type of land compared to the lands in the slopes and ridges. With land redistribution following land reforms, the non-traditional landowning community benefitted; however, the size of holdings reduced through property division. Besides, the holdings of the majority of working farmers are too small to provide adequate returns from farming and the small plots are fragmented in ways that reduce the effectiveness of technological and environmental interventions if managed separately.

Fall in Operated Area

The area under operational holdings shows a fluctuating trend in Kerala. The maximum operational area reported from 1970-71 was 18.05 lakh hectares enumerated in 1980-81. Incidentally, it may be noted that in 1966-67 operational area was still higher at 18.28 lakh hectares. Continuous diversion of agricultural land to accommodate other uses is a matter of serious concern at present; however, the trend was marked even before 1970. Growth in the operational area has not only been slowed down but also was negative in some periods. The operational area declined from 17.96 lakh ha in 1990-91 to 15.70 lakh ha in 2000-01 and further down to 15.10 lakh ha in 2010-11 (Government of Kerala, 2013b). This indicates that the 2.86 lakh ha area has gone out of operation in a twenty years' period. The average declining rate was much higher (22,600 ha/year) during the first ten years since 1990-91 compared to the next decade of 2000-01 to 2010-11 when the annual rate declined to 6000 ha. The higher declining rate is largely attributable to the real estate boom and expansion of non-agricultural activities linked to economic liberalization/globalization initiated in the early 1990s. This steady decline in the operated area is a matter of serious concern.

In recent years, agricultural land, specifically paddy growing lands, is being kept fallow for various reasons, as a result, the operational area shows a declining trend. There is a geographical limit to which operational area can be expanded and it has almost reached a saturation level, whereas the number of holdings is growing primarily due to population increase and division of property. Therefore, the holding size is getting more and more fragmented. Since 1980-81, the number of operational holdings decreased under all size classes except the size class of <1 hectare, and in the case of >10 ha size class, the number increased for one-time point between 1980-81 and 1985-86. By 2010-11, there were only 1854 holdings in the size class of >10 ha. Of these, only 1261 holdings were in individual possession. The operational area under this class of large holding size declined from 12.6 per cent in 1970-71 to 7.92 per cent in 2010-11.

Spatial Variations of Landholdings

Data available for the year 2010-2011 shows that among 14 districts in Kerala, Wayanad has 1,74,416 operational holdings or 2.4 per cent of total holdings in the state which is the lowest and Thiruvananthapuram has 7,50,903 operational holdings or 12.4 per cent of total holdings in the state, which is the highest (Table 5.4). The average holding size varies from 0.11 ha in Thiruvananthapuram district to 0.64 ha in the Idukki district. The spatial distribution of the number of operational holdings broadly follows the trend of population density and relief character. It is found that the districts of Idukki, Wayanad, and Pathanamthitta dominated by high lands and rugged relief and low population density, together, accounted for 10.7 per cent of the total number of holdings and 21.32 per cent of the total operated area in 2010-11.

While the proportion of the number of holdings reported by these three districts together hovered around 11 per cent to 12 per cent from 1985-86 onwards, the proportion of operated area increased from 18.7 per cent in 1985-86 to 21.3 per cent in 2010-11. This indicates that operated area in the hilly districts is relatively higher compared to other districts and the rate of reduction in the operated area is also less in these hilly districts. Nevertheless, an increase in the proportion of the number of holdings and a decrease in the proportion of area, although within 1 per cent, indicates that the fragmentation of holdings is taking place in all districts in Kerala irrespective of their geographical location.

Table 5.4: Distribution of Numbers of Operational Holdings and Area Operated by Districts in Kerala (2010-2011)

Districts	*Operational Holdings (2010-2011)*		
	Number	*Area (ha)*	*Average size (ha)*
Triruvananthapuram	750903	86352	0.114
Kollam	580297	90187	0.155
Pathanamtitta	296818	66771	0.225
Alappuzha	477343	67161	0.141
Kottayam	418736	123984	0.296
Idukki	259615	164859	0.635
Ernakulam	647395	120052	0.185
Thrissur	684740	99369	0.145
Palakkad	499524	145484	0.291
Malappuram	690472	132858	0.192
Kozhikode	638351	110586	0.173
Wayanad	174416	90456	0.518
Kannur	485221	127115	0.262
Kasaragod	226958	85592	0.377
Kerala	6830789	1510816	0.221

Data Source: Government of Kerala, 2013b.

In a ten years period from 1990-91 to 2000-01, the number of operational holdings grew more than 25 per cent in Malappuram, Kasaragod, Thiruvananthapuram, and Ernakulam districts. However, the underlying reasons for this increase are not the same in all the districts. In the case of Thiruvananthapuram and Ernakulam districts, which are traditionally urbanized, the growth of operational holdings is related to urban expansion and intradistrict migration from rural to urban areas. In fact, the number of holdings in the size class below 0.02 ha in Thiruvananthapuram and Ernakulam districts, increased by 59 per cent and 91 per cent respectively in a ten years' period from 1985-86 to 1995-96. This clearly indicates that residential sectors experienced tremendous growth. In the case of Malappuram and Kasaragod districts, the growth is mainly related to natural population growth, which is evident from the demographic data. Decadal (1991-2001) growth of population in Malappuram and Kasaragod districts were 17.2 per cent and 12.3 per cent respectively.

The number of operational holdings is decreasing in recent years. Data for the years 2005-06 and 2010-11 indicated that the number of operational holdings in Kerala came down from 6.90 million to 6.83 million. The spatial trend shows that all districts in south Kerala barring Pathanamthitta, and Palakkad district in central Kerala recorded a reduction in the number of operational holdings, whereas all Malabar districts except Kannur experienced an increase in the number of operational holdings (Fig. 5.4).

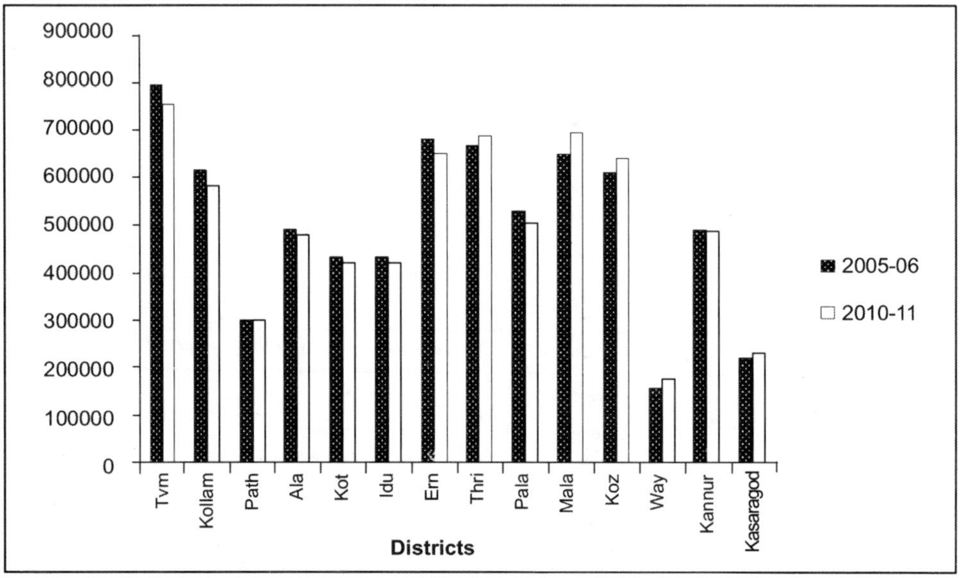

Fig. 5.4: Distribution of Number of Operational Holdings by Districts in Kerala (2005-06, 2010-11).
Data Source: Government of Kerala, 2006a, 2013b.

STAKEHOLDERS

Landowners, forest dwellers, and tribal groups are all stakeholders in land resource management. The 2010-11 agricultural Census indicated that 92 per cent area were operated by individuals followed by institutional holders (7.5%) and joint holders (0.5%). Considering the fact that individuals operate more than 90 per cent of agricultural land in Kerala, it is a challenge to intervene in the private land and facilitate the improvement of soil quality, strengthen land resource base and introduce conservation practices. Forest dwellers and tribal groups, original stakeholders of forest lands have lost their entitlement due to rules and regulations controlling forest areas. These rules were framed during the colonial period and still continue with certain amendments. Other stakeholders are national and state governments for strategic issues like setting up defence establishments, resettlements to honour national and international treaties, and establishing institutions of regional and national importance. The state government, districts, municipalities, and panchayats require land for building infrastructure, local institutions, community requirements, promotion of tourism, settlement, waste dumping, and facilitate human welfare. NGOs spearheading conservation movements are also stakeholders for land resource management. All these issues deserve due attention while preparing a land resource management plan for Kerala.

LAND MARKET

The land market in Kerala is growing. Foreign remittance plays a very important role in demand and supply. A comprehensive assessment is necessary for land markets. It should include socio-economic characteristics of the buyers and sellers, spatial variations of the market, the purpose of buying (productive, speculation, real estate, hedge against inflation, residence) and selling (emergency, cash-in on investment, moving out, immediate survival) and cultural values. All this information along with data on land quality should be part of land bank information. This will help in proper valuation and decision making in devising appropriate land use.

LAND DATA BANK

Kerala needs a data bank on land resources. Data are being generated by various agencies. There are problems with compatibility, besides, information is also partial. The problem of proper records in land resource assessment is an impediment noted even by the Comptroller and Auditor General of India (CAG). It is important to initiate a programme to develop a land resource data bank for the state covering all attributes of land and its spatial dimension. A comprehensive data base is essential for evolving strategic land resource management plan. Fortunately, there are some initiatives under the Panchayat Resource Mapping programme. It is a participatory exercise for documenting data in a cadastral scale for all Panchayats in the state. This programme initiated by the Centre for Earth Science Studies is now being coordinated by Kerala State Land Use Board. Spatial data on the cadastral scale are generated. These data are now being digitized and consolidated at watershed and district levels. A comprehensive programme may be devised to use these data, update and refine them, and feed into the data bank.

CHALLENGES OF LAND MANAGEMENT

High population density, corridor type settlement system, urban sprawl on agricultural land, landslides associated with high rainfall, soil erosion, reclamation of wetlands, occupation of flood plains, encroachment, increase in fallow land, intervention in water bodies, rock quarry, and sand and soil mining, and land shaping are important issues impacting land management in Kerala. Besides, commercialisation of agriculture, out-migration, labour shortage, the spread of modern education, the emergence of salaried class, the disintegration of joint families, economic liberalisation and globalisation are impinging factors in land resource management of Kerala.

It is necessary to design integrated land use plan considering production and conservation components together and bring balance between present demand for land and future developments. The FAO (1999) postulation for integrated land use plan can provide a guideline for Kerala. This plan proposes:

(i) " to introduce mechanisms and incentives for changes in the allocation of land to uses for which it is suitable in biophysical and socio-economic terms, and prescribe appropriate management practices and options to ensure that land resources are conserved;

(ii) (it is) essentially a mechanism for decision support to guide the stakeholders in selecting the best sustainable landuse options which are consistent with their objectives. This approach is participatory and recognises the rights of all stakeholders, including those such as indigenous groups or rural women who may be traditionally marginalised in some societies, to negotiate on the use and management of land resources;

(iii) (to use) negotiation as the basis for resolving landuse conflicts and agreeing mutually attainable objectives. Essentially, the approach embodies an interactive partnership between government and people to address their common concern to manage land sustainably for the benefit of present and future generations" (FAO, 1999: iv).

A state land use policy is the most important instrument to introduce integrated land use planning. This policy should address land tenure, create a conducive environment for sustainable land use practices, and provide fora for free and effective negotiation among stakeholders and link national policy with that of local initiatives. Key issues for strategic land use planning in Kerala should be knowledge-based, customised and capable of meeting demands for the future development of the smart city, transport corridor, and environment-sensitive zones. Issues related to climate change and disaster management should also be taken into account. The present land use practices in Kerala can be broadly grouped into 10 land use zones (Table 5.5). However, these zones are not spatially contiguous. Inventory of land for each of these categories across the state is necessary to draw a realistic picture. It may be prudent to identify these zones according to river basins and consolidated at the district level. The participation of stakeholders is an essential component.

Table 5.5: Land Use Zones in Kerala

Zones	*Characteristics*
LUZ-1	Ecologically sensitive areas—mostly for conservation, no development zones
LUZ-2	Areas for strategic development of national importance
LUZ-3	Areas of strategic mineral deposits, rare earth deposits and other valuable minerals
LUZ-4	Urban centres and surrounding areas
LUZ-5	Industrial areas and industrial parks
LUZ-6	Rural settlements and habitations
LUZ-7	Agricultural areas—food crops, cash crops, irrigation infrastructures, command areas
LUZ-8	Transport corridors and related infrastructures
LUZ-9	Areas of tourist interest, and heritage and religious sites
LUZ-10	Hazard prone areas: natural and manmade

Source: Modified after Nukala and Mutz, 2015.

The pressure on land is very high in Kerala. In the one hand, lands are intensely used and there are growing demands for land from different sectors. On the other hand, there are problems of land fragmentation, land degradation, declining productivity, and loss of land. It is important to devise a proper land resource management plan for Kerala. The existing land governance may be revisited and the state should strive to work out a strategic land policy to address all the relevant issues and to promote sustainable land resource management.

CHAPTER

6

Primary Sector-I: Agriculture and Performance of Crops

SPECIALITY OF KERALA'S AGRICULTURE

The agricultural scenario in Kerala differs from the rest of India. Tree crops or perennial crops dominate agricultural land use in Kerala. Lands suitable for rice or other seasonal crops raised in alluvium soil are limited in the state. Homestead garden is a unique phenomenon here. Varieties of fruit trees along with coconut, banana, and tapioca are cultivated around the dwelling places. Even rubber plantations are raised beside settlements. Per hectare agricultural output in Kerala in terms of money is well above the all-India average due to the dominance of cash crops. Agro-climatic conditions and micro-level landscape diversity facilitate the cultivation of seasonal crops, annual crops, and plantation crops nearby throughout Kerala. Hardly, there is any exclusive crop zone, although the nature of crops changes with altitude and distance from the coastline. Multi-tier cropping in garden lands and double and triple cropping of rice fields are common features.

Kerala opted for extensive as well as intensive agriculture quite early to cater to the need of high population pressure and prevailing favourable agro-climatic conditions. Land use data indicate that spatial expansion of agriculture has reached a saturation point. Whatever expansion was possible had already been taken place. This extensive use of land often leads to situations reaching beyond the carrying capacity of the ecosystem, the manifestation of which can be observed in a substantial drop in productivity of some of the crops cultivated in marginal areas and also in ecologically unsuitable areas. There are some agricultural expansions within the forest boundaries, which are not generally accounted for in the published agricultural statistics. Analysis of remote sensing data products may help in this context.

FACTORS GOVERNING KERALA'S AGRICULTURE

In the search for an explanation of the distinctiveness of Kerala's agriculture, it is possible to single out four main factors. These are:

- Agro-ecological set up: High diversity both at the macro and micro level, long sun shine, limited temperature variations and high and well-distributed rainfall conducive to growing a wide range of crops,

- Trade tradition: The long tradition of the spice trade through sea routes since the early days,
- Demographic pressure: High population density, dispersed settlements, a narrow gap between rural and urban areas, out-migration of labour force mainly to gulf countries, and
- Government policy: Land reforms, land use regulation, wage rate, and local level initiatives/decentralisation.

These factors have certain important salient effects on Kerala's agricultural system. Some of those are: (i) high out-turn from the farmland due to dominance of plantation agriculture and cash crops, (ii) intense use of land in rural, sub-urban, and urban areas, (iii) high price of farm labour compared to other states, particularly adjoining states, and (iv) influence of foreign remittance on land use. This phenomenon of foreign remittance is relatively recent; nevertheless it is emerging as an important determining factor of agricultural practices in Kerala.

LOCATIONAL FACTORS: THEORETICAL ANALYSIS

Apart from the factors mentioned above, Kerala's agriculture may also be viewed in terms of its complex and continually evolving relationship to the market and economics of production. Farmers' decisions to use their lands for a particular use or to raise a particular crop reflect more than an assessment of potentials and limitations of the physical resources base. Selection of crops, even varieties of crops is also influenced by socio-economic factors, such as land value, taxes, accessibility to market, perishability, cost of labour, transport cost, and availability of wholesalers. Farmers in Kerala use their land intensively due to high population pressure and the suitability of land for subsistence crops and also cash crops. Relatively large inputs of labour, capital, and energy are required in intensive land-use systems. Rising land values in the vicinity of urban centres and settlement areas mean that output from agricultural practices must be very high to compete with potential non-agricultural uses. The use of land for housing development, industrial park, shopping centres, auditoriums, or amenities like airports, etc. can be far more profitable than raising crops.

Proximity to expanding urban centres can be a negative factor to continue with agriculture, but the superior accessibility to large markets for agricultural produces can be a tremendous advantage. Location is an important geographic asset in the matters of changing agricultural land use in Kerala. Historically, Kerala was in the middle of the spice route between east and west and the emporium of Mussiris (Kodangalur) impacted agricultural practices in the surroundings. Expansion of coconut cultivation in the coastal tract and pepper cultivation in the hinterlands of series of small ports developed along the river mouths were linked to trade and market since the fifteenth century. Rubber introduced in the early 20th century has expanded throughout Kerala and now it is the second-ranking crop after coconut in area coverage. Rice cultivation is fast losing its ground, and in many districts of Kerala, the position of rice is well down in the list of crops. Kerala farmers are highly innovative, open-minded entrepreneurs and are capable of shifting land use to respond to changes in the market, technologies, and new sources of competition.

It may be interesting to examine Von Thunen's theory (O. Kelly and Bryan, 1996) of agricultural land use in the context of Kerala's agricultural transformation. Von Thunen's theory, proposed before industrialisation, was based on a set of assumptions like an isolated country, and constant bio-physical variables and transport efficiency. The idea proposing a concentric land use pattern with the city in the centre followed by intensive farming/dairy, forest, extensive field crops (grains), and ranching/grazing/forest in the outer periphery appeared to have been modified in case a river passing through the area. Although not similar, Kerala exhibits a type of zoning both from the coast to the Western Ghats and also from the valleys to the ridges. The bid-rent theory based on Von Thunen is significant. The economic pressures on farms within the suburban orbit, following road network, which seems to be expanding continuously is significantly high.

AGRICULTURAL INPUTS AND MANAGEMENT

Water is a critical input for agricultural development. Kerala is blessed with rainfall in two seasons—south-west and north-east monsoons. North-east monsoon is relatively weak towards north Kerala. The Palghat Gap is a dry area with high temperatures during the summer months. There are several other drought-affected areas distributed in different parts of Kerala. Besides, the behaviour of the monsoon is increasingly erratic. To address these issues and for unhindered agricultural development, the supply of irrigation water is necessary. It is estimated that about 16 lakh ha of land could be brought under irrigation through major and medium irrigation projects and nine lakh ha area can be catered through minor irrigation projects (Planning Commission of India, 2008a). However, coverage under irrigation is less than 25 per cent. We provide a brief discussion here on irrigation.

IRRIGATION

Canal, tank, well, and tube well are primary sources of irrigation. Lift irrigation from rivers, pond irrigation and tapping of spring water to irrigate crops are also practiced. All these sources are clubbed together under the category of other sources. By 2016-17, the net irrigated area in Kerala was 3.78 lakh ha and the gross area irrigated was 4.58 lakh ha (Table 6.1). Irrigated areas had fallen compared to 2000-01. However, when the irrigated area is computed in terms of the net sown area, there was a marginal increase in 2016-17 compared to 2000-01, which was due to a reduction in net sown area over the years. Fallow lands are increasing in Kerala.

Table 6.1: Net Area Irrigated from Various Sources in Kerala (2000-01, 2016-17)

Sl. No.	*Source*	*2000-01*		*2016-17*		*Change*
		Area in ha	*Per cent*	*Area in ha*	*Per cent*	*(per cent)*
1.	Government canals	100926	26.5	63403	16.8	-37.20
2.	Private canals	4041	1.1	932	0.2	-77.00
3.	Tanks	49972	13.1	49657	13.2	-0.63
4.	Wells	115703	30.3	122478	32.4	5.86
5.	Other sources	110399	29	141291	37.4	28.00
6.	**Total**	**381041**	**100**	**377761**	**100**	**-0.86**
7.	Area irrigated more than once	76824		79812		3.98
8.	Gross irrigated area	457865		457573		-0.06
9.	Net area irrigated to net area sown (per cent)	17.27		18.74		1.47
10.	Gross irrigated area to gross cropped area (per cent)	15.15		17.7		1.95
11.	Irrigated area under rice to total irrigated area (per cent)	45.44		28.64		-16.80

Data Source: Government of Kerala, 2002; 2018.

Irrigated area in per cent to the net sown area at the district level varies from 3.2 per cent in Kozhikode to 46 per cent in Alappuzha (Fig. 6.1). Alappuzha, Thrissur, Palakkad, and Kasaragod districts record more than 30 per cent of their net sown area under irrigation. There are several river valley irrigation projects developed in Palakkad and Thrissur districts. Ernakulam is served by the Pariyar canals. Alappuzha is catered by the Pamba irrigation project. Initially, Palakkad, Thrissur, and Ernakulam districts in Central Kerala were the focus of irrigated agriculture in Kerala due to the high concentration of irrigable land and the dominance of seasonal crops in this tract. However, their combined share in the net irrigated area of Kerala dwindled from 59 per cent in 1991-92 to 40 per cent in 2016-17. This is due to the expansion of the irrigated area in other parts of Kerala and also a reduction in net area under agriculture, particularly rice, which needs irrigation. Ernakulam district experienced a major shift due to an increase in urban-industrial land use.

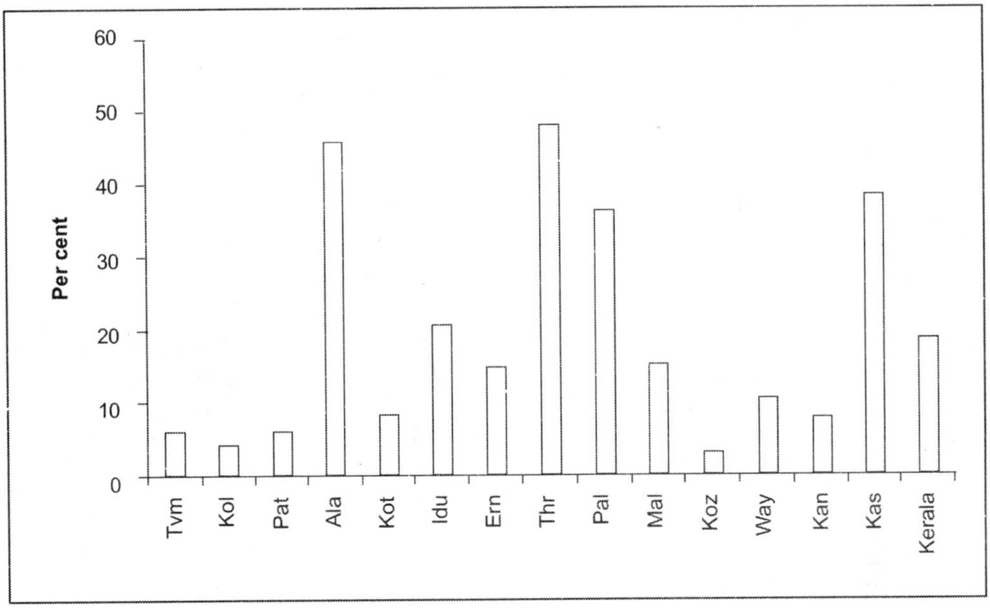

Fig. 6.1: Net Irrigated Area in Per Cent to Net Sown Area by Districts and State Average of Kerala (2016-17).
Data Source: Government of Kerala. 2018.

Change in Mode of Irrigation

Nature of irrigation changed perceptibly. The area under canal irrigation has fallen drastically by 37 per cent between 2000-01 and 2016-17. The decline in irrigated area under the Government canal and private canal indicate that canal irrigation is losing importance. Under these circumstances, the question arises about the wisdom of further investment in canal irrigation projects. It is also noted that the private sector is gaining importance in the irrigation sector, which mostly depends on groundwater. As a whole, there is a growing dependence on groundwater. In 2016-17, 43.5 per cent of irrigation water was supplied from wells and tube wells/bore wells, which was 38 per cent in 2010-11, 30.3 per cent in 2000-01, and 19.4 per cent in 1991-92. There was 154 per cent growth in the area under irrigation from wells, bore wells, and tube wells, (Fig. 6.2). The impact of increasing use of well water is reflected in declining groundwater level. Several areas in Kerala reported over-exploitation of groundwater.

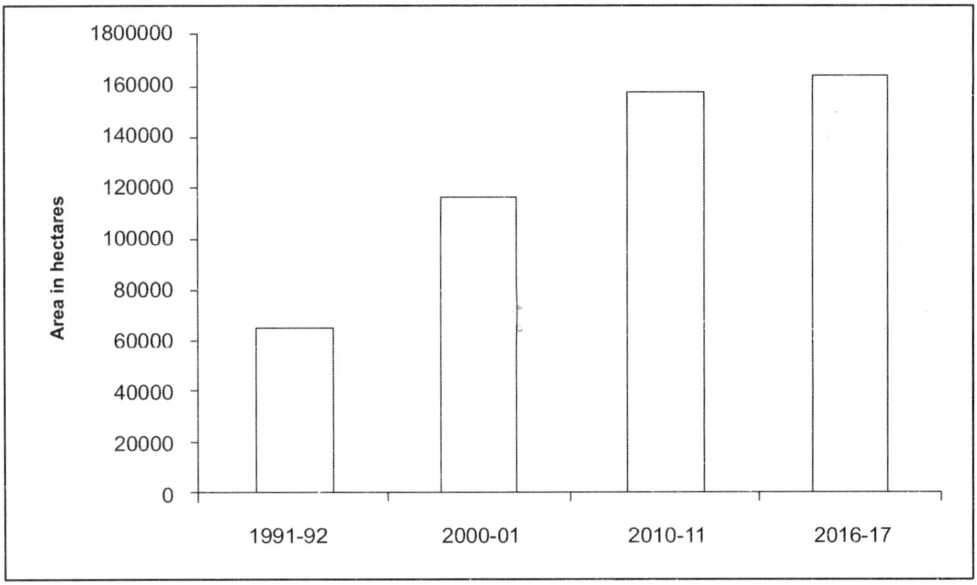

Fig. 6.2: Temporal Variations of Area Irrigated by Groundwater (Wells, Tube Wells and Bore Wells) in Kerala (1991-92 to 2016-17).

Data Source: Government of Kerala, 1993, 2002, 2012a, 2018.

Irrigated Crops

Irrigation facilities are extended to all types of crops in Kerala—seasonal, annual and perennial. Among the crops, coconut is the major crop recording 35 per cent of its area under irrigation followed by rice (29%), banana (10%), arecanut (7%), and vegetables (5.5%) (Table 6.2). During 2016-17, irrigated coconut and irrigated rice accounted for 1.6 lakh ha and 1.31 lakh ha respectively. The share of these two crops together in the gross irrigated area was 63.6 per cent in 2016-17 which was 67 per cent as of 2010-11 and 81.6 per cent during 2000-01. Till recently, the maximum irrigated area was reported by rice. By the turn of the century, rice accounted for a little over 45 per cent of gross irrigated area, which is now only 29 per cent. Drastic fall in rice cultivation has resulted in this changed scenario. The irrigated area under vegetables has increased by 198 per cent between 2000-01 and 2016-17. Compared to 2000-01, the areas irrigated under some of the crops, namely banana, arecanut, vegetables, and nutmeg have increased during 2016-17.

Most of the irrigated rice fields are concentrated in Palakkad, Thrissur, and Alappuzha districts. Thrissur reported the maximum area under irrigated coconut followed by rice, banana, and arecanut.

FERTILISER USE

Soil fertility of Kerala has been categorized as low nitrogen and medium in the content of phosphorus and potassium in the all-India context. Deficiency has also been reported for micro-nutrients. Proper application of fertilizer is necessary to increase agricultural output. Kerala's fertilizer consumption is well below the national average. Since 2000-01, Kerala's fertiliser

Table 6.2: Temporal Variations of Gross Area Irrigated under Different Crops in Kerala (2000-01, 2016-17)

Sl. No.	Crops	2000-01		2016-17		Change (Per cent)
		Area in ha	Per cent	Area in ha	Per cent	
1.	Rice	208047	45.4	131063	28.7	-37.0
2.	Vegetables	8529	1.9	25407	5.6	198.0
3.	Coconut	165957	36.2	159949	35.0	-3.6
4.	Arecanut	30501	6.7	32489	7.1	6.5
5.	Nutmeg/clove	2466	0.5	19211	4.2	679.0
6.	Banana	19448	4.3	44841	9.8	131.0
7.	Betel leaves	990	0.2	261	0.1	-74.0
8.	Sugarcane	3367	0.7	1063	0.2	-86.0
9.	Others	18560	4.1	43289	9.5	128.0
	Total	**457865**	**100**	**457373**	**100**	**-0.10**

Data Source: Government of Kerala, 2001; 2017a.

consumption increased from 58kg/ha in 2000-01 to 100kg/ha in 2015-16. The corresponding figures for the all-India average were 86.3kg/ha and 134.9kg/ha. The increasing cost of fertilizer is one of the hindrances to increase consumption, as most of the crops raised in Kerala are not under the purview of procurement prices administered by the government of India. Moreover, given low use efficiency, questions have been raised about continuing increased use of fertilizer.

OTHER MANAGEMENT PRACTICES

There are several management initiatives to improve the agricultural situation in Kerala. They include collective farming under Kudumbashree, Horticulture Mission, Vegetable and fruit promotion council, Rashtriya Krishi Vikas Yojana (RKVY), crop insurance, expansion of hi-tech agriculture, farm mechanisation, and strengthening of plant health management. Kudumbashree is a unique initiative in Kerala for women self-help. It has introduced collective farming to achieve food security both at the household level and community level. In 2011-12, there were 6116 collective farming groups known as Joint Liability Groups (JLGs) which had cultivated 9912 hectares of land. The major crops cultivated are rice, vegetables, banana, and tubers. Fallow rice fields can be gradually brought under cultivation at the instance of Kudumbashree (Fig. 6.3). However, it needs further facilitation and technology support.

FLUCTUATING AGRICULTURAL GROWTH

Agriculture as an economic sector has passed through successive high and low growth paths since the 1960s. In the first phase (1960-61 to 1969-70), the growth rate of the agriculture sector was 3.39 per cent followed by a stagnating second phase (1968-69 to 1982-83), that was succeeded by a revived third phase (1983-84 to 1993-94), when the growth rate of agriculture sector touched 3.85 per cent, and in the fourth phase (1994-94 to 2001-02), it fell to 1.91 (Pushpangadan and Parameswaran, 2006). In subsequent years, the agricultural sector faced problems due to a low growth rate, in addition to increased input costs and a fall in farm income. The problems turned to a severe crisis as reflected in farmers' suicide in those areas where farmers cultivated export-oriented high-value crops (Nair and Menon, 2007). Apart from fall in prices due to volatility of the international market, export oriented cash crops like

Fig. 6.3: Kudumbashree Members Taking Part in Land Management Activities in Kanjikuzhy Panchayat of Alappuzha District in Kerala. Photo by V. Dhanya.

pepper, coffee, tea, and cardamom are suffering from low yield due to ecological stress of low rainfall, the incidence of drought, fall in the groundwater table, and decrease in soil moisture. Mono-cropping and deforestation also contributed to the loss of soil moisture.

Two broad issues emerge from this brief discussion. These are: (i) Further spatial expansion of agriculture is not possible, in fact, it has reached a saturation point and area under agriculture is declining, and (ii) agriculture has again fallen into the low growth path and is passing through a crisis. We would like to discuss the agricultural scenario of Kerala keeping these two broad issues in view.

TOTAL CROPPED AREA AND IMPORTANT CROPS

The total cropped area (TCA) in Kerala was 25.8 lakh hectares in 2016-17, which was 22.11 lakh hectares in 1957-58. It recorded an increase of 16.9 per cent during this 58 years' period. However, this increasing trend was not uniform (Table 6.3). The period from 1957-58 to 1975-76 witnessed a steady growth, thereafter, it was fluctuating. The TCA reached the maximum of 30.67 lakh hectare in 1995-96 and afterwards it declined by 15.8 per cent till 2016-17. There were 14 important crops, which together covered 89 per cent of the total cropped area in 1957-58. By 2016-17, these 14 crops together covered 80 per cent of the total cropped area. There was readjustment in area coverage among the crops and also new crops like cocoa, vanilla, and new varieties of vegetables were introduced. Out of the 14 crops considered here, only two crops, namely, sugarcane and tea recorded negative growth between 1957-58 and 1975-76 (Table 6.3). However, the period between 1975-76 and 1995-96 witnessed a fall in coverage under six more crops. Most of the food crops had lost their area coverage. Even coconut, which showed an increasing trend in all these years, recorded a fall in coverage of 14.6 per cent between 1995-96 and 2016-17. There were 10 crops that recorded negative growth between 1995-96 and 2016-17.

Compared to 1975-76 percentage share of food crops like rice and tapioca decreased by 80 per cent and 79 per cent respectively by 2016-17. The area under non-food crops like rubber, coffee, areca nut, and pepper recorded a steady increase. Rubber and coffee gained the maximum by 450 per cent and 400 per cent respectively between 1957-58 and 2016-17. By 2016-17, coconut covered around 30.2 per cent of the total cropped area followed by rubber and rice. Coconut cultivation is expanding at the expanse of rice cultivation in coastal plain and low lands, where-as rubber is replacing degraded forests, tree crops, and tapioca. Land use change has serious implications on water quality and overall hydrological conditions. Agriculture scenario warrants urgent attention.

Table 6.3: Changes in Cropping Pattern in Kerala, (1957-58 to 2016-17)

Crops	*Area in thousand hectares*				*Change in per cent*		
	1957-58	1975-76	1995-96	2016-17	1957-58 to 1975-76	1975-76 to 1995-96	1995-96 to 2016-17
Rice	767	876	471	171	14.2	–46.2	–63.7
Tapioca	214	327	114	69	52.8	–65.1	–39.5
Other cereals and pulses	28	45	24	2	60.7	–46.7	–91.7
Sugarcane	9	8	6	1	–11.1	–25.0	–83.3
Banana and plantain	42	52	73	114	23.8	40.4	56.2
Coconut	463	693	914	781	49.7	31.9	–14.6
Areacanut	50	77	71	98	54.0	–7.8	38.0
Pepper	91	108	192	85	18.7	77.8	–55.7
Ginger	9	12	13	5	33.5	8.3	–61.5
Cashewnut	44	109	103	42	147.7	–5.5	–59.2
Cardamom	28	54	44	39	92.9	–18.5	–11.4
Tea	40	38	35	30	–5.0	–7.9	–14.3
Coffee	17	42	82	85	147.0	95.2	3.7
Rubber	100	207	449	551	107.0	116.9	22.7
Others	309	333	476	511	7.76	42.9	7.4
Total cropped area	2211	2981	3067	2584	32.6	3.2	–15.7

Data Source: Chattopadhyay and Franke, 2006; Government of Kerala, 2017a.

FOOD AND NON-FOOD CROPS

All crops in Kerala are categorized into two broad groups: food crops and non-food crops. Crop husbandry witnessed a gradual change in favour of non-food crops. The ratio between area coverage of food crops and non-food crops was 67:33 in 1960-61, which turned to 50:50 by 1990-91, 39:61 in 2010-11, and, by 2016-17, it stood at 37: 63 (Fig. 6.4). In absolute terms, the area under food crops came down from 15.65 lakh hectares in 1960-61 to 9.45 lakh hectares in 2016-17, whereas the area under total cropped area increased from 23.49 lakh hectares to 25.84 lakh hectares during the same period. This signifies that not only the area under food crops is decreasing but also the expansion of agriculture is taking place mainly in the non-food crop sector. In a food deficit state like Kerala, this trend has serious socio-economic implications.

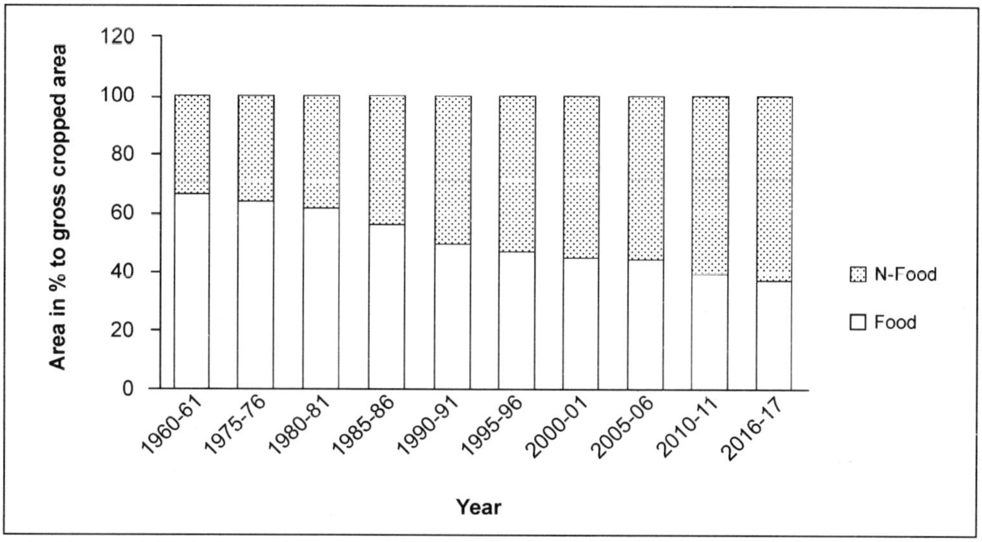

Fig. 6.4: Temporal Variations of Area under Food and Non-Food (N-Food) Crops in Kerala (1960-61 to 2016-17).

Data Source: Agriculutal Statistics of Various Years, Government of Kerala.

SEASONAL, ANNUAL AND PERENNIAL CROPS

Based on the time duration of cultivation and nature of crops, the major crops in Kerala can be grouped into two: (i) seasonal and annual crops consisting of rice, other cereals, tapioca, other tubers, sesamum, banana and plantain, sugarcane, ginger, groundnut and vegetables, and (ii) perennial crops comprising of coconut, arecanut, pepper, rubber, cashew, cardamom, tea, coffee, and all other tree crops. Some of the seasonal and annual crops are raised more than once within a year as evident from sowing and harvesting seasons (Table 6.4). The area under seasonal crops declined from 8.9 lakh hectares or 38 per cent of total cropped area in 1960-61 to 4.31 lakh hectares or 16.7 per cent of total cropped area in 2016-17. As rice constitutes the bulk of the seasonal crops, this decline is mainly on account of a fall in the area covered by rice. Other seasonal crops like tapioca, groundnut, sesamum, and pulses also recorded a substantial reduction in their coverage during the last 45 years. There are biophysical limitations to extensively grow all types of pulses in Kerala, however, certain varieties like black gram, green gram, and cowpea can be raised. Area coverage under pulses has drastically reduced in spite of higher prices. It covered only 1738 hectares in 2016-17. This shift towards plantation crops or commercial crops is a structural change in agriculture observed since the mid-seventies. In fact, in the year of 1975-76, the area under seasonal crops covering 9.29 lakh hectares touched the all-time peak. There was a fluctuating trend in coverage of seasonal crops during the period from 1960-61 to 1975-76, however, the subsequent period, particularly from 1982-83 onwards, witnessed a sharp decline in coverage of seasonal crops (Fig. 6.5). There is a steady fall in area under rice and tapioca and high growth of area under rubber plantation. Coconut records fluctuating but increasing trend. Except for rubber, the area under all important crops has declined since 1995-96.

Table 6.4: Sowing, Harvesting and Peak Marketing Seasons of Selected Crops in Kerala

	Crop	Season	Sowing	Harvesting	Peak Marketing
	(1)	(2)	(3)	(4)	(5)
1.	Rice	Autumn 1st crop Winter 2nd Crop Summer 3rd Crop 4th crop	April-June August-October November-December January-March	August-October December-February February-March April-May	September-October January-February March-April May-June
2.	Tapioca	1st Crop 2nd Crop 3rd Crop	October-November March-May July-September	August-September November-January May-July	August-September December-January June-July
3.	Other Pulses	1st Crop 2nd Crop	May-June October	August-September December-January	August-September January
4.	Sugarcane	1st Crop 2nd Crop	November-February January-March	October December December-February	November-December February
5.	Ginger (Raw)	---	April-May	November-January	December-January
6.	Pepper	---	—	November-January	December-January
7.	Sesamum	1st Crop 2nd Crop 3rd Crop	February-March August-October December-January	June-July December-January March-April	July-August December-January April-May
8.	Cotton	---	August-September	February-March	February-March
9.	Sweet Potatoes	1st Crop 2nd Crop 3rd Crop	June-July September-October November-December	September-October December-January February-March	September-October December-January February-March
10.	Turmeric	---	April-May	December-January	January-February
11.	Lemongrass	---	---	June-September	September

Source: Department of Economics and Statistics, Government of Kerala.

CROPPING INTENSITY

Cropping intensity is an index to assess how intensely a cultivable land is used or how many crops are raised from land in a year. It is commonly measured as a ratio of gross cropped area (GCA) and net sown area (NSA). In 1960-61, cropping intensity in Kerala was 122 per cent. It steadily increased to 136 per cent by 1975-76, and since then there were fluctuations and the highest cropping intensity (140 per cent) was recorded in the year 2005-06. However, the gross cropped area was the highest in the year 1995-96. This trend indicates that agriculture in Kerala is undergoing various structural adjustments in the matter of crop cultivation. By 2016-17, cropping intensity in Kerala was 128.2 per cent. It signifies that 28.2 per cent of Kerala's land was cultivated more than once. District-wise distribution of cropping intensity in 2016-17 varied from 107.5 per cent in Kasaragod district to 147.8 per cent in Wayanad district closely followed by Palakkad (Table 6.5). Seasonal crops played important role in increasing cropping intensity. Multitier cropping, which is a unique phenomenon in Kerala, also has an important role in enhancing cropping intensity. Computation of cropping intensity in Kerala has a problem as it includes perennial crops also.

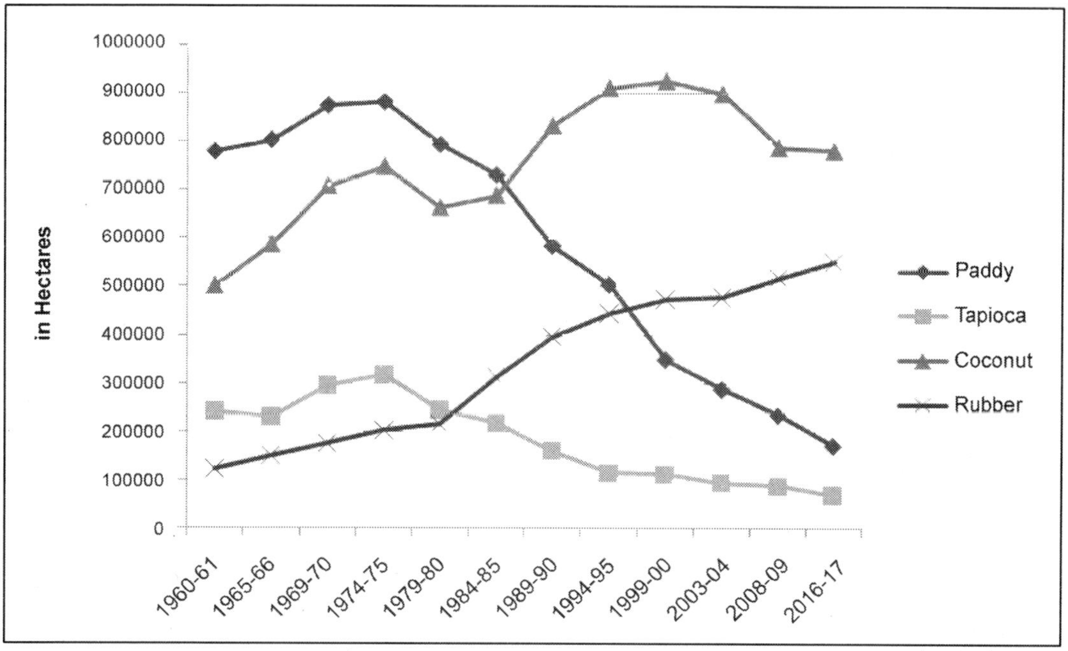

Fig. 6.5: Temporal Trend in Area Coverage of Selected Crops in Kerala (1960-61 to 2016-17).
Data Source: Agricultural Statistics of various years, Government of Kerala.

PERFORMANCE OF CROPS

The performance of selected crops has been discussed to provide an in-depth picture of the agricultural scenario in Kerala.

RICE

Rice is the staple food of people in Kerala. It is grown throughout the year. Three traditional seasons are *Virippu* (*Kharif*/Autumn), *Mundakan* (*Rabi*/Winter), and *Puncha* (*Boro*/Summer). The term *Puncha* also refers to the low-lying seasonally waterlogged areas, where cultivation is possible only during summer months and sometimes dewatering is necessary from these lowlands to sow rice seedlings. Kuttanad spreading over Alappuzha, Kottayam, and Pathanamthitta is the main area of *Puncha* cultivation. Based on micro-relief, the extent of water-logging, and water quality, the other types of rice cultivations are *Kulappala* (floating varieties of rice), *Kole* (similar to Kuttanad in Thrissur and Malappuram districts), and *Pokkali/ Kaipad* (tidal inundated saline areas). Large rice fields in blocks are known as *Padasekharams* (Figs. 6.6 and 6.7). Cultivations in these *Padasekharams* are done through co-operatives formed among the shareholders of land within *Padasekharams* or members of *Padasekharam* committees. Settlements have developed within these *Padasekharams* in recent years. The crop calendar in the Kuttanad area is slightly different as *Puncha* cultivation dominates (Fig. 6.8). A second rice crop is also raised in recent years. There are four common systems of rice cultivation in Kerala: dry seeded upland rice, dry seeded lowland rice, wet seeded lowland rice, and transplanted lowland rice (Mathew and Dev, 1997).

Table 6.5: Distribution of Area under Selected Crops, Total Cropped Area, Gross Cropped Area and Crop Intensity by Districts in Kerala (2016-17) *(In Hectares)*

Districts/crops	*Rice*	*Tapioca*	*Coconut*	*Rubber*	*Total non-food crops*	*Gross cropped area*	*Crop intensity (per cent)*
Thiruvananthapuram	1392	14628	70467	32200	109178	158979	123.2
Kollam	1302	14421	50983	37260	94690	147942	120.4
Pathanamthitta	2640	5025	15877	50900	75981	103972	129.2
Alappuzha	32453	2559	33670	4500	46420	102621	121.3
Kottayam	17216	5956	25610	114410	153695	204337	127.2
Idukki	695	6998	16122	40590	135943	270356	131.5
Ernakulam	4730	5415	43079	60170	113122	166086	111.9
Thrissur	21100	1172	80504	15660	103965	170978	132.8
Palakkad	65513	1900	59547	37870	133709	278455	147.4
Malappuram	7140	5283	102836	42770	161032	237860	137.4
Kozhikode	1987	1477	119064	21930	147085	198389	134.3
Wayanad	7822	1726	10322	10800	101179	166875	147.8
Kannur	4671	1661	88217	48070	151508	221367	117.8
Kasaragod	2737	443	65234	33920	110892	155789	107.5
Kerala	171398	68664	781496	551050	1638400	2584007	128.2

Data Source: Government of Kerala 2017a.

Changing Trend of Rice Cultivation

India witnessed a food shortage during the 1960s and started importing food. There were all-out efforts by the Government of India to produce food. This had culminated in the green revolution, and by the 1970s, India become self-sufficient in food grain production. Kerala was not in the belt of the Green revolution, however, it recorded the introduction of high yielding varieties of rice and a steady increase in area under rice during the 1970s. The total area under rice in Kerala peaked with coverage of 8.81 lakh hectares in 1974-75 followed by 8.76 lakh hectares in 1975-76. This high coverage of rice of more than eight lakh hectares continued for a decade since 1967-68. Production of rice also increased. It touched a peak in 1974-75 and could meet 50 per cent of total demand in that year (Government of Kerala, 1999). Rising production in the seventies inspired the policy-makers, administrators, and planners to take initiatives for attaining near-self sufficiency in rice production. However, subsequent trends belied all expectations as Kerala witnessed a sharp decline in area under rice.

There was a steady fall in area coverage of rice noted from 1977-78, and by 2016-17, the total area under rice in Kerala was only1.71 lakh hectares or 6.6 per cent of the total cropped area. The reduction in area under rice in a period of 42 years, from 1974-75 to 2016-17, was 80.5 per cent. The average annual rate of reduction during 1980-81 to 1990-91 was 3 per cent, followed by 3.8 per cent during 1990-91 to 2000-01 and in the subsequent period from 2000-01 to 2010-11, it was 3.9 per cent, and from 2010-11 to 2016-17, it was 3.8 per cent. Rice was the 1st ranking crop in terms of area coverage till 1980-81. By 1990-91, it was the second-ranking crop after coconut and by 2000-01 rice ranked third below coconut and rubber. The impact of the reduction in area under rice on total production can be partly offset by enhanced productivity.

Change in Seasonal Coverage of Rice

Disaggregating data at the seasonal level, it emerged that, the maximum coverage of area under

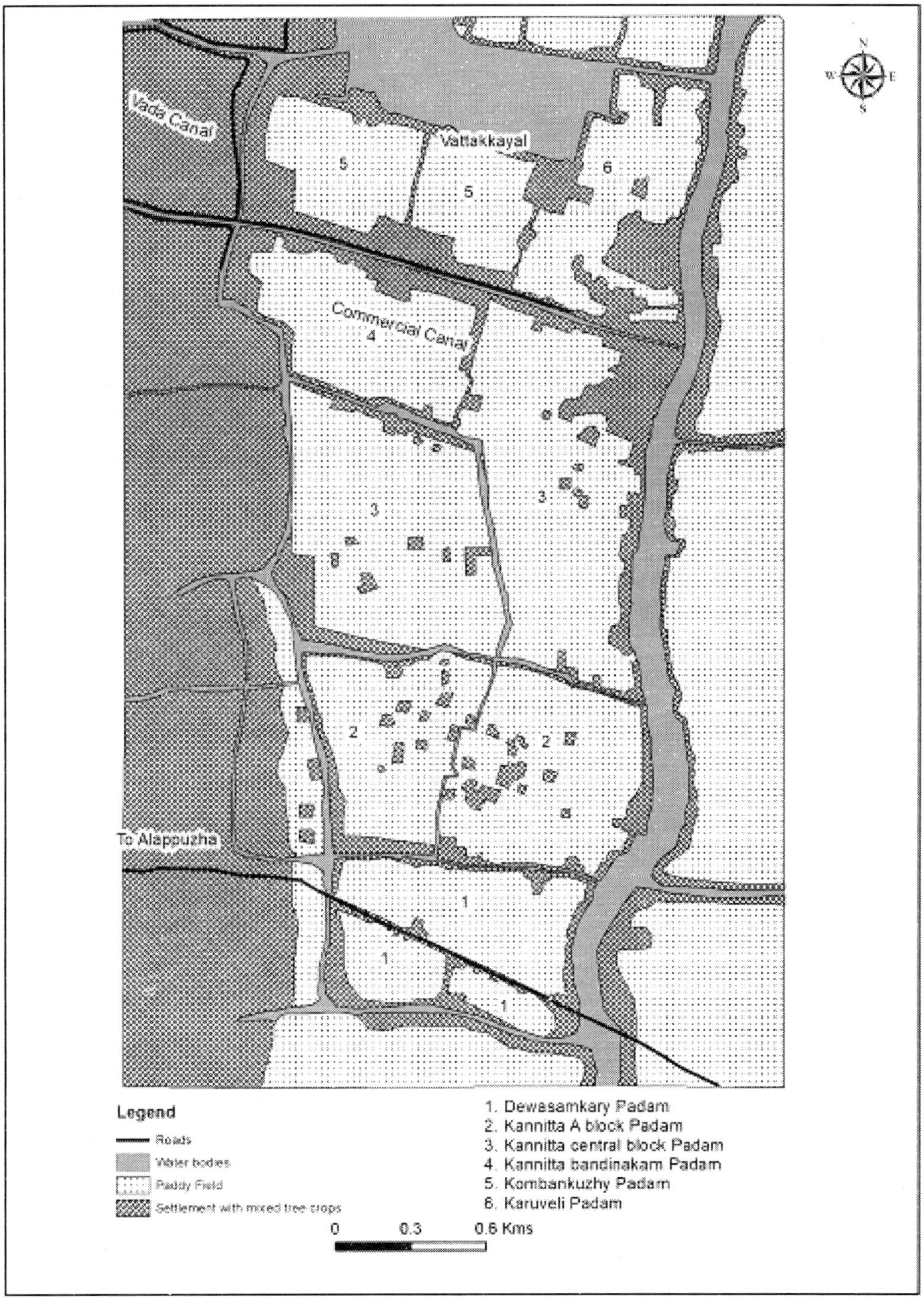

Fig. 6.6: Padasekharams in Kuttanad Area of Alappuzha District in Kerala.

Data Source: Field Mapping.

Fig. 6.7: View of a Rice Field in a Padasekharam, Kuttanad of Alappuzha District in Kerala. A Farmer Describes Problems of Rice Cultivation in Kuttanad. Photo by S. Sajna.

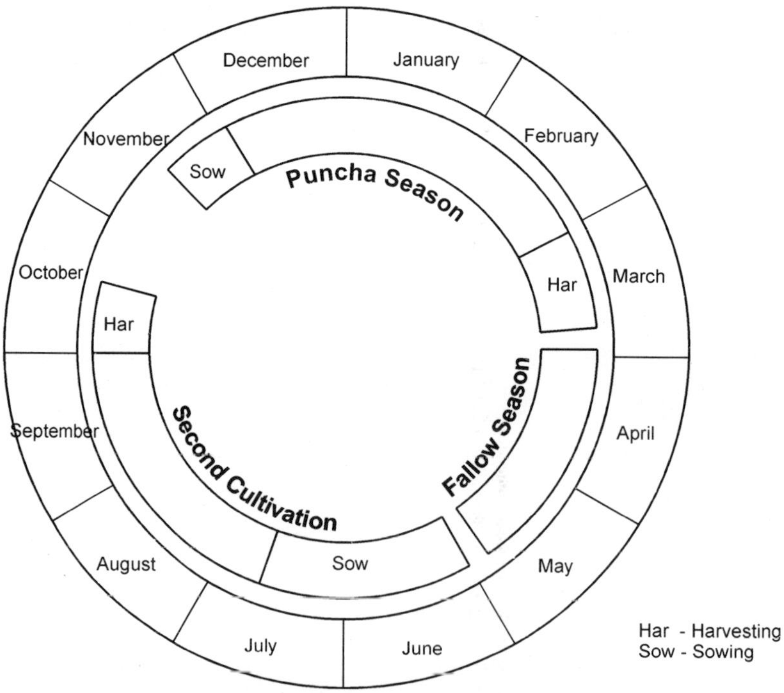

Fig. 6.8: Crop Calendar in Kuttanad of Alappuzha District in Kerala.
Data Source: Field Survey.

rice was during the winter season (42 per cent of the total area under rice) followed by autumn (33 per cent) and summer (25 per cent) in the year 2016-17. This coverage was 47 per cent, 37 per cent, and 16 per cent for the cropping seasons of winter, autumn, and summer respectively in 2000-01. These figures suggest that the relative share of seasonal coverage changed over the years. While both winter and autumn seasons witnessed a decline in their respective shares, the proportion of coverage increased in the case of the summer season during 2016-17. It can be reasonably inferred from this distribution that the land areas used for raising winter and autumn rice are declining at a higher rate compared to land area under summer rice. Summer rice is cultivated in the relatively low-lying areas, which remain waterlogged during other seasons. These lands are less prone to conversion due to seasonal flooding and the high cost of conversion. Micro relief plays an important role in land conversion from rice to non-rice use.

HYV and Local Varieties of Rice

High Yielding Varieties (HYV) dominated in Kerala by covering 95.4 per cent of the total area under rice in 2016-17. Little over 76.5 per cent of the total rice area was irrigated with HYV counting for a major share. Around 23 per cent of the area under HYV and 33 per cent of the area under local varieties remained un-irrigated. Cultivation of HYV, a product of Green Revolution, warrants proper water management, and in most states in India, HYV is restricted to the irrigated areas only. This high coverage of the HYV rice area in Kerala under un-irrigated condition is primarily attributable to rainfall distribution and the nature of rice fields. Almost all rice fields in Kerala are wetlands and a sizable portion of rice lands are located in the valleys surrounded by sloping uplands. Some of these valleys are spring-fed and therefore remain under the wet condition for a long time. Besides, these areas receive moisture from the surrounding uplands. High moisture availability in the valleys is ecologically linked to the adjoining land units. The area under HYV in the un-irrigated sector dominates during the autumn season. The maximum coverage under HYV is reported during winter in the irrigated sector (39.5 per cent of the total area under HYV). So far, more than 80 high yielding varieties have been released in Kerala.

Traditional rice varieties in Kerala counted more than 2000 (Mathew and Dev, 1997: 211). There are separate varieties for *virippu, mundakan*, and *puncha* seasons. Some of the traditional varieties have definite periods of maturity and some have particular harvesting season. Traditional varieties are well known for withstanding drought conditions and other vagaries like excessive rain, waterlogging, and salinity incursion, although the yield rate of traditional varieties could not be improved much.

Distribution of Area under Rice Cultivation by Districts

Lands suitable for rice cultivation are not uniformly distributed due to the physiographic peculiarities of Kerala. Flood plains, flat bottom valleys, and terraces across the altitudes and outer coastal plain are given for rice cultivation. Palakkad, Thrissur, and Ernakulam belt within Bharatapuzha and Periyar basins and Kuttanad area in Pamba and adjoining river basins spreading over Alappuzha and Kottayam districts are the main rice growing areas. These arrangements are reflected in district level variations in coverage under rice. Palakkad and Thrissur together accounted for 50.5 per cent of total rice land in Kerala during 2016-17 (Table 6.5). Alappuzha-Kottayam belt (Kuttanad) followed with 29 per cent coverage. Besides, the

Wayanad plateau with broad valleys at higher altitudes also has lands suitable for rice cultivation. Around 4.6 per cent of rice fields in Kerala is in the Wayanad district.

Among three seasons, the winter season reported the maximum coverage (72,000 ha) followed by Autumn (56,601 ha) and Summer (2,544 ha). However, these seasons are not equally important for all the districts. The districts of Kollam, Idukki, Ernakulam, Thrissur, Malappuram, Kozhikode, and Wayanad reported the maximum coverage under rice during the winter season. Autumn is the most important season for rice cultivation in Thiruvananthapuram, Palakkad, Kannur, and Kasaragod. Summer is the main rice-growing season in Alappuzha, Kottayam, and Pathanamthitta districts sharing areas under the Kuttanad belt. Thrissur district also has a sizable area under summer rice raised in *kole* lands. Palakkad is noted for both Autumn and Winter rice due to irrigation facilities. Area coverage under these two seasons in Palakkad district accounted for 38 per cent of total rice lands in Kerala during 2016-17.

Decline in Area under Rice Cultivation

Time series data indicates that area coverage under rice declined in all the districts (Fig. 6.9). However, the rate of reduction is relatively less in the traditional rice-growing areas like Palakkad and Alappuzha. Ernakulam district, although part of the traditional rice-growing area, has experienced considerable loss due to the expansion of industrial-urban land use. The Spatial dimension of the nature of the decrease of rice fields deserves attention while planning for rejuvenation of rice cultivation in Kerala. There is an all-round effort to revive rice cultivation in the state particularly since 2007. These initiatives have yielded some results. The proposed Kuttanad package (KSPB, 2019) also envisaged the revival of rice cultivation in this traditional rice-growing area.

Production and Productivity of Rice

Kerala produced 4.36 lakh tonnes of rice by 2016-17. The year 1981-82 recorded the maximum production of 13.39 lakh tonnes although the area under rice cultivation reached a peak in the year 1974-75 when Kerala produced 13.31 lakh tonnes of rice. Since 1981-82, rice production continued to decline. By 1990-91, the production of rice was 10.6 lakh tonnes, which declined to 7.03 lakh tonnes in 2001-02 and further down to 5.22 lakh tonnes during the year 2010-11. Total production fell by 38 per cent in a span of 15 years from 2001-02 to 2016-17. Winter rice recorded the highest production in all the years as this is the main harvesting season. It is also noted that the reduction of production is high during the autumn season.

The average productivity of rice in Kerala was 2205 kg/ha in 2016-17, which was 2182 kg/ha in 2001-02 and 1959 kg/ha in the year 1991-92. Among three seasons, autumn rice recorded the highest (2877 kg./ha) productivity. The seasonal difference in productivity is also low in the main rice-producing areas. An increase in productivity indicates better management and increasing use of HYV seeds. Decline in production due to shortfall in the area can be partially recovered by raising productivity and perhaps Kerala should focus on this aspect of production management as the area under rice cultivation is continuously declining.

COCONUT

Kerala is the land of coconut. People immediately recognize Kerala when they look at a picture depicting a coastline with leaned coconut trees, or a stretch of back-water lined by coconut

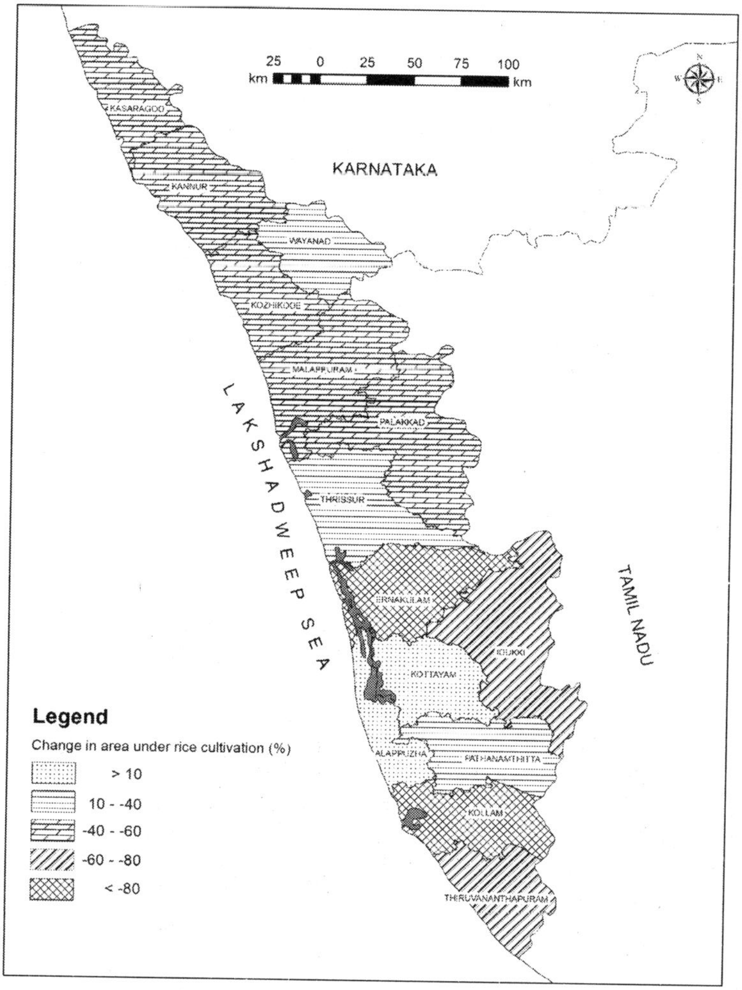

Fig. 6.9: Spatial Trend of Change in Area under Rice Cultivation by Districts in Kerala (2005-06 to 2016-17).
Data Source: Government of Kerala, 2006a; 2017a.

trees. Coconut grows in almost all countries in the Tropics; however, it is not yet possible to trace its origin to any particular country or region. Although coconut is mainly concentrated in the low lands, it is found across the higher altitudes also, possibly up to an elevation of 1000 m near the equator. Coconut palm grows best in areas receiving around 2000 mm of annual rainfall well distributed throughout the year (Markose, 1997). It is grown in Kerala in almost all types of lands ranging from coastal plain/river valleys to high altitude hilly areas, although all these lands may not be ecologically suitable for coconut cultivation. This is reflected in wide variations in productivity.

By 2016-17, Kerala had a 7.81 lakh hectare area under coconut, which is declining over the years. The peak coverage since 1960-61, was 9.26 lakh hectares attained in the year 2000-01. Afterward, there was a steady fall. Kerala's share in area coverage under coconut plantation at the all-India level has also declined over the years. Available data indicate that while in

1960-61, Kerala accounted for 69.58 per cent of the area and 69.52 per cent of production in the country, by 2011-12, Kerala's share declined to 40.2 per cent and 42.12 per cent respectively. Two neighbouring states, Karnataka and Tamil Nadu witnessed a huge expansion of area under coconut cultivation. Their combined share increased from 29 per cent in 1991-92 to 45.5 per cent in 2011-12 (Government of Kerala, 2014).

Varieties of Coconut

Coconut palms are of two varieties: Tall and Dwarf. West Coast Tall is an important variety. The tall varieties have a life span of 80-100 years. They start bearing fruits in 6-8 years and attain steady bearing by 12-15 years. The dwarf varieties, which are available in three colours of green, yellow, and orange, survive for 40-50 years and bear fruit by the 4th year. Copra and oil contents are less in dwarf varieties compared to tall varieties. The coconut research stations in Nileshwar and Kasaragod developed hybrids between tall and dwarf varieties in the 1930s. Two varieties developed are T x D (tall female parent) and D x T (dwarf female parent). Both the hybrids are early bearing (3-4 years) and yield relatively more fruits, provided management conditions are good. High breed varieties perform poorly than local varieties in case management is poor.

Methods of Coconut Cultivation

The selection of seed-nuts is very important as the performance of the plant can be learnt only several years after planting. Planting pits are normally of one cubic metre in size with 7.5 m to 9 m spacing. In one hectare there will be 175 palms (Markose, 1997). Well-drained, moist, and alternate layers of sand and clay provide appropriate soil conditions for growing coconut palms. "The advantage of coconut-based farming systems are efficient utilization of solar energy filtering through the canopy of coconut palms; efficient use of water and land; elimination of weeds; generation of employment opportunities; ensuring increased net income per unit area; deduction of risk on market glut due to overproduction of a single commodity and protection of environment and soil health." (Markose, 1997: 222)

Coconut gardens are significant for intercropping. High biomass production due to intercropping raises land productivity and also the productivity of the coconut crop. In a well-managed coconut garden, the multi-storeyed cropping systems are composed of coconut in the top tier; other tree crops like cocoa, papaya, etc. in the second tier; tapioca, yam, ginger, turmeric, etc. in the third tier. Pepper in the coconut trunk forms one more tier. Pineapple forms another tier either as fencing crop or ground cover in some places. Banana and vegetables also grow under coconut.

Distribution of Area under Coconut by Districts

Coconut in Kerala is grown in all types of soil and therefore its distribution is not restricted to anyone place. However, spatial variations in productivity indicate ecological limitations to cultivate this crop to its full potential. Among the districts, Kozhikode reported the maximum area under coconut of 1.2 lakh ha or 15.23 per cent of the total area under coconut in Kerala in the year 2016-17 followed by Malappuram (13.2 per cent) and Kannur (11.3 per cent). Coconut is the first ranking crop in eight districts. Hilly districts of Wayanad, Idukki, and Pathanamthitta all together accounted for 5.4 per cent of the total area under coconut in Kerala. Coverage under coconut records spatio-temporal variations over the years. The fall in coverage in southern districts like Thiruvananthapuram, Kollam, Pathanamthitta, Alappuzha, and Kottayam may be due to the incidence of root wilt disease. In fact, Alappuzha is the worst affected district. Almost all coconut trees in this district need replacement.

Production and Productivity of Coconut

In 2016-17, Kerala produced 5384 million nuts, which was 5479 million nuts in 2001-02. Although reduction in area under coconut between 2001-02 and 2016-17 was 14 per cent, production of nuts declined by 2 per cent. Expansion of area under high yielding varieties might have contributed to this scenario. In 2001-02, the average yield rate in Kerala was 6049 nuts/ha which rose to 6889 nuts per hectare in 2016-17. Coconut productivity has increased over the years in Kerala, however, other states like Tamil Nadu and Andhra Pradesh are performing better in the matter of productivity. Data for the year 2004-05 showed that while Kerala's average yield rate was 6379 nuts/ha, Maharashtra, Andhra Pradesh, and Tamil Nadu recorded 15189 nuts/ha, 11532 nuts/ha, and 9083 nuts/ha in the same order. Karnataka stands next to Kerala in terms of area, however, it records productivity of only 3139 nuts/ha.

Malappuram and Kozhikode are leading producers of coconut in Kerala by contributing 16 per cent and 15 per cent of total production respectively. These two districts along with Kannur and Kasaragod districts produce more than 50 per cent of total production in Kerala. District level yield rate varied from 3907nuts/ha in Idukki district to 8538 nuts/ha in Malappuram district during 2016-17. Wayanad has also reported low productivity. Ecological limitations perhaps play an important role. Idukki and Wayanad districts are located within the Western Ghats with a sizable area lying above 1000 m altitude.

TAPIOCA

Tapioca, also known as cassava, is a tuber crop. It is a staple food for a sizable section of people in the tropics, especially in Africa and Latin America. Tapioca is consumed as a cereal substitute in Kerala. Mexico is considered as the native place of tapioca, which was taken to the Philippines by the Spanish. Tapioca was introduced in India by the Portuguese during the early 18th century and it was actively promoted in south Kerala by the erstwhile king of Travancore kingdom (Nayar and Nayar, 1997). Kerala is the leading producer of tapioca in India both in terms of area and production.

Tapioca is a tropical crop. It cannot withstand waterlogging, high salt content, and alkalinity. It grows throughout Kerala. Lateritic soils, red loam, coastal alluvium, and alluvium soils are suitable. Most varieties of this crop can be harvested by 8-10 months after planting. Some short duration crops are ready to be harvested by the 7th month after planting. Tapioca used to be considered as a non-responsive crop, as it was raised in virgin soils without the application of fertilizers. However, now it is grown with proper administration of fertilizers. As exhaustive crop tapioca removes N, P, and K from soils. It has been estimated that 30 tonnes of fresh tubers/ha remove 180-200 kg of N, 15-22 kg of P and 140-160 kg of K. Adequate quantities of N, P_2O_5, and K_2O have to be applied to sustain the yield.

The total area under tapioca cultivation in three seasons of summer, autumn, and winter (based on harvesting season) was 2.37 lakh ha in 1961-62, which rose to 3.26 lakh ha in 1975-

76. Thereafter, the area under tapioca started dwindling and by 2001-02 it was 1.11 lakh ha and by 2016-17, it was only 0.69 lakh ha. The progressive decline in area under tapioca is mainly attributable to land use shift from food crop to non-food crop, especially, to accommodate rubber in place of tapioca in most of the areas.

Distribution of Area under Tapioca by Districts

Tapioca grows in all districts of Kerala, however, in varying proportions. In 2016-17, Thiruvananthapuram headed the list of the districts by having around 14,628 ha of the area under tapioca closely followed by Kollam (14,421 ha). These two districts together accounted for 42 per cent of the total area under tapioca in Kerala during 2016-17 (Table 6.5). Kasaragod district reported the least coverage of the 443 ha area. District-wise position changed significantly over the years.

Production and Productivity of Tapioca

The total production of tapioca in Kerala was 16.2 lakh tonnes in 1961-62. A peak production of 53.90 lakh tonnes was attained in 1975-76. Thereafter there was a sharp decline and by 2001-02, it was only 24.55 lakh tonnes. Production started increasing again and by 2016-17, it was 25.29 lakh tonnes with a yield rate of 36842 kg/ha. Pathanamthitta district recorded the highest yield rate of 45,285 kg/ha followed by Idukki (42,565 kg/ha) and Ernakulam (42,455 kg/ha). Idukki, Ernakulam, Thrissur, and Kottayam formed a block with a yield rate >40,000kg/ha.

PEPPER

Kerala is historically well known as the land of black pepper. International trade in ancient days was centred on spices, and pepper was the principal commodity. There are historical evidence indicating travels of merchants from Phoenicia and Greece way back in 8th Century CE and later Arabs to do barter trade for buying spices from Malabar. The sea route to Kerala from other countries is known as the 'spice route'. Search for the sea route to India for spice trade led Christopher Columbus in the 16th Century to discover America or the new World and in 1498 Vasco de Gama landed in Kozhikode (Ravindran *et al.*, 1997). Traders came from Arabs, Persians, and Europeans. European communities fought with each other to capture the pepper trade, and pepper growing hinterlands in Kerala.

Black pepper, a herb spices, requires a warm humid tropical climate with well distributed rainfall of 2000-3000 mm and temperature between 10-40 °C. It can grow from sea level up to an altitude of 1200 m. Humus rich well-drained soils having a pH of 4.5 to 6.0 are well suitable for growing pepper. Over 100 varieties have been reported from Kerala, however there are 32 varieties most common (*ibid*). Varietal diversity is most significant for the pepper crop. It is harvested during December-February in plains and January to April in hills.

Out of 11 crops listed under spices and condiments, pepper covered 32 per cent of the total area under spices and condiments during 2016-17. The total area under pepper was 85,207 ha or 3.3 per cent of the gross cropped area in Kerala. There is a fluctuating trend in pepper cultivation since 1960-61 when the coverage under pepper was little less than one lakh hectare. The rate of growth slowed down in the seventies and by 1980-81 it was negative growth. In the span of the last fifty years, the highest growth (+47 per cent) was recorded from 1985-86 to

1990-91. Steady growth continued till 2005-06 when 2.38 lakh hectares were brought under pepper cultivation. Since then, coverage declined and compared to 2005-06 the reduction in coverage was around 64 per cent by 2016-17.

Distribution of Area under Pepper by Districts

Although Kerala as a whole enjoys favourable climate for pepper cultivation, the main coverage is restricted to the hilly areas. Idukki district accounted for a little over 51 per cent of the total area under pepper in Kerala followed by Wayanad (12%) and Kannur (5%). Pepper was first and third ranking crops in Idukki and Wayanad districts respectively during 2016-17. Alappuzha district reported the lowest coverage.

Production and Productivity of Pepper

India contributed 16 per cent of the World's pepper production and Kerala produces around 80 per cent of India's production. It is one of the important commercial crops highly dependent on the export market. In 2016-17, Kerala produced little over 34 thousand tonnes of pepper. It fell from a whopping 87 thousand tonnes produced in 2005-06. The production declined by 61 per cent, whereas, area reduced by 64 per cent. The present yield rate is 400 kg/ha ranging from 221kg/ha in Malappuram to 596 kg/ha in Kasaragod. Idukki district reported the highest productivity of 618kg/ha in the year 2005-06, and since then, productivity declined and by 2016-17, it came down to 428 kg/ha. Export earnings from pepper declined since 2000-01 as a result of a shortfall in exported quantities. There is strong competition from Vietnam in the International market. The Pepper economy also suffered as India is now importing pepper from Sri Lanka under Indo-Sri Lanka free trade agreement and Vietnam under WTO provision.

Pepper is a shallow root crop. Soil moisture is declining due to deforestation and a gradual increase in atmospheric temperature. The major pepper-producing areas are affected due to ecological change and as a result, productivity is coming down. Price fluctuation due to the international market contributes to a reduction in coverage. There is a need for a special package to revitalize pepper cultivation and to safeguard pepper growers. As pepper cultivators consist of mostly small and marginal landholders, revitalization of pepper cultivation will have a larger impact on livelihood security.

RUBBER

The introduction of rubber (*Hevea Brasiliensis*) as a commercial plantation in the first decade of the 20th century at the instance of the Periyar Syndicate in Travancore triggered the large-scale land use change in Kerala. Rubber plantation has emerged as an important contributor to the agro-economy of Kerala during the past one century. It has brought economic prosperity to many small and marginal farmers. Rubber is an industrial raw material and supports several large and small scale industries. However, prolonged rubber cultivation has some reportedly adverse ecological impact, which is yet to be properly studied (Chattopadhyay, 2015).

India ranks fourth in the World by producing 8 per cent of total natural rubber production. Kerala accounts for more than 90 per cent of rubber production in India. Kerala along with adjoining areas of Karnataka and Tamil Nadu are traditional rubber growing areas in the country.

Kerala's share at the all-India level is coming down with the expansion of rubber growing areas in other states.

By 2010-11, the area under rubber covered 20 per cent of the gross cropped area of Kerala and ranked second among the crops in area coverage next only to coconut. Rubber replaced many of the garden land crops and degraded forest areas under private ownership. It has even replaced rice in valley heads. Although there is a steady growth in the absolute area, the rate of growth shows a downward trend in the case of area, production, and yield. Rubber prices in Kerala used to be determined locally, however, with the introduction of WTO market price in Kerala and liberalisation of the economy, the rubber market in Kerala is gradually getting integrated with the international market. Rubber growers in Kerala are now exposed to international competition. The exclusion of rubber from agricultural commodities has added another problem of dumping of rubbers by major countries producing natural rubber.

Warm tropical climate with a temperature of 21 °C to 35 °C and evenly distributed rainfall of about 2000 mm provide an optimum condition for the growth of rubber plants. Although the rubber plant shows wider adaptation, it is best suited in the altitudinal range from 20 m to 500 m above mean sea level. Productivity reduces at higher altitudes. Rubber plants are sturdy, tall, and fast-growing and have a life of 30 to 35 years. After planting rubber saplings, there is a gestation period of 5/6 years prior to the tapping of latex.

Distribution of Area under Rubber Plantations by Districts

By 1961-62, rubber occupied 1.33 lakh hectares of area, which increased to 5.51 lakh ha in 2016-17. Starting from south Kerala, rubber has now spread over all districts including highland districts like Wayanad and Idukki in the Western Ghats. Area coverage of rubber varied from little over 4.5 thousand hectares in Alappuzha to more than 114 thousand hectares in Kottayam, which is the main centre of rubber plantation. Kottayam district accounted for 21 per cent of the total area under rubber plantation in Kerala followed by Ernakulam (11%) and Pathanamthitta (9.4%). Rubber plantations are mostly concentrated along foothills. Trivandrum-Kottayam stretch is the most densely covered zone. Rubber plantation is aggressively promoted by Rubber Board. Even small and marginal farmers raise rubber plantations in their meagre landholdings.

Production and Productivity of Rubber

Total rubber production in Kerala was 5.4 lakh tonnes in 2016-17. Kottayam district occupied the preeminent position by producing 20 per cent of all rubber production in Kerala. Alappuzha district contributed only 0.88 per cent of state production corresponding to its share of the area of 0.83 per cent. Wayanad district accounted for 1.9 per cent of the area under rubber in Kerala; however, it contributed only 1.4 per cent of rubber production in the state. This is possibly due to the comparatively low productivity of rubber in the Wayanad district. District-wise figures indicated that productivity varied from 722 kg/ha in Wayanad to 1057kg/ha in Pathanamthitta. Kozhikode, Kannur, Kollam, and Alappuzha districts recorded a yield rate of > 1000 kg/ha. Productivity in Kerala averaged at 981 kg/ha in 2016-17 which was 1442 kg/ha in 2010-11. Kerala is witnessing a fall in the productivity of rubber. Perhaps, rubber cultivation in Kerala has reached its peak and it is now spreading in areas, not ecologically suitable, thereby the average productivity of the state is coming down.

TEA, COFFEE, COCOA AND CARDAMOM

Tea, coffee, and cocoa are the other three plantation crops listed under non-food crops in 2016-17. Among these three, coffee covered the maximum area of 85,000 ha in 2016-17 which was around 21 per cent of the total area under coffee in the country. Karnataka accounts for the highest coverage (56%) in the country. Wayanad is the major coffee producing area by having 79 per cent of the total area under coffee in Kerala by 2016-17. Coffee is grown even in smallholdings at the household level. Idukki and Palakkad are the other two districts growing coffee. Robusta is the major variety of coffee planted in Kerala. The productivity of coffee in Kerala (808 kg/ha) is lower than that of the national average (852 kg/ha).

Kerala accounted for 6.4 per cent of the total area under tea plantations in the country and produced 6.9 per cent of total national tea production. Tea plantations covered 30,205 ha of area in Kerala by 2016-17. They were mostly concentrated in the Idukki district (72.7%) followed by Wayanad (17.6%). Apart from Idukki and Wayanad districts, other districts that have some coverage under tea plantations include Thiruvananthapuram, Kollam, Palakkad, and Thrissur. Cocoa is raised only in 14,404 ha of area distributed in all districts of Kerala. Idukki has the maximum concentration of 65.6 per cent of the area under Cocoa in the state.

Cardamom, listed under spices and condiments is a high-value crop. Kerala produces 79 per cent of total cardamom production in the country. A little over 39,080 ha of area are given to raise cardamom plantations in Kerala. Idukki district has the highest concentration of cardamom gardens (79.7%). Wayanad and Palakkad are the other two districts that reported more than 2000 ha each under this crop. Pathanamthitta, Kottayam, Malappuram, and Kozhikode are another four districts, which reported some areas under the cardamom garden. Cardamom accounted for 3.7 per cent of Kerala's total value of Rs. 31,983 crore from major agricultural crops in 2014-15.

The distribution of these crops indicates agro-ecological conditions. Cocoa can be raised throughout Kerala, whereas cardamom, coffee, and tea, each has specific agro-climatic requirements limited to certain altitudes.

HOMESTEAD FARMING

Homestead farming is a sub-system of land use systems. It is practiced in several countries across the World principally to provide food security and other livelihood requirements at the household level. This unique agro-ecologic system tends to optimise utilisation of land resources, solar energy, and recycling of household wastes (Nair and Sreedharan, 1986). Kerala is, historically, well known for homestead farming system and biodiversity of homestead gardens. Traditional homestead farming system with highly diverse crop composition is a low input sustainable agro-ecosystem, which has co-evolved with native socio-economic, ecological, and cultural systems (Sasidhar and Lakshmi, 1997). This agro-ecosystem is often compared to the tropical rainforest due to structure, species diversity, energy flow, biogeochemical cycling, nutrient cycling, water cycling, and other ecological characteristics (*ibid*). The high biodiversity of the homestead gardens is a conscious decision of the farmers to grow plants and trees for meeting various household needs. Traditionally, the composition of home gardens includes spices and fruit trees. Besides, plants having medicinal qualities are also grown. Some households plant high-value timber plants like teak near plot boundary. There are structural and functional diversity of homestead crops. The structural configuration of the home garden consists of tall trees like coconut, and arecanut in the top most layers. Mango, Jack fruit, breadfruit, sapota, tamarind etc. comprise the second tier. The third tier is occupied by banana, papaya, guava, custard apple, etc. and there are pineapple, ginger, lemon, turmeric, yam, tapioca, sweet potato, etc. in the fourth tier. Black pepper is raised as a climber. Production from homestead gardens meets a sizable portion of household demand of food consumption. The homestead trees, particularly jack fruit, mango, and guava shed litter round the year and thereby return a considerable amount of nutrients to the soil. It has been estimated that the annual leaf-litter productions by jackfruit, wild jack, mango, and guava are 3.37, 3.95, 2.42, and 1 tonne ha^{-1} $year^{-1}$ respectively (John, 2014). These litters coupled with kitchen wastes and animal wastes, in case of households rearing cows, goats, and poultry, provide nutrient needs of the crops and plants within the homestead garden. There is a complete recycling of organic matters on the household scale. Cashew is also found as homestead crop, especially in Kasargod, Kannur, Kozhikode, Malappuram, Kollam, and Thiruvananthapuram districts.

The introduction of rubber has changed the homestead cropping pattern. Due to high economic return and other socio-economic conditions, rubber is preferred as homestead crops in many places across Kerala. In initial years, when rubber plants are small, there are local varieties of banana and pineapple grown side by side in these plots. Once the rubber plants are well established and the canopy develops, the entire plot is occupied by mono-cropping of rubber plantation.

Homestead cultivation is under stress due to various factors including real estate boom and fragmentation of holdings. Land use change, shortage of agricultural labour, and falling commodity prices are major constraints in homestead farming (John, 2014). There are also problems of low yield, inadequate marketing facilities of surplus products from homestead gardens, and lack of awareness among the new generation people about the nutritional and medicinal values of the homestead crops. The structural diversity that used to be the primary characteristics of the traditional homestead farming is gradually disappearing as observed in the recent village-level analysis (Andrews and Kannan, 2016). Homestead farming is usually practiced in isolation. Homestead farmers seldom discuss among themselves about crops and technology use. There is a need for clustering approach in optimum utilisation of homestead farms, introduction of technology, which is at low level, and market household products. Rejuvenation of homestead cultivation is necessary for long term benefit of Kerala. It will help in economic-ecologic capacity building, empowerment and food and livelihood security of the local people.

VALUE OF MAJOR CROPS

Twenty four crops were considered to compute the value of crops in Kerala for the year 2014-15. The total value was estimated at INR 31,983 crores. This gave an average value of INR 123,424 per hectare of the gross cropped area. Eight crops contributed 88 per cent of the total value of crops. Coconut contributed the maximum value accounting for 25.9 per cent followed by rubber (18.2%), tapioca (11.5%), banana and plantain (8.59%), black pepper (8.2%), arecanut (7.1%), rice (4.95%) and cardamom (3.68%). Although rice ranked 3rd in area coverage, its position was 7th in value addition. The contribution of districts in the total value of crops in Kerala varied from 4.31 per cent in the Pathanamthitta district to 13.73 per cent in the Idukki district (Fig. 6.10).

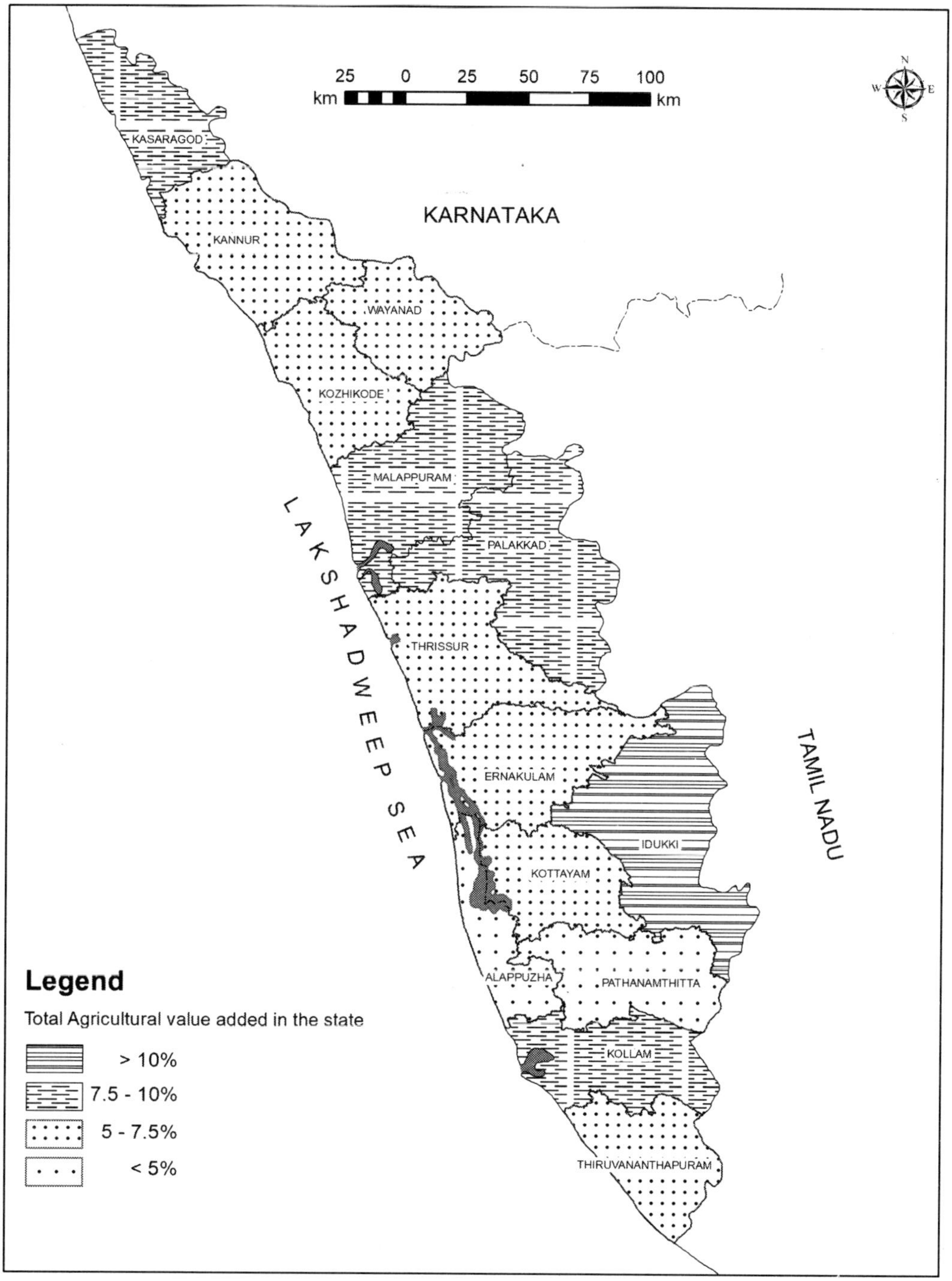

Fig. 6.10: Contribution of Districts to Total Value of Crops in Kerala.
Data Source: Government of Kerala, 2018.

The concentration of high-value plantation crops in the Idukki district had contributed to this trend. There were district-level variations in generating agricultural income. The importance of crops varied from district to district. Coconut contributed 61.7 per cent of total income from the agricultural sector in Kozhikode district and 41.6 per cent in Malappuram district whereas rubber contributed 55 per cent of agricultural income in Kottayam and black pepper and cardamom together accounted for 60 per cent of Idukki district's income from agriculture. The hill districts like Wayanad and Idukki received more income from spices and plantations of tea, coffee, and cardamom. Change in the importance of crops is reported. Cardamom, ginger, and pepper are gradually losing their importance. This is partly attributable to environmental changes experienced in the areas where these crops are usually raised.

CHAPTER

7

Primary Sector-II: Animal Husbandry, Fisheries and Forest Resources

ANIMAL HUSBANDRY

Livestock plays an important role in the rural economy, and traditionally, it maintained a symbiotic relationship with agriculture. It is an asset, and often it is the main source of income for people with limited or no access to agricultural land. Apart from milk, meat, and egg, the contribution of livestock by way of draught power and manure is significant. There are also other income-earning by-products like skin, hide, blood and bone. The livestock sector generates livelihood opportunities, employment, and entrepreneurship, and provides protein security. We briefly discuss here some of the salient features of this sector.

LIVESTOCK

The livestock sector contributed 3 per cent of Gross State Value Added (GSVA) from all sectors and 29 per cent of GSVA coming from agricultural and allied sectors in Kerala during 2015-16. This sector witnessed a 4.3 per cent growth in recent years. It is significant as agricultural and allied sectors are experiencing negative growth. Gross capital formation in livestock is relatively high. In spite of all these, Kerala experienced a drastic reduction in various categories of livestock over the last couple of decades (Government of Kerala, 2017b). Available data since the formation of Kerala in 1956 till 2012 indicated that there were two distinct phases of development of this sector. The first phase from 1956 to 1987 recorded steady growth and the second phase from 1987 to 2012 was a declining phase (Table 7.1). In 2012, the total livestock population in Kerala was 27.33 lakh heads, which was 54.71 lakh heads in 1987. As reported in 2012, cattle comprised 48.6 per cent of total livestock followed by goats (45.6%), buffaloes (3.7%), and pigs (2.1%). Marginal farmers possessing less than 1 hectare of land owned around 87.7 per cent of cattle, 86.6 per cent of buffalo, 92.6 per cent of goats, and 57.5 per cent of pigs in Kerala. The economic linkage between livestock and poultry and resource-poor farmers is well evident. Distribution of population and distribution of livestock population at the district level is strongly correlated with the level of significance at > 99 per cent (Fig. 7.1).

Table 7.1: Livestock Population in Kerala (in Thousands) (1956-2012)

Type/Year	1956	1961	1966	1977	1987	1996	2003	2007	2012
Cattle	2510	2753	2857	3006	3424	3396	2122	1740	1329
Buffalo	488	485	471	454	329	165	65	58	102
Goats	NA	NA	NA	NA	1581	1861	1213	1729	1246
Pigs	NA	NA	NA	NA	137	143	76	59	56
Total	2998	3238	3328	3460	5471	4555	3476	3586	2733
Poultry	NA	NA	NA	NA	17938	26833	12876	13959	24282

Source: Planning Commission, 2008; Government of Kerala, 2015a.

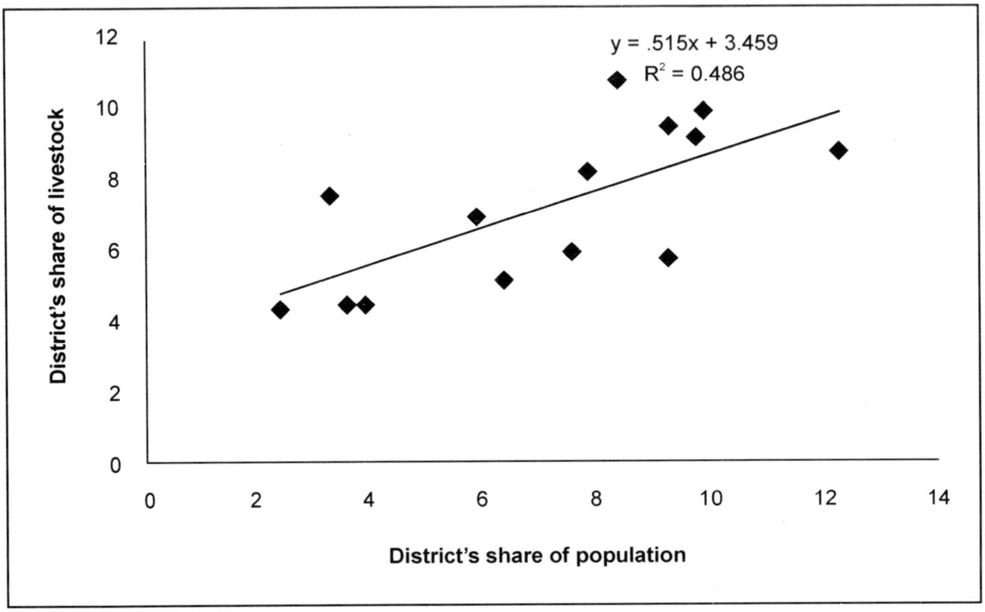

Fig. 7.1: Relation between District's Share of Population and Livestock Population in Kerala.

Rearing of livestock is common to all districts. Palakkad district recorded the highest concentration (10.7%) of the livestock population followed by Thiruvananthapuram, Thrissur, and Ernakulam districts (Fig. 7.2). These four districts together house around 39 per cent of all live-stocks in Kerala. This pattern of distribution changes when the number of animals is considered in relation to households in each district. It is found that Kerala as a whole has 309 livestock animals for every 1000 households and this number varies from 209 animals in Kozhikode to 617 animals in Idukki (Fig. 7.3). Wayanad and Palakkad are other two districts possessing 528 and 450 animal heads for every 1000 households respectively.

Cattle and Buffalo

Traditionally, cattle and buffalo are an integral part of agriculture and the rural economy. It is a popular supplementary avocation for small and marginal farmers. In 1956, there was 29.9 lakh bovine population comprised of 25.10 lakh cattle and 4.88 lakh buffaloes (Table 7.1). The total number of cattle and buffalo population touched 37.53 lakh in 1987 recording a growth of 25 per

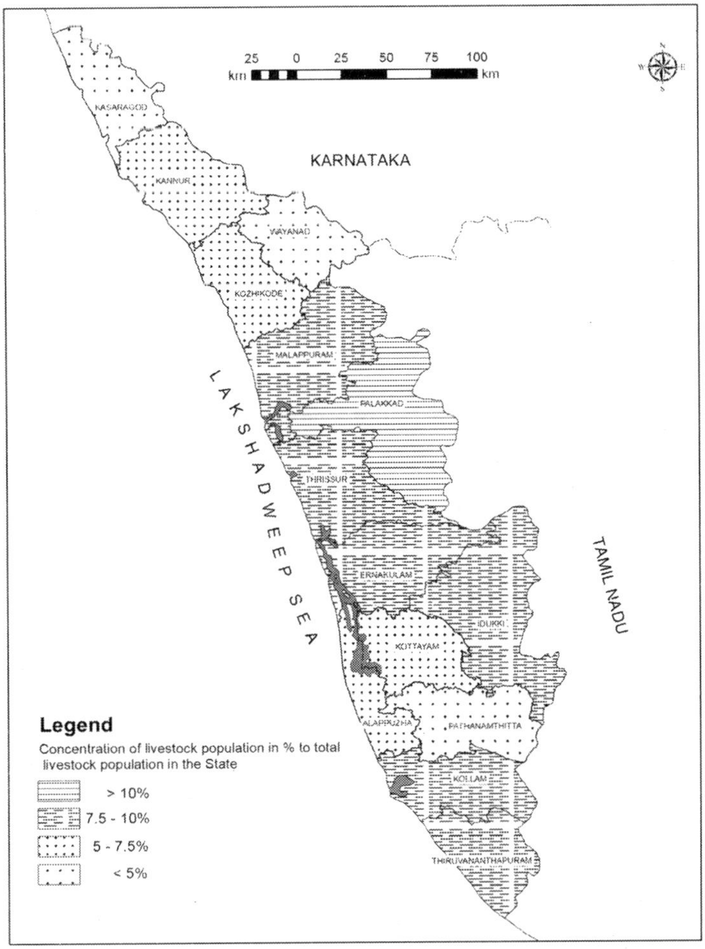

Fig. 7.2: Concentration of Livestock Population by Districts in Kerala, 2012.
Data Source: Government of Kerala, 2015a.

cent over a span of three decades. The bovine population declined since 1987 and by 2012, the number came down drastically to 14.31 lakh heads (Government of Kerala, 2015a). The fall was 62 per cent in a span of 25 years. Cattle accounting for 13.29 lakh constituted 93 per cent of the total bovine population in 2012. Female cattle were 11.8 lakh. While the cattle population increased during the first phase and started declining during the second phase, buffaloes declined steadily from 1956 to 2007 and recorded an increase of 44 thousand between 2007 and 2012. Attempts to introduce cross-breed varieties among the cattle population started in 1956. It picked up slowly and by 1987 cross-breed varieties overtook the indigenous varieties. Thus growth in the first phase was partly triggered by improvement in the genetic qualities of indigenous cattle through cross-breeding technology introduced under the Indo-Swiss project (George and Nair, 1990). The data for 2012 indicated that as much as 94 per cent of the total cattle population in Kerala was cross-breed. Female cross-breed constituted 83.9 per cent in 2012. Indigenous varieties still have a sizable presence only in the Kasaragod district. Little over six lakh households or 6.85 per

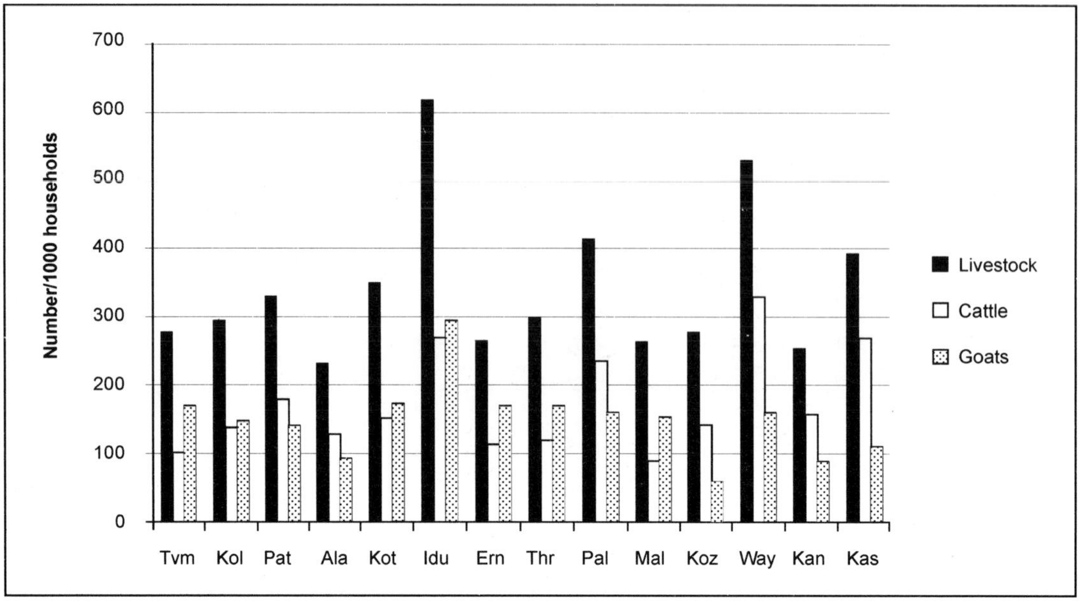

Fig 7.3: Distribution of Livestock, Cattle and Goats by Districts in Kerala, 2012.
Data Source: Government of Kerala, 2015a.

cent of total households in Kerala own cattle. Besides, there are 1488 non-household enterprises and institutions possessing cattle. Around 44 per cent of all cattle in Kerala are concentrated in five districts of Palakkad, Ernakulam, Thrissur, Kozhikode and Kollam.

The number of cattle per 1000 households varies from 90 in Malappuram district to 329 in Wayanad district (Fig. 7.3). Considering district-wise households rearing cattle it is found that 15.2 per cent of total households in Wayanad district possess cattle followed by Idukki (13%) and Kasaragod (13%). All three districts lag behind economically and house a sizable number of marginal farmers.

Goats

According to the 2012 livestock census, Kerala had 12.46 lakh goats which were 15.81 lakh in 2007. Considering the last 25 years, it is found that Kerala husbanded the highest number of goats of 18.61 lakh in 1996 (Table 7.1). Since then, the goat population decreased with an intermediate high during 2007. Both male and female varieties decreased during the inter census period between 2007 and 2012. Around 4.5 lakh households or 5.1 per cent of the total households in Kerala domesticate goats. The number of goats per 1000 households varies from 59 in Kozhiode district to 295 in the Idukki district (Fig. 7.3). The proportion of households rearing goats varies from 3.6 per cent in Palakkad and Kasaragod districts to 10.8 per cent in the Idukki district. Special attention is necessary for goat rearing as it has a multiplying impact on the economy of poor people, particularly landless people. Goat development is taking place in Kerala with little policy guidelines.

POULTRY

Poultry is an important sub-sector of agriculture and is growing at a faster rate. It provides cheap sources of animal protein and with proper technological intervention, this sector evolved from a backyard practice to an industrial-commercial venture. Poultry uses substantial quantities of non-edible agricultural and industrial bi-products. Fowls constitute the lion share of poultry birds in Kerala. Besides, there are duck, turkey, and others. Kerala had 242.82 lakh poultry birds in 2012 with fowls accounting for 89.4 per cent. While the livestock population showed a declining trend, there was significant growth in the poultry sector with fluctuations from time to time (Table 7.1). Poultry birds increased by 88.6 per cent between 2003 and 2012, however, compared to 1996, there was a decline of 9.5 per cent during 2012. Around 22 per cent of households in Kerala maintain poultry and the majority of them are marginal farmers who own 93.5 per cent of poultry birds. On average, there are 2747 poultry birds for every 1000 households in Kerala (Fig. 7.4).

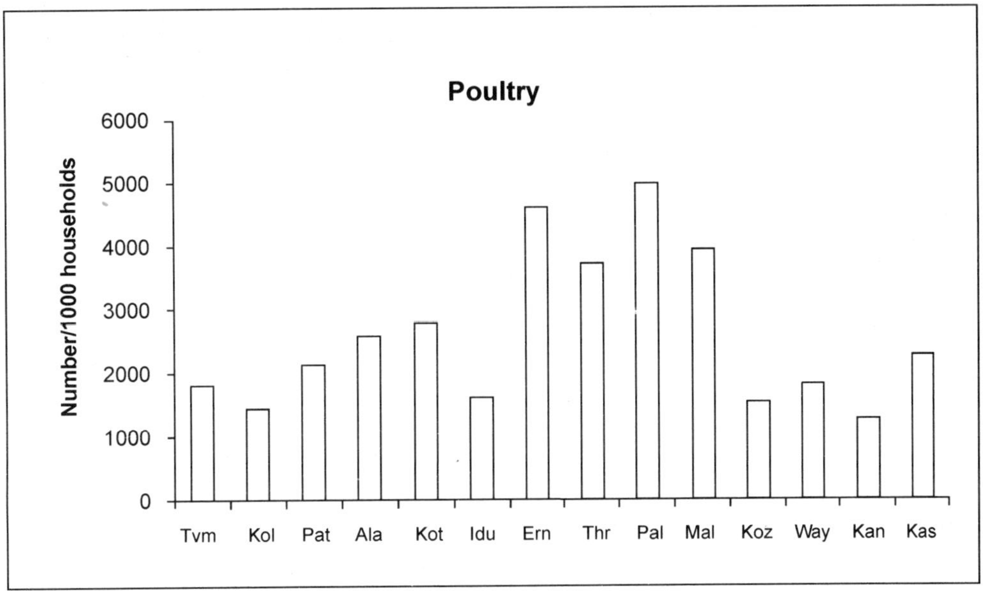

Fig. 7.4: Distribution of Poultry Population by Districts in Kerala, 2012.
Data Source: Government of Kerala, 2015a.

Palakkad ranks first by possessing 4975 birds for every 1000 households followed by Ernakulam (4591) and Malappuram (3916). District level distribution of households owning poultry varies from 14.1 per cent in Kannur to 49.8 per cent in Pathanamthitta. This indicates that every second household in the Pathanamthitta district has backyard poultry. This district caters to the need of adjoining urban centres. In the case of Malappuram, poultry is found in every third household, and in Wayanad and Idukki it is found in every fourth household.

PRODUCTION OF MILK, EGG, AND MEAT

Kerala has a 6.7 lakh female adult cattle population, which is 0.5 per cent of the national total. By 2016-17, Kerala produced 25.20 lakh tonnes of milk or 1.5 per cent of total milk production in India. Cross-breed cattle produced 93.5 per cent of total milk production in Kerala; goats produced 5 per cent, and the rest came from buffaloes and indigenous cows. Milk production in Kerala increased gradually from 7.78 lakh tonnes in 1977-78 to 27.17 lakh tonnes in 2001-02. There were fluctuations since then and the highest production of 27.92 lakh tonnes was reported in 2012-13. More than 90 per cent of milk comes from cross-breed cows. The rapid increase in cross-breed varieties in Kerala has been facilitated by the state government through the provisioning of infrastructure facilities for the production and distribution of materials for artificial insemination. Kerala has 2900 Artificial Insemination centres and 7 regional semen banks to provide services for cross-breeding. At present, the average milk yield is 9.2 litres/day, which is second only to the state of Punjab.

Per capita availability of milk in Kerala is 189 gm/day against all-India average of 352 gm/day. While per capita milk availability is increasing at the all-India level, Kerala is witnessing a declining trend. However, the demand for milk is increasing in Kerala. The Kerala Co-operative Milk Marketing Federation (KCMMF) responsible for the marketing of milk procured 3834.58 lakh litres of milk against the sale of 4774.20 lakh litres in 2016-17 through 12 dairies spread throughout Kerala (Fig. 7.5).

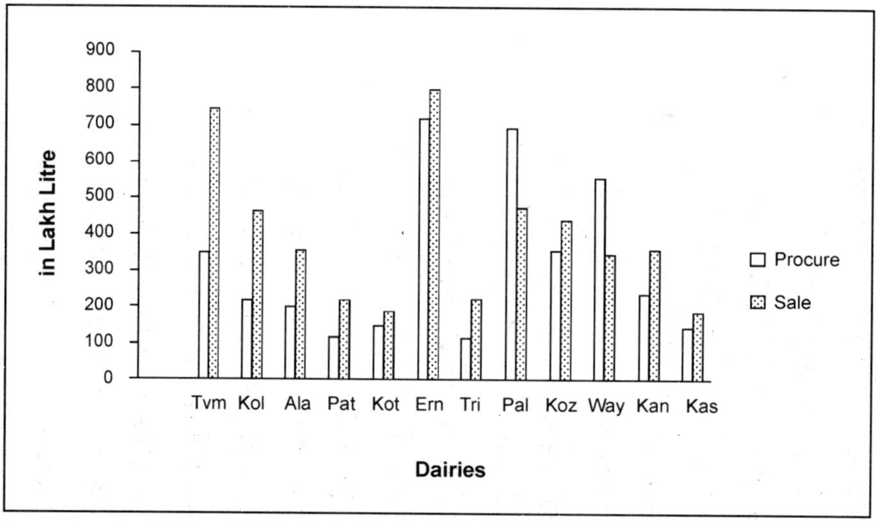

Fig. 7.5: Procurement and Sale of Milk by Dairies in Kerala, 2016-17.
Data Source: Government of Kerala, 2018.

Kerala experiences a shortfall of around 20 per cent of liquid milk. This is being met by procuring milk from Tamil Nadu, Karnataka, and skimmed milk powder. The annual growth of milk production in Kerala is 2 per cent, well below the national average of 4.8 per cent. There are 2891 functional societies with a total member of 9.4 lakh of which women member count for 2.21 lakh. Although women members constitute 23.5 per cent of total members, only 5.7 per cent of the dairy co-operatives have women as presidents, and women dairy co-operatives accounted for 2.2 per cent of all dairy co-operatives in Kerala.

About two-thirds of dairy animals are concentrated in seven districts forming three zones, namely the northern zone covering Kannur and Kozhikode districts, the central zone covering Palakkad, Thrissur, and Ernakulam districts, and the southern zone consisting of Kollam and

Thiruvananthapuram districts. Zone specific approach is necessary to address the problems of feed, technology, productivity enhancement, and enterprise development.

Kerala produced 2.44 billion eggs during 2016-17. Per capita availability has been 64 eggs per annum. Egg production declined over the years. Around 99 per cent of total egg production came from backyard poultry. Improved fowls contributed 68.8 per cent of total egg production and local *(desi)* fowls produced 25.7 per cent. Meat production in Kerala was around 4.69 lakh tonnes during 2016-17. Poultry meat constituted 40.3 per cent of total meat production followed by cattle (31.2%) and Buffalo (22.6%). Goat contributed only 4.5 per cent of total meat production in Kerala. During a span of 10 years from 2006-7 to 2016-17, egg production in Kerala grew by 105 per cent against all-India average growth of 64 per cent whereas in the case of meat production Kerala's growth (135%) was well below all-India average (205%).

Kerala's internal production does not match the demand. Kerala produced 45 per cent of her egg requirement and 40 per cent of her meat requirement, domestically, during 2014-15. The shortfall was met from neighbouring states of Karnataka and Tamil Nadu, and also from Andhra Pradesh. Kerala's dependence on external sources is increasing over the years both for egg and all types of meats. It is important to note that Kerala used to export eggs during the 1970s. The decline in livestock and poultry, changing food habits and preference for chicken meat in place of red meat perhaps have fuelled high demand in the poultry sector.

CHALLENGES OF LIVESTOCK REARING

Continued reduction of the bovine stock is a major issue warranting the immediate attention of policy-makers and livestock resource managers. There are several reasons underlying this trend. The cropping pattern in Kerala is changing fast. Cultivable lands are increasingly being converted to provide space for fast-growing urbanization. Plantation crops and tree crops are replacing livestock friendly seasonal crops. Crop wastes, particularly hay and straw traditionally used as dry fodder for the bovine population have reduced with the decline in rice-growing areas. Kerala does not have pasture lands like other states. Most of the livestock rearer at present depend on stall feeding. However, the availability of cattle feed is a major problem as Kerala does not produce sufficient fodder crops. Fodder availability is estimated to be around 36 per cent of total requirements (Government of Kerala, 2017b). Marginal and small farmers owning more than 90 per cent of livestock are not in a position to maintain cattle or buffalo like their ancestors. In this context, goats are important for poor people in rural areas.

Apart from the shortage of fodder, other reasons contributing to this scenario are marginalisation of agricultural holdings, the declining trend in family participation, in particular among youth, high cost of production as a result of increasing reliance on externally sourced purchased inputs, uneconomic price level due to inflow of cheaper livestock products from neighbouring states and change in favour of urban way of life. Technological change in farming operations like the use of tractors, thrashers, and harvesters, and the development of modern transport facilities for carrying agricultural goods have reduced the role of drought cattle in favour of milch cattle. As milk production per animal has increased with the adoption of cross-breed varieties, there is little need to maintain a large number of cattle for small farmers to cater to the need of households.

Due to all these issues doubts are raised about the viability of homestead based livestock husbandry with technological support as is being done now. There are various suggestions to establish large cattle farms for the supply of milk and meat. However, livestock provides and supplement livelihood opportunities for large sections of landless households and small and marginal farmers. It is the only source of income for many poor people, who will find it difficult to survive in absence of adequate support for maintaining livestock.

MANAGEMENT OF LIVESTOCK SECTOR

The livestock sector in Kerala is beset with various problems, many of which are management-related, although, Kerala has made significant progress in the dairy production system in recent years. In fact, the 2019 study of the National Dairy Research Institute (NDRI) has placed Kerala in the top among 20 milk-producing states in India (*The Hindu*, 2020a). Kerala outshone all other states in the country in the matter of animal breeding, resource availability, and policies and regulations. It ranks second in veterinary infrastructure, seventh in dairy production, and eighth in value addition and marketing. The willingness of highly literate dairy farmers in Kerala to adopt scientific dairy farming practices, and to acquire technical knowledge about the sector and favourable Government policy favoured the development of the dairy sector in Kerala.

Nevertheless, there are challenges of self-sufficiency, productivity, and management. New opportunities are opening up with the change in lifestyle, people's willingness to spend more on food, increasing demand for meat and milk products, demand-supply gap, increasing demand for processed food, and more and more thrust on organic farming and locally grown food. It is important to take a holistic view and practice integrated agriculture by linking organic farming and livestock development. Around 84 per cent of urban households and 97 per cent of rural households are non-vegetarian in Kerala. There is a vast internal market, which may be tapped with proper intervention measures. Animals, agriculture, and integrated farming will be indispensable for farm sector development and transition to sustainable livelihood strategy. Agro-ecological zone and location-specific innovative intervention measures are required to address the challenges and realize the potential.

The dairy sector offers ample opportunities for women empowerment. The 2019 Survey Report of Women Dairy Farmers highlighted that in spite of the massive presence of women in the dairy sector their representation in dairy co-operatives is low. The report also brought out that women dairy farmers work for more hours (4 to 4.5 hours a day) than their male counterparts. Around 77 per cent of women dairy farmers own 1 to 2 cows and 51.5 per cent of them belong to the BPL category. The government can design specific projects to support women dairy farmers. It will help addressing not only the issue of dairy management but also the overall improvement of the household economy.

FISHERIES

Traditionally, fish is an important component of food for a large section of the population in the world. Besides, being the source of high-quality animal protein, it supplies essential micronutrients, minerals, and fatty acids required for human beings. Kerala with a continental shelf area of 40,000 km^2 has a marine fishery potential of 0.57 million tonnes in the nearshore water, which is around 26 per cent of national potential and 38 per cent of the potential of the west coast although Kerala possesses only 7 per cent of all-India coastline and 7.8 per cent of continental shelf area of the country. These figures indicate the high productivity of the coastal waters of Kerala. Inland water spreading over 4,000 km^2 consisting of 44 rivers, 30 major reservoirs, freshwater ponds and brackish coastal water bodies (lagoons and estuaries) provides enough scope for inland fishing. By 2016-17, Fisheries and aquaculture contributed Rs. 4051 crore to Gross State Value

Addition (GSVA), which was around 8.5 per cent of GSVA from the primary sector. There are 222 villages in Kerala with the dominance of marine fishing and 113 villages with inland fishing communities. The active fishing population is a little over one fifth of the total fishing population over a million. Only 18 per cent of the active fishing population operates in the inland sector.

FISH PRODUCTION

Based on the hydrodynamic condition and water quality, the coastal water ecosystem of Kerala can be divided into two broad fishing zones: South of Kollam (Neendakara/Asthamudi mouth), and North of Kollam. The development of mud banks has further modified the coastal water ecosystem, particularly between Alappuzha and Kochi. Due to the remarkable damping effect on the incident waves, mud banks provide a congenial environment for fishing even during the high monsoon when marine fishing in the open sea becomes highly risky. Mud banks, locally known as '*Chagara*' even inspired novelists to compose novels based on fishermen's life.

This division of two zones lying to the north and south of Neendakara (Kollam) is well reflected in the fishing methods, fish catch, and other fishery-related activities comparable to all-India zonation. During the year 2015-16, Kerala's share of marine fish landing in India was 14.2 per cent, which was 19 per cent in the year 1997-98. Total marine fish catch in Kerala had increased from 0.28 million tonnes in 1980 to 0.66 million tonnes in 1990. The phenomenal increase was mainly due to modernisation like an increase in trawling operation, the motorization of country craft, the introduction of purse seines, the emergence of ring seines, and increased operation of multi-day trawlers and extension of fishing area.

The catch crossed the mark of 0.60 million tonnes only in 1989. It peaked in 1990 and then slid down. It was always below 0.60 million tonnes from 1991 to 2001 with the lowest catch (0.53 million tonnes) recorded in the year 1995. The catch again crossed the mark of 0.60 million tonnes in 2002. A study on Cochin Fishery Harbour indicated that from 1999 onwards, deep-sea trawlers started to harvest the depth zone up to 400 m to exploit deep sea resources (Ammini and Augustine, 2003). Deep sea prawn fishing also started in November 1999, and it was a successful operation (Nandakumar *et al.*, 2001). However, the catch started declining again and by 2011-12, it was 0.55 million tonnes and by 2016-17, it further slipped down to 0.49 million tonnes. There was near stagnation in production during the last two decades with average of around 0.57 million tonnes. Initial growth in the 1980s underpinned by technological up-gradation of fishing gears and enlargement of fishing ground could not be sustained. Substantial increase of motorized fishing crafts from the late 1980s to the early years of this century has not contributed to any significant increase in fish production. However, an increase in the number of mechanized crafts has resulted in a decrease in the share of fish landings from traditional crafts. This has led to frequent confrontations between traditional fishermen and operators of mechanized crafts. The indiscriminate use of fishing gear and nets in violation of the Marine Fisheries Regulation Act has also aggravated the problem (Planning Commission of India, 2008a).

There are three main groups of fishes harvested in Kerala: Pelagic, Demersal, and Molluscs and crustaceans. The pelagic group accounts for more than 65 per cent of all catches. It is reported that the demersal group of fishes is showing a declining trend over the years (Chattoapdhyay and Shravankumar, 2007). Although there are large varieties of species harvested, only 16 species together account for more than 85 per cent of total catch as noted since 1980 (Table 7.2). Oil sardine is the dominant species, however, its share in total catch has come down over the years. The relative share of species changes over the years. The fish catch varies seasonally. The monsoon season spreading over July, August and September witness the highest landings of fishes accounting for 31 per cent of the total catch. Spawning takes place during June-July. A trawling ban is also imposed during this period to avoid disturbance to fish growth. In some parts of the Kerala coast, a unique system, known as '*Kara Nila*' is prevalent among marine fishermen (Chattopadhyay and Franke, 2006). It is a traditional community group formed by marine fishermen, where a fisherman can take part in fishing by rotation and the catch is always equally divided among the group members. This locally evolved system intends to provide sustained livelihood opportunities among the fishing population by sharing fish catches.

The inland fish production in Kerala is steadily increasing. It crossed one lakh tonnes by 2008-09. In 1980-81, the catch was only 0.26 lakh tonnes, which had increased to 1.4 lakh tonnes in 2011-12 and 1.88 lakh tonnes in 2016-17. There is scope for expanding inland fisheries. In the coastal tracts and lowlands, inland fishing provides occupational choice to the fishermen during rough weather season, when fishing in the sea becomes difficult.

Table 7.2: Temporal Variations in Marine Fish Catch by Species in Kerala (1980-2016/17) (in per cent to total catch)

Species	*1980*	*1990*	*2000*	*2010-11*	*2016-17*
Sardine (oil and lesser)	28.90	29.00	40.70	42.20	13.41
Penaeid prawns and non-penaeid prawns	19.45	6.90	11.30	10.56	11.65
Mackerel	6.60	11.90	5.60	8.04	12.26
Perches	6.40	10.20	8.40	5.41	10.91
Cat fish	5.00	0.40	Neg	0.03	0.30
Ribbon fish	4.60	1.50	1.40	2.74	3.44
Tunneis	3.80	4.90	2.60	2.15	5.53
Crabs	2.60	3.30	1.00	0.82	2.79
Soles	1.60	2.30	2.80	2.49	1.38
Seer fish	1.30	0.80	0.80	0.45	3.49
Cephalopods	1.50	3.70	5.10	2.54	11.58
Anchoviella	2.80	4.05	3.70	5.79	3.69
Elasmobranches	2.40	1.05	Neg	0.53	1.44
Saurida and sarus	2.50	1.70	1.30	0.99	1.66
Caranx and related species	1.70	10.40	7.80	8.72	4.75
Others	9.85	7.90	7.50	9.07	11.72
Total catch (in tonnes)	279,543	662,890	608,467	560,398	488,336

Data Source: Government of Kerala, 1981; 1991; 2012a; 2017c.

Fish Production by Districts

District-wise distribution of fish production is based on the fish landing. By 2016-17, Kerala harvested 0.676 million tonnes of fish comprising both marine (72%) and inland (28%) fishes. There are nine districts along the coastline. District level distribution of marine fish landings indicated that Kollam recorded the highest landing of marine fishes (37%) followed by Ernakulam (22.5%) (Table 7.3). Neendakara in Kollam district is the main fishing harbour in Kerala. The Vembanad lake is the major water body for inland fishing in Kerala. Three districts of Ernakulam, Alappuzha, and Kottayam surrounding this lake together produced 43.5 per cent of total inland fish catch. The fishing operation was conducted through 17 fishing harbours and 10 fish landing centres as of 2016-17. Besides, there are several other fish landing centres for traditional fishermen. Alappuzha and Ernakulam are also noteworthy for shrimp production. Malappuram-Kozhikode stretch of Malabar coast accounted for 14 per cent of total marine fish landings in Kerala.

Table 7.3: Distribution of Total Fish Catch by Districts in Kerala (in lakh tonnes), 2016-17

District	*Marine*		*Inland*		*Total*	
	Catch	*Per cent*	*Catch*	*Per cent*	*Catch*	*Per cent*
Thiruvananthapuram	0.70	14.3	0.05	2.7	0.75	11.1
Kollam	1.80	36.9	0.19	10.1	1.99	29.4
Pathanamthitta	-		0.03	1.6	0.03	0.4
Alappuzha	0.30	6.1	0.32	17.0	0.62	9.2
Kottayam	-		0.12	6.4	0.12	1.8
Idukki	-		0.03	1.6	0.03	0.4
Ernakulam	1.10	22.5	0.38	20.2	1.48	21.9
Thrissur	0.04	0.8	0.26	13.8	0.30	4.4
Palakkad	-		0.22	11.7	0.22	3.3
Malappuram	0.20	4.1	0.06	3.2	0.26	3.9
Kozhikode	0.46	9.4	0.05	2.7	0.51	7.5
Wayanad	-		0.02	1.1	0.02	0.3
Kannur	0.10	2.1	0.04	2.1	0.14	2.1
Kasaragod	0.18	3.8	0.11	5.8	0.29	4.3
Kerala	4.88	100	1.88	100	6.76	100.0

Data Source: Government of Kerala, 2018.

EXPORT OF FISH PRODUCTS

The export of marine products from India grew significantly over the years. However, Kerala's share is declining. In 1981, Kerala's share at the national level was 38 per cent, which had slid down to 13 per cent in 2016-17. From 2012-13 to 2016-17, the growth of export from India was 22 per cent, and that from Kerala, it was negative (-4%). In value terms, the growth was 101 per cent and 45.7 per cent for India and Kerala respectively during the same period. This indicates that Kerala's fisheries sector is under stress and other fish producing states are performing better than Kerala. Frozen shrimp accounted for 36 per cent of export earnings of marine products in 2011-12 followed by frozen squid (20%). Kerala's annual productivity of shrimps was 0.63 tonnes/ha in 2011-12. This was well below the national average. Aquaculture in Kerala gained momentum under the *Matysakeralam* programme launched during 2008-09 to promote inland/brackish water aquaculture.

CHALLENGES OF FISHERY MANAGEMENT

In spite of modernization, marine fish production in Kerala appears to be stagnating or declining. As a result, per capita availability of fish is declining in one hand, and on the other hand income of people depending on fishing and the related sector is reducing over the years as population growth rate among fishing population continues to be high. While the cause of stagnation is a subject matter of debate, there is hardly any difference of opinion about the multiplicity of problems encountered by this sector and the people associated with it. The traditional fishermen communities in Kerala are socially and economically backward. They depend mainly on the fisheries sector which is overcapitalized and low income generating. Fishermen have little alternative employment prospects. As the catch is stagnant and the fishermen population is increasing, there is a declining trend in per capita income. Stagnation in the production level can only deteriorate their living standard. Catch per trip is also decreasing as fishing crafts are increasing over the years.

The marine fisheries sector is almost saturated. The problems in this sector affect the larger society in Kerala as fish is part of the regular diet of a major section of people in the state, and it is also a foreign exchange earner. There is scope for expansion of inland fisheries, culture fisheries and ornamental fisheries. People depending on fisheries resources need special attention. There is a need for all-round policy intervention to overcome the problems.

FOREST RESOURCE

Forest is a unique ecosystem, which has been acknowledged for a long. It plays multiple roles at global and also at the local level by providing various ecosystem services beneficial to human beings like maintaining hydrological cycle, soil generation, soil and water conservation, purification of air and water, nutrient recycling, maintaining biodiversity, mitigation of climate change, and carbon sequestration. As a source of economic goods, it supplies industrial raw materials, fuelwood, non-timber forest product like food, fodder, fibre, medicine, and supports employment generation, recreation, and protection of cultural and natural heritage. The 1972 Stockholm Conference recognised forests as the largest, most complex, and self-perpetuating of all ecosystems, and emphasized the need for sound land and forest use policies, ongoing monitoring of the status of the world's forests, and the introduction of forest management planning. Kerala is well endowed with forest resources, which is rich in biodiversity and maintains a high level of productivity compared to the rest of India. A comprehensive forest policy is needed to address several challenges in forest management, which now stresses on conservation and ecosystem services.

DISTRIBUTION OF FOREST

Land use data of Kerala show that the legal area under forests is 11309 km^2, which is 29 per cent of the total geographical area of the state, well above the all-India average of 23.8 per cent. Out of the total forest area of 11309 km^2, it is reported that 8853 km^2 is in the category of dense and degraded forest, 1550 km^2 are under plantations, and the remaining 906 km^2 area is given to various agencies, such as the Kerala State Electricity Board (KSEB) and Public Works Department (PWD) for the river valley projects, the Forest Development Corporation, the Kerala Livestock Development Corporation, and other Government agencies (Govt. of Kerala, 2017d). Around 141 km^2 has been declared as an ecologically fragile area (EFA). Little over one-third of EFA is located in the Palakkad district. There are disputes about the figures for the dense forest cover. One of the early estimates indicated that intact forest vegetation in Kerala was little over 6630 sq. km or 17.06 per cent of the total geographical area in 1973 (Chattopadhyay, 1985). A detailed resurvey of the forest area is overdue. Field observation and satellite image analysis have brought out that there is considerable thinning of vegetation in many areas. Forests are found in all districts except Alappuzha. Kottayam district has only 101 km^2 of the area under forest. Idukki district has the maximum forest area of 2991 km^2, followed by Palakkad with 1604 km^2. Wayanad district has 884 km^2 of forest area (Fig. 7.6).

Forest Type

The climatic variability and physiographic diversity are well manifested in the forest types, its floristic composition, physiognomy, and life form. The forests in Kerala have been divided

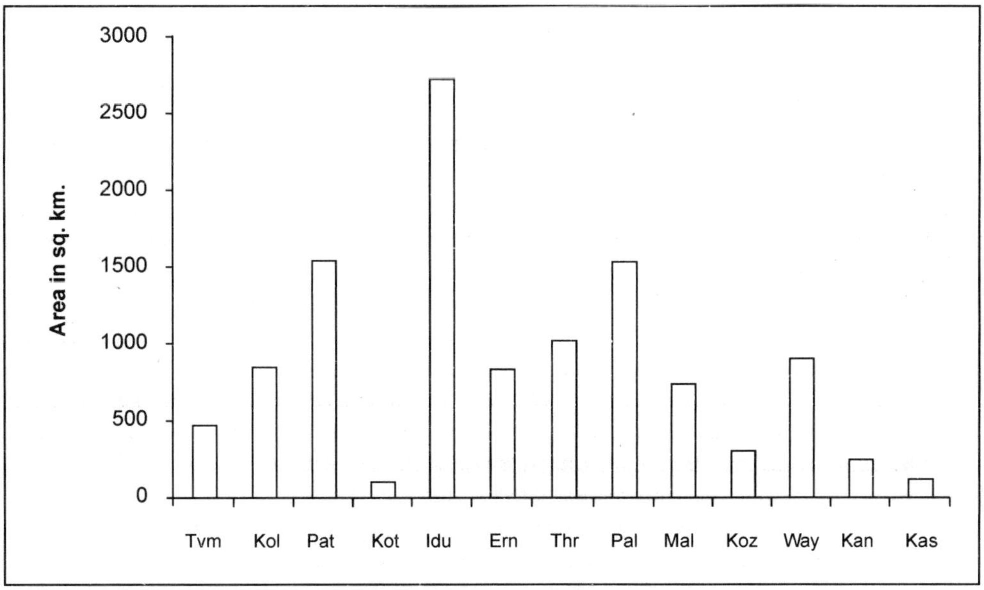

Fig. 7.6: Distribution of Forest Area by Districts in Kerala, 2016.

Data Source: Government of Kerala, 2017.

into seven major types (Chand Basha, 1997). There are 20 subtypes and many further subdivisions depending upon floristic composition and other local factors. Forest Department classified main vegetation types into the categories of wet evergreen and semi-evergreen (3877 km^2), moist deciduous (3616 km^2), dry deciduous (391 km^2), sub-tropical montane and temperate shola forests (286 km^2), and grasslands (501 km^2). Dry deciduous and sub-tropical montane type of vegetation can be marked in patches having low temperatures and meagre rainfall.

The West coast tropical wet evergreen and semi-evergreen forests cover around 34.3 per cent of the total forest area in Kerala. This type of forest is mostly seen between 600 m and 1100 m altitude having rainfall of more than 200 cm, temperature between 15°C to 30°C, and humidity between 70 per cent and 100 per cent. It is the climax vegetation in Kerala characterized by dense vegetation and at least three tiers with undergrowth of numerous ferns and tall herbs (Fig. 7.7).

The highest tier often attains the height of 40 to 45 m, the second and third tiers are in the range of 15 to 30 m and below 15 m respectively. Many species develop plank buttresses and the trees are heavily infested with orchids, aroids, mosses, and ferns (Fig. 7.8). Climbers are common. These forests are the storehouse of medicinal plants hosting more than 180 species and many wild relatives of cultivated plants (*ibid*). Ecologically this forest is in the most advanced stage with high floristic richness and provides much of tangible and intangible benefits. More than 50 per cent of this forest was under the system of selection felling. Degraded areas require 50 to 80 years to get back to the original stage, and in many instances, it may not be possible to return to pristine condition.

The West coast tropical semi-evergreen forest is considered transitional between evergreen and moist deciduous forest types. This type of forest is found to develop in the altitudinal range between 600 m to 800 m and in some places, it extends up to 900 m. Semi-evergreen

Fig. 7.7: View of Tropical Evergreen Forest in Kallar Catchment in Kerala. Note the dense closed canopy coverage, characteristics of tropical evergreen forest. Photo by Mahamaya Chattopadhyay.

Fig.7.8: Interior View of Tropical Evergreen Forest along the Slopes of Wayanad Plateau, Kerala. Note vegetation density, climbers and buttress roots. A small stream is seen flowing cutting across the forest. Photo by G. Sankar.

forests are also found associated with highly disturbed evergreen forests. The floral composition of this forest is an admixture of evergreen and moist deciduous species. Vegetation in the lower storey is similar to that of the evergreen forests. There are 37 medicinal plants exclusively found in this type of forest and it also houses many commercially valuable trees. More than 70 per cent of this forest was subject to selection felling in the past. If left undisturbed deciduous species will gradually disappear with the increasing dominance of evergreen species (*ibid*).

Some of the common species of tropical evergreen forests comprise of *kunthirikkam* (*Canarium strictum*), *vellakil* (*Dysoxylum malabaricum*), *churuli* (*Mesua ferrea*), *vellayini* (*Dipterocarpus indicus*), *kambagam* (*Hopea parviflora*), *chempunna* (*Calophyllum elatum*), *vedipla* (*Cullenia excelsa*), *pali* (*Palaquium ellipticum*), *payini* (*Vateria indica*), etc. Evergreen and semi-evergreen forests are the major sources of raw materials for plywood and match industries. Reeds that occur in this belt are traditionally used for mat making, basket making, and other cottage industries in addition to their use in the manufacturing of paper and pulp.

South Indian moist deciduous forest is a seral (intermediate) stage in the ecosystem succession towards climatic climax vegetation, which is a wet evergreen forest. A regular forest fire and anthropogenic disturbances have deterred advances of this vegetation towards the climax stage. This type of forest covers 32 per cent of the total forest area in Kerala. Trees reach 30 to 35 m height. The tires are not so distinct like evergreen forests. Buttressed trees are rare but fluting is common. This type of forest thrives well in the altitudinal range of 200 to 1250 m within a temperature regime of 20 °C to 35 °C. Rainfall of 150 cm to 200 cm and humidity ranging from 40 per cent to 70 per cent provide preferable conditions. Although species diversity is low and vegetation is less dense, compared to evergreen forests, this forest is commercially very rich with valuable species like *teak* (*Tectona grandis*), *karimarudu* (*Terminalia tomentosa*), *poomarudu* (*Terminalia paniculata*), *veeti* (*Dalbergia latifolia*), *venteak* (*Lagerstroemia lanceolata*), *irul* (*Xylia xylocarpa*), *venga* (*Pterocarpus marsupium*), etc. The major feature of this forest is that trees remain leafless from December to June. Bamboos and reeds are common. Ground flora consists of many species of herbs with medicinal properties. Around 300 medicinal plants have been reported. Plantations of teak and softwood are raised by clearing moist deciduous forests. In lower elevations, eucalyptus plantations are also found. In the areas of selection felling, there is hardly any possibility to recuperate to the original vegetation. The evergreen, semi-evergreen and moist deciduous forest types are located in the rainfall zone of 250 cm to 350 cm and above and at a temperature of more than 20 °C. They are mostly located above 300 m altitude. These forests are found at lower levels in the north-western flank of the Wayanad Plateau in north Kerala.

Southern subtropical broad leaved hill forests are inferior to the wet evergreen forests of lower altitudes. Dry savanna areas predominate between the watercourses. It covers only 0.5 per cent of the forest area in Kerala. The height of the trees rarely crosses 20 m. Trees are heavily clothed with mosses, lichens, aroids, orchids, and ferns. This type of forest is mostly found in places where the altitude ranges between 1300 m to1800 m, temperature varies between 10 °C to 20 °C, rainfall is between 300 cm to 550 cm, and even up to 650 cm and relative humidity is 80 per cent to 100 per cent. This forest is not rich in species diversity, houses only 4 per cent of medicinal plants but ecologically important due to its contribution to the hydrological cycle.

Southern montane wet temperate forests, known as Shola forest are found along streams in the valleys within high ranges, mostly above 1900 m elevation. It constitutes only 3.4 per cent of the total forest area. Rainfall is more than 400 cm and temperature varies between 10 °C to 15°C. The relative humidity is between 80 to 100 per cent. The height of the trees seldom reaches 10 m.

Shola forests in the valleys are surrounded by grasslands along the ridges. Around thirty species of medicinal plants are found in this forest. Streams surrounded by this forest are perennial. The fire spread into the Shola forest from burnt grasslands results in shrinkage of area under Shola forest in most areas. Some common species of these forests are *vellachampakam* (*Michelia nilagirica*), *kattukarana* (*Gordonia obtusa*), *and cherukalyavi* (*Meliosma simplicifolia*).

The southern tropical dry deciduous forests, which account for 3.5 per cent of the total forest area are found in the leeward eastern slopes of the Western Ghats in the Anjanad valley, Pambar valley to the north-east of the Anamalai and Attapadi hills. Sandalwood (*Santalum album*) is the most important species in this forest. These are rain shadow areas with an annual rainfall of less than 100 cm. The winter temperature is below 20°C and humidity varies between 30 per cent and 60 per cent. Elevation ranges from 200 m to 1100 m. This type of forest is open with trees reaching the height of 15 m to 20 m and commercially less valuable. Deciduous trees remain leafless during the period from December to June. Woody climbers are associated with deciduous forests. Frequently affected by fire, this forest is highly degraded due to heavy pressure from grazing. This restricts regeneration, and in places, this forest turns to scrub vegetation with the dominance of xerophytic plants. Around 115 medicinal plants are reported from this forest.

Grasslands develop as the progressive seral stage and also as the retrogression stage of the forest due to regular annual fire. They are found in isolated patches throughout Kerala mostly occupying the ridges. Some of the grasslands are due to climatic climax and some others are due to human intervention. Grasslands also developed in stretches of forest clearance along the Western Ghats crest where roads were laid for the purpose of plantation development in the nineteenth century. As a whole, grasslands cover around 4.40 per cent of the total forest area and are seen at various altitudinal levels. Grasslands found at an elevation of above 1500 m are considered as high-level grasslands and those developed below 1500 m are low-level grass lands. High-level grasslands are extensive in Munnar (Fig. 7.9). Attapady and upper reaches of Silent Valley are other important areas of high-level grasslands. These grasslands are interspersed by Shola forests. High rainfall and consequent heavy run-off, strong wind, and shallow soil constitute the biophysical setup.

MANGROVES

Mangroves grow in the intertidal zone where freshwater mixes with saline water like estuaries, river mouths, and coastal wetlands with formations of tidal flats/mudflats (Fig. 7.10). At present, mangrove vegetation is confined to the confluences of some of the major rivers in Kerala. There are remnants of mangrove vegetation all along with the coastal wetlands in the state. Two sites of dense mangrove vegetations are found in the mudflats around Vallapatnam river mouth in Kannur district and recently accreted land in Puduvipin to the north of Cochin mouth in Ernakulam district. Common mangrove species found in Kerala are *Avicennia officinalis, Avicennia marina, Rhizophora apiculata,* and *Acanthus ilicifolius.* One of the reasons contributing to the destruction of mangrove vegetation is that majority of mangrove areas are privately owned. Transformation of mangrove sites into prawn farms are noted in many parts of Kerala, particularly along the coastal stretch of Cherthala-Kochi.

SACRED GROVES

Scared groves by definition are groves of plants associated with gods or goddesses under any

Fig. 7.9: High Level Grasslands in Munnar Plateau in Kerala. Note long slopes and scar of slope failures along the road. Photo by G.K. Suchindan.

Fig. 7.10: Mangrove Vegetation, Kannur District in Kerala. Note waterlogged ground surface and the stilt roots. Photo by Mahamaya Chattopadhyay.

culture. They may be considered as the religio-cultural natural landscape outside the main domain of the forest. Sacred groves are found in many parts of the world, particularly in areas dominated by ancient culture and traditional societies. India is rich in scared groves, although, there is little documentation about this unique feature manifesting humanity's reverence to plants and animals. Kerala is well known for sacred groves associated with *Naga* (Serpent God), *Sastha-Durga,* and a number of gods and goddesses from the primitive pantheon. *Sarpa Kavu* or a grove associated with the serpent god '*Naga*' is most common in Kerala. These *Kavus* are maintained by individual families, groups of families (*Tharavedu*), and statutory agencies of temple management (Chandrasekhara and Sankar, 1998). Sacred groves are home to varieties of plants including medicinal plants, reptiles, birds, butterflies, and are associated with water bodies like ponds, wells, springs, and streams. In some sacred groves, there is a *sthala vriksha* or a tree of the place, which is worshipped and protected due to socio-ecological and medicinal importance (Menon, 1997). Some of the common species of *sthala vriksha* are peepal (*Ficus religiosa*), neem (*Azadirachta indica)*, maruthu (*Terminalia paniculata*), kanikonna (*Cassia fistula*), and bilva (*Aegle marmelos*) (*ibid*).

A comprehensive sacred of sacred groves is yet to be materialised in Kerala. A preliminary survey indicates that there are more than 2000 sacred groves in Kerala distributed mostly in the lowlands and midlands extending up to foothills. The size of the sacred groves varies widely. There are some groves with few trees covering an area of one cent or less, whereas there are also groves with luxurious vegetation spreading over more than 20 hectares like that of Iringole *Kavu* (Menon, 1997). Mannarasala is one of the famous *kavus* in south Kerala. People preserve sacred groves and refrain from felling trees due to religious belief. As a result, thick native flora remained protected. Plant species found in these groves represent the indigenous species prevalent in the locality. The general floristic composition resembles low-level evergreen forest vegetation. Sacred groves are also associated with mangrove swamps in poorly drained regions of south Travancore experiencing a long rainy season. *Myristica magnifica* with prominent dense stilt roots are seen in these swampy groves. Sacred groves, which serve as nursery and store-house of rare species and many of the local ayurvedic, tribal, and folk medicines, are now threatened with changing socio-cultural conditions and increasing population pressure. There is also a change in perception. Socio-religious and socio-economic entanglements revolve around the core issue of mixed benefits from large sacred groves with multiple stakeholders and there is also de-natured alienation (Notermans, *et al.*, 2016). Emphasis is more on intensified religious rituals than conservation of naturalness of the groves. It is important to preserve these relics of original forests as part of nature conservation and ecosystem services.

FOREST PLANTATIONS

Organised exploitation of forests started by the British with the introduction of plantation forests and agricultural plantations. Nilambur forest plantation was set up in 1842. The first European coffee plantation was reported in the year 1878. Forest plantations are now an important component of forest resources in Kerala (Fig. 7.11). There are 34 types of plantations with significant coverage. In 2016-17, the area under forest plantation was 1550 km^2, of which 50 per cent area was under Teak plantations followed by Softwood plantations (9%), Acacia (6%), Eucalyptus (5%), Cane and Bamboo (5%), and Medicinal plants (1.3%). Other plantations covered less than 1 per cent each. The forest divisions of Wildlife Parambikulam, Konni and Wildlife Wayanad, each accounting for more than 8000 ha under hardwood are important plantation areas. Other significant forest divisions include Malayattoor, Thiruvananthapuram, Kothamangalam, Nilambur south, Nilambur north, Chalakudy, and Kottayam, each of which raises more than 4000 ha area under forest plantations.

Fig. 7.11: Close View of Teak Plantations in Kolathupuzha Catchment, Kerala. Note that plantation floor is devoid of any basal vegetation cover. Photo by Srikumar Chattoadhyay.

FOREST PRODUCE

Major forest products include timber, reeds, bamboo, sandalwood, and firewood. By 2016-17, the quantity of round log and squared log timber production was 31134 cubic metre. Other products included little over 5201 metric tons of reeds, 52102 kg of sandalwood, and 4796 metric tons of firewood (Government of Kerala, 2017d). Minor forest products consist of ayurvedic herbs, spices, fibres, grass other than fodder, incensive plants, honey, and bee's wax. Production of ayurvedic herbs was more than 10 lakh kilograms. Net revenue earned from the forest products was Rs. 239 crores during 2016-2017.

WILDLIFE CONSERVATION

By 2016, an area of 3211 km^2 or 28.4 per cent of the total forest area has been earmarked as protected forest under National Parks and Wildlife sanctuaries to protect wildlife, and to promote forest conservation. The six wildlife divisions managing the wildlife protected forest areas are: Thiruvananthapuram (Thiruvananthapuram district), Thekkady (Pathanamthitta and Idukki districts), Eravikulam (Idukki district), Parambikulam (Palakkad district), Silent Valley (Palakkad district), and Wayanad (Wayanad district). There are five National Parks, 17 wild-life sanctuaries, two bird sanctuaries, one tiger reserve, and two biosphere reserves, namely

(i) Nilgiri (parts of Wayanad, Nilambur, and Silent valley) and (ii) Agasthamalai (parts of Thiruvananthapuram, Kollam, and Pathanamthitta) (Fig. 7.12). These areas are distributed throughout Kerala which is significant for decentralized management, and also to promote habitat-specific development. The Silent Valley National Park is nationally and internationally well known. The environmental movement in Kerala was centred on the Silent Valley during the 1970s. This 89 km^2 virgin tropical rainforest is still pristine. There was a proposal to construct a hydel power project in this area. The movement mobilized against this proposal. The site is now protected as a national park. The wildlife sanctuaries and national parks attract tourists from various parts of the country and abroad. The Periyar Wildlife Sanctuary is a well-known site, globally, with a four-star hotel and a steady stream of visitors eager to view the wild animals live in their natural habitat.

FOREST MANAGEMENT AND INSTITUTIONS

Institutionalised forest management began by mid 19th century. Regular forest establishment was organised in the line of British model and working plans were meticulously prepared. Indian Forest Act, 1865, 1878, Madras Forest Act, 1882, Travancore Forest Act 1887, and Cochin Forest Act, 1905 were promulgated and forests were declared Reserve forests, thereby free access of people traditionally depending on the forest was restricted. In 1806, Captain Watson of the police department was appointed conservator of Malabar—the first forest conservator in India. Teak in the Malabar area was the main attraction. The export of teak was reported as early as 1796. The demand for teak timber for Military purposes was the main thrust for large scale exploitation of forest in the Malabar area. Revenue collection in Travancore and Cochin encouraged the expansion of agriculture in the forest areas. Before the reorganization of the state in 1956, parts of the forests in the Travancore-Cochin area were privately owned. Following the reorganization, all forests were taken over by the Government. Forest administration in Kerala is operated through eight forest circles, 35 forest divisions, and 105forest ranges. Thirty-five co-operative societies are functioning for the collection of minor forest produce. Timbers are traded through 29 timber depots functioning under six sales divisions distributed in five circles.

PARTICIPATORY FOREST MANAGEMENT

Kerala has introduced Participatory Forest Management (PFM) since 1998 for the conservation of bio-diversity and to improve the livelihood of forest-dependent people by forming partnership institutions at the grassroot level (Government of Kerala, 2012b; 2017d). There are two types of institutions: (i) Vana Samrakshana Samity (VSS) for the territorial forest divisions. There were 387 members under VSS for 24 territorial divisions in 2010-11 and (ii) Eco-Development Committees (EDC) for sanctuaries and national parks. There were 187 members of EDC for 11 Wildlife Divisions. Altogether, 66,713 families were involved in VSS, and 12,316 families participated in EDC. While more than 50 per cent of families under VSS belonged to other than SC and ST communities, in the case of EDC, more than 50 per cent of families are of SC and ST communities. Besides, there were 446 Joint Forest Management Committees (JFMCs) functioned under National Afforestation Programme (NAP) during 2007-2008 to 2011-2012 and generated more than 20 lakh labour days' work.

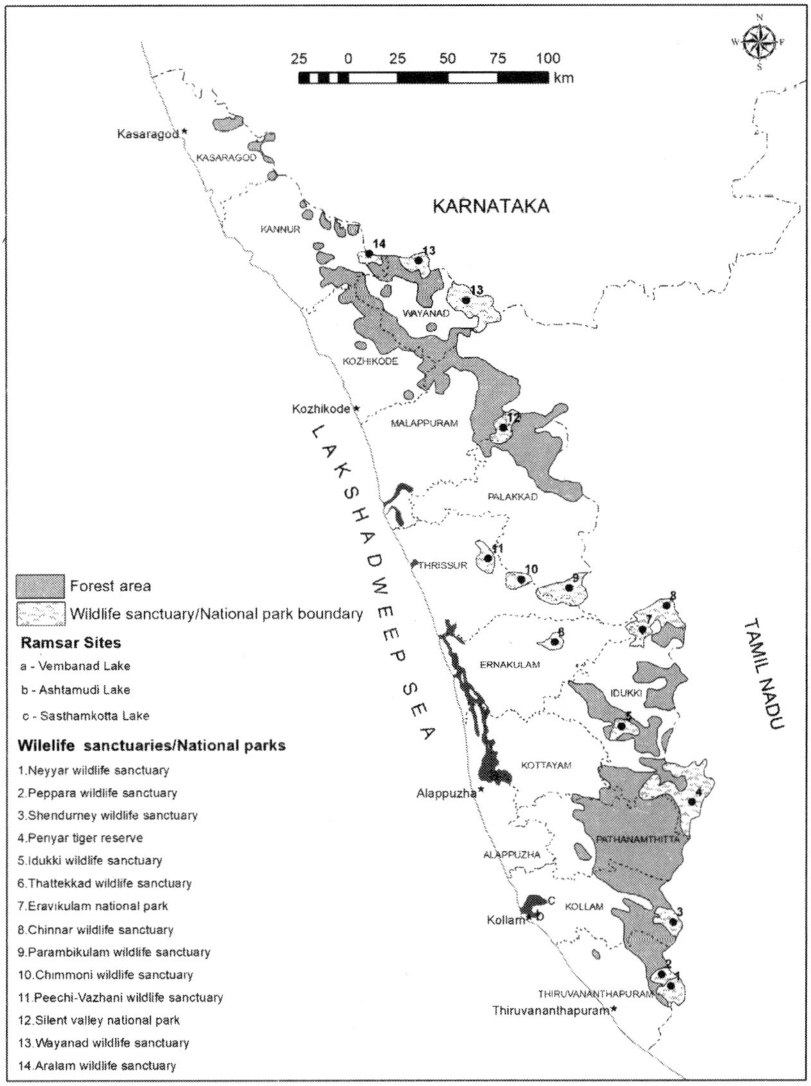

Fig. 7.12: National parks, Wildlife Sanctuaries and Ramsar Sites in Kerala. Note that this map shows selected wildlife sanctuaries only.

Data Source: Government of Kerala, 2012b.

ECOTOURISM

Ecotourism is one of the major activities promoted jointly by the Forest Department and Tourism Department of the Government of Kerala. Sixty ecotourism destinations have been developed in 14 districts by 2010-11. The districts of Kozhikode and Palakkad each has 9 ecotourism sites followed by Wayanad (7 sites) and Idukki district (7 sites). The impacts of these initiatives are yet to be fully assessed.

EMERGING CHALLENGES OF FOREST MANAGEMENT

Traditionally, the main objective of forest management has been sustained yield of commercial timber and to address problems in forestry. However, over the year, forest management in the tropics has evolved in response to social, economic, and political factors associated with growing population pressure and the need for land for agriculture and other uses. This has enlarged the role of forest management practices from simply supplying industrial raw materials like timber to address much larger issues like ecosystem services and other forest produces providing sustenance to the people living within and outside the forest area. Multiple-use forest management is significant in this context. It envisages a forest policy integrating agriculture, energy, industry, and other sectors, a thorough site evaluation, forest land capability classification, and zone specific approach for primary and secondary use (FAO, 1984). Conservation acquired centre stage of forest management in Kerala supported by the Wildlife (Protection) Act, 1972, the Forest Conservation Act, 1980, and New Forest Policy (1988) of the Government of India (Chundamannil, 1993). There is a clear departure from earlier practices.

Forests in Kerala are fragmented and are now mostly restricted to rugged slopes. Road construction across forests is the principal driver leading to fragmentation. This has increased human intervention within forest areas, besides, intervening in the paths of wildlife movement. Loss of forest vegetation has impacted the hydrological behaviour of river catchments. Water holding capacity of provenance of rivers has changed with an increase in instantaneous flow and reduction in base flow. Landslides have increased due to the removal of forest vegetation. There is also an increase in temperature in the cleared area within forests. A forest fire and human animal conflict are some other problems reported almost every year. Around 17 km^2 area was destroyed due to the incidence of fire during 2015-16. Loss of biodiversity and the growth of invasive species are other issues warranting attention. Forest management has to consider all the ecologic issues apart from multiple uses, and develop plans accordingly. The strategies for forest development should envisage environmental stability, reduction of degraded forests, conservation of biodiversity, eco-regeneration, and increase of productivity, enhancement of forest cover, and promotion of participatory forest management. A comprehensive approach is required to accomplish all these objectives.

CHAPTER

8

Industrial Development

INDUSTRIAL DEVELOPMENT: A BRIEF OVERVIEW

Industrial development in Kerala lags behind the level of human development that the state has achieved and globally acclaimed for. This is a matter of serious concern because prior to the planned development era, the Travancore part of Kerala had a reasonably diversified industrial base and its position was well above the national average (Subramanian, 1994). Now, with several decades of development planning, Kerala's position is much below the national average and that of the neighbouring states. The State Planning Board (2005) in its approach paper expressed their concern about the inadequate development of the productive sector to sustain human development in Kerala. Like all other sectors, the industrial structure of Kerala differs from the rest of the country. Kerala does not have mineral deposits like many other states in India, which have developed industrially. Industrial development in Kerala began largely based on primary resources. These industries are labour intensive, seasonal, and can hardly provide employment throughout the year. They are capable of providing sustenance but are not potent enough to propel the economy in a high growth trajectory. Modern industrial development in India overshadowing traditional industries began during the colonial period. While new industries were established, there were hardly any attempts or little attention was paid to the technological innovation or up-gradation of the traditional industries. As a result, most traditional industries depend on primitive technology with little or no transformation over the years. There is a lack of large private capital investment in the industrial sector. Central investment is also limited. All these have played their role in low industrial development in Kerala. Even the industrial growth witnessed during 1980-90 could not be sustained. Globalisation and the Government of India policy of disinvestment of the public sector impacted Kerala's industrialization. Small scale industries could not withstand competition in the market.

Although there are constraints like scarcity of land, lack of mineral deposits, and a dearth of large capital investment, Kerala enjoys distinct benefits like geo-strategic location, availability of the educated human resource, infrastructure, internal market for consumer goods due to a high level of consumption compared to all-India average, the dominance of cash crop and large base of traditional industries. Key industries in Kerala include agro-based and food processing, coir, textiles and garments, seafoods and other marine products, chemicals, and information technology/information technology-enabled services (IBEF, undated).

One of the notable features of the secondary sector of Kerala is that employment in the secondary sector was 31.8 per cent in Kerala against all-India average of 24.3 per cent (Sanitha and Singla, 2016). However, the contribution of the secondary sector to Gross Value Addition was 26 per cent in Kerala against all-India average of 29 per cent. This apparently lopsided development of low contribution and high employment in Kerala in contrast to all-India average can possibly be attributed to the nature of resource base, the spatial distribution of resources, and the process of industrial development that has been followed in the state.

INDUSTRIAL RESOURCE BASE

The industrial resource base of Kerala is different from the rest of India. Kerala is not rich in mineral resources like coal or iron ore, however, it can boast of water resources generating hydropower, heavy mineral (Zircon, Thorium, and Monazite), and glass sand deposits in some parts of its coast plain. There are good china clay deposits and limited quantities of bauxite deposits. Lime shell deposits in the Vembanad lake provide raw materials for white cement and limestone deposits in Walayar, Palakkad are used to produce cement. Kerala has a long tradition of agro-based industries. The development of the coir and cashew industry is well known. Both the crops grow throughout Kerala, particularly in the lowland and midland sections. The plantation crops like coffee, tea, and cardamom promoted processing industries in selected areas of the Western Ghats. Rubber plantation, which is next to coconut in area coverage, is providing raw materials for the rubber factory. The extraction of rubber from latex is a household industry in Kerala that provides much of the economic benefits to the small rubber growers. The forest resources spreading all along the Western Ghats provide a resource base for timber and wood-based industry, bamboo industry, and industries based on non-timber minor forest products. Fish, both marine and inland, is another important resource base for the fisheries industry. In recent years, a knowledge-based industry like IT has made a significant impact on industrial development. Kerala's speciality is oriented to local resources and availability of the market at the local, national, and international levels.

ENERGY RESOURCE

Energy is the main driving force for economic development and contributes to enhance the quality of life. Electric power generation in Kerala is from three sources: hydel, thermal, and wind. In sharp contrast to all-India picture, Kerala's energy sector is dominated by hydropower, which accounts for 71 per cent of total installed capacity in the state. Thermal energy contributes 24.3 per cent of the total installed capacity. National Thermal Power Corporation (NTPC) at Kayamkulam is Kerala's dedicated thermal station. It has an installed capacity of 360 MW. KSEB started producing wind energy in the Kanjikode wind farm however its share in total installed capacity is marginal. One wind energy project under the private sector at Agali with 6 MW installed capacity has been commissioned. Wind farm produces 59.27 MW, and 75.42 MW comes from solar power.

The first hydroelectric power generating station was started by Kannan Devan Hill Produce Co. in 1940 at Munnar. The Pallivasal project, commissioned in 1940-41 was the first public sector power project in Kerala. The expansion of the Pallivasal project, and commissioning of the Sengulam scheme were carried out during the First Five-Year Plan. The Kerala State Electricity Board (KSEB) was established in 1957 after reorganisation of the state in 1956 to oversee power development and its distribution. There was an addition to installed capacity during five year plan periods; however, Kerala started experiencing power shortage since the beginning of the

Third Plan Period (1961-66) itself. With the commissioning of Sabarigiri in 1966 and Idukki (Stage I) in 1976, the power situation in Kerala improved and the generating capacity rose to 1012 MW during the Fifth Plan period. There were 24 hydel power projects with a total installed capacity of 1921 MW in 2009. By 2016-17, the total installed capacity was 2961.11 MW (Government of Kerala, 2018). The installed capacity was 86 MW in 1956. Besides, there are 21 hydropower schemes at various stages of implementation. On completion, they are expected to add nearly 380 MW in installed capacity under hydel power. KSEB continues to invest in hydel power and tries to maintain the ratio of 80:20 between hydel power and thermal power.

Attention has also been given to develop small hydro power projects which have the potential to emerge as renewable energy sources for remote areas particularly in the Ghats section having isolated human habitations. Energy availability in the villages can be enhanced by tapping local sources. A survey conducted by Total Energy Security Mission (TSEM) and Agency for Non-Conventional Energy Research and Training (ANERT) identified 875 sites with various capacities. The Meenvallam small hydro project in Palakkad district is a pioneering attempt. One of the problems of developing SHPs in Kerala is the high fluctuation of drainage discharge in monsoon and non-monsoon periods. The creation of storage capacity for all the sites may not be feasible. A cluster approach may sometimes help in this situation. Non-conventional energy sources are being developed. Activities include the installation of biogas plants, improved *chulha*, wind power plants and solar power plants.

Electrical energy consumption in Kerala increased from 350 million units in 1956 to 20453 million units in 2016-17 (Government of Kerala, 2018). Per capita consumption increased from 18.5 units to more than 600 units during the same period. Around 51 per cent of consumption was in the domestic sector, followed by high tension and extra high tension (20.6%), commercial (14.8%), industrial (5.7%), and bulk supply (3.05%). Public lighting consumes 1.88 per cent of total consumption. Kerala is now a power deficit and has entered into a power purchase agreement with various Central agencies to procure 1481.15mw. Kerala State Electricity Board Ltd. is also purchasing power from wind farms and small hydro projects installed within Kerala. Consumption in Kerala is expected to grow by 54 per cent in the coming 10 years. The domestic sector, commercial sector, and public lighting are expected to witness more than 60 per cent growth in each case. With economic growth and climate change, there is a possibility of increased domestic consumption well above the projected growth. It is a big challenge for Kerala to meet this demand. Conscious efforts are required specially to intensely develop non-conventional sources like wind, solar, and micro/medium hydro projects to transform Kerala from a power deficit to a power sufficient state. Demand management, institutional changes, control of transmission loss, and energy conservation are some of the areas warranting proper attention.

LOCATIONAL FACTORS

Locational factors are important determinants for setting up industries. An optimum location is decided based on the lowest cost of raw materials, manufacturing, and delivery of the product to the market. Direct cost includes the cost of raw materials, labour, transport, energy, water supply, and waste disposal. Indirect costs are those needed for the acquisition of land and construction of buildings, movable equipments within the building, local taxes, maintenance, utilities, and administration. The cost varies with industries and in the case of the manufacturing industry the direct costs are more important. The main direct cost involves raw materials, power, and fuel, labour and marketing. All these are critically related to transportation in various measures. The transport costs are to cover (i) quantity of materials handled, (ii) perishability, (iii) cost per unit of transportation, and (iv) distance between producing and consuming centres or their degree of concentration (Stansfield Jr., 1983). If there is a great loss in bulk as in the case of separation of minerals from black sand, which are extracted from beach sand the industrial location will be at the site of the raw materials, like Indian Rare Earths (IRE) and Kerala Minerals and Metals Limited (KMML) units in Chavara of Kollam district. Brick and tile factories in Ernakulam-Thrissur stretch bear testimony of raw material based locations as good quality clays are available here in Periyar-Chalakudy valley. When raw materials are perishable like fish the processing units move towards the source and around the port as in the case of an item like prawn which commands high export value. In the case of the coir industry, where retting is needed for extracting fibre from copra, processing units move towards the retting sites. The high concentration of coir industry in Alappuzha district in the coastal plain, characterized by brackish water bodies, bears testimony of this arrangement. The location of bamboo industries in the foothills of certain pockets of the Western Ghats is also a manifestation of raw material based industry.

One of the important characteristics of traditional industries in Kerala is that the raw materials, based on which the traditional industries have developed, are mostly available throughout the state, thereby providing a foundation for decentralized industrial development. Ports are important locations for industries, particularly for those which move long distances mainly due to cheap water transport. The natural port Kochi emerged as the focal point of many industries including petro-chemicals and exportable marine products. Inland waterways, road ways, and railways running close to each other from Kollam to Ernakulam provide unique conditions for corridor type industrial development. The combination of railways and road-ways connecting Palakkad and Coimbatore provides a congenial environment for corridor type industrial development spreading over Tamil Nadu and Kerala. In fact, this industrial corridor is further expanding to Thrissur and Ernakulam and joins with Kochi industrial belt. Both Government of India and the Government of Kerala are giving special attention to develop this corridor.

The locational factors for newly emerging IT industries are different from that of other industries. It requires skilled human resources, accessibility, and electronic communication facilities besides uninterrupted power supply. The IT industry in Kerala is developing under private entrepreneurship housed in a specially created infrastructural set up by the Government. The availability of skilled labours and institutions to impart technical education are important facilitators for this industry.

DEVELOPMENT OF MODERN INDUSTRY: A HISTORICAL BACKGROUND

Modern industrial development in Kerala was initiated in the late 19th century. It was in the year 1881 when the first manufacturing unit was established by an American at Kollam to manufacture clothes (Sankaranarayanan and Bhai, 1994). A coir factory started in Alappuzha shortly after that and by the end of the 19th century, there were industrial establishments started functioning in Palakkad and Kozhikode under the then Madras presidency. The decade preceding independence in 1947 witnessed the establishment of a good number of industries under the *Diwanship* of C.P. Ramaswamy Iyer. Some historians consider this period as the golden era of Kerala's industrial development (*ibid*). The Fertiliser and Chemical Travancore Ltd. (FACT), Indian Rare Earths Ltd. (IRE), Travancore Titanium Products (TTP), Punalur Paper Mills,

Travancore Sugar and Chemicals, Kerala Ceramics, Forest Industries (Travancore) Ltd., Travancore Electrochemical Industries, Travancore Rayons, Travancore Cements were some of the major units established before 1947.

During the post-independence period, particularly in the 1960s some of the important industries established in Kerala were Travancore Cochin Chemical Limited, Hindustan Insecticides Ltd., Cooperation Spinning Mills, Premier Tyres, Hindustan Machine Tools Ltd., Cochin Shipyard, Cochin Refineries, Hindustan Paper Corporation, Apollo Tyres, KELTRON, Kerala Minerals and Metals Ltd., Hindustan Latex Ltd, Malabar Cements, and Modern Bakeries Ltd. Majority of these industries are under public sector.

SPATIAL TREND OF INDUSTRIAL DEVELOPMENT

The spatial trend of industrial development in Kerala is not uniform. Ernakulum is the leading district in Kerala in industrial development. Four districts, namely, Ernakulum, Thrissur, Kollam, and Palakkad together account for 50 per cent of all registered working factories and 62 per cent of industrial employment in Kerala. Due to the high concentration of cashew factories, Kollam leads among the districts in the number of employment. Kozhikode and Kannur districts in the north have a certain concentration of industries. Wayanad, Kasaragod, Idukki, and Pathanamthitta are industrially backward districts in Kerala. Apart from Kasaragod the rest three districts of Wayanad, Idukki, and Pathanamthitta are located within the hilly Western Ghats and lack proper infrastructure facilities.

Kochi-Kalamassery-Alwaye emerged as an important industrial belt in Ernakulam. Several modern industries have been established in this area, which is adjacent to the Kochi port, one of the important natural harbours on the west coast of India. Kochi port enjoys waterways, railways, and roadways connection with its hinterlands. This belt is also well connected with the rest of the country through railways, roadways, and by Air. Similarly, Palakkad-Coimbatore belt has a sizable concentration of industrial units. District-wise contribution to gross value addition (GVA) in the secondary sector of Kerala, as a whole, varies with a coefficient of variation (CV) of 43.4 per cent, comparable to the CV of population density (43.7 per cent). It may be inferred from these figures that the spatial trend of secondary sector development is similar to that of population density distribution. In the secondary sector GVA of Kerala, Wayanad district contributes 2 per cent whereas Ernakulam district contributes 12.3 per cent (Fig. 8.1).

MANUFACTURING INDUSTRY

According to the annual survey of Industries by the Government of India organized sector comprised of 6890 operating factories providing employment to 3.39 lakh persons by 2015-16 (Government of Kerala, 2019). The unorganised manufacturing sector engages 14.2 lakh persons (Government of Kerala, 2017c). Two major traditional industries, coir and cashew processing employed 3.8 lakh and 2.5 lakh workers respectively (Government of Kerala, 2019). The majority of these workers are women and are paid low wages. The average wage of factory workers a in Kerala is less than the national average due to the dominance of traditional sectors in manufacturing. The growth of manufacturing industries in Kerala is relatively slow compared to the neighbouring states. The contribution of the manufacturing sector to the gross state domestic product has decreased over the years. In 1990-91, it was around 15.44 per cent, which was comparable to the national average but it fell down to 7.5 per cent in 2011-12, and

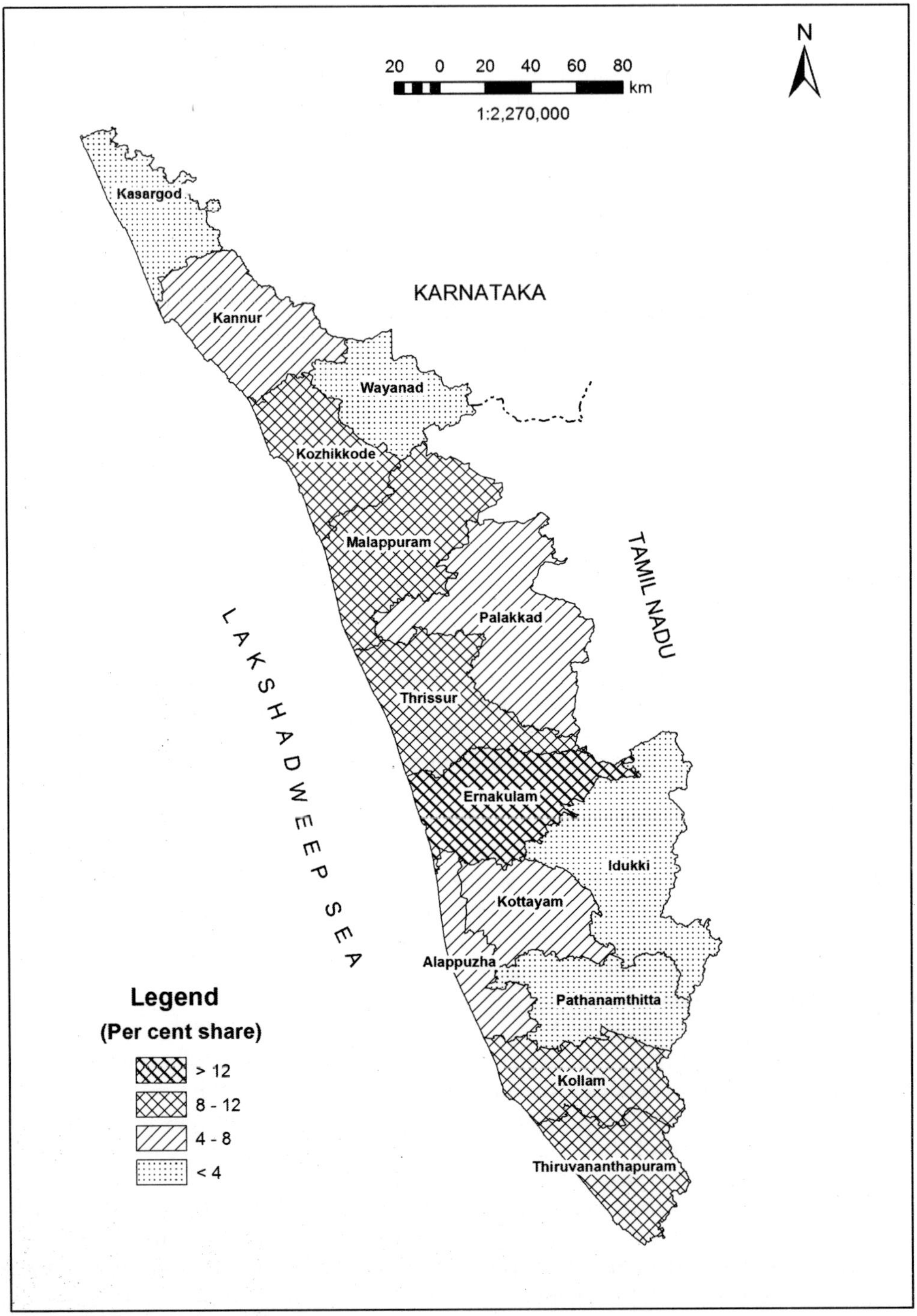

Fig. 8.1: Contribution of Districts to Gross Value Addition (GAV) in Secondary Sector in Kerala, 2017.

Data Source: Government of Kerala, 2018.

by 2014-15, although the share of the manufacturing sector increased to 9.10 per cent, it was well below the national average (17.2%). Income from the manufacturing sector registered a growth of 12.5 per cent in 2012-13, however, it declined subsequently and in 2016-17. It was only 2.2 per cent. During 2011-12, the manufacturing sector contributed 7.5 per cent of GDP and provided 14 per cent of employment, whereas, at all-India level these figures were 16 per cent and 13 per cent respectively (Government of Kerala, 2017c). The proportion of employment in the manufacturing sector is comparatively high in Kerala and this is mainly due to the dominance of labour intensive traditional manufacturing sector.

Among all the districts in Kerala, Alappuzha district receives the highest share (15.8 per cent) from the manufacturing sector in its domestic product. The coir industry is the main contributor in this district. There are seven districts receiving more than 10 per cent contribution from the manufacturing sector in their domestic products (Table 8.1). However, when the contribution of individual districts to the total manufacturing sector in Kerala is considered, Ernakulam stands out followed by Alappuzha, Thrissur, and Kollam. Ernakulam district, which is the most industrialized district in Kerala, received an 11 per cent contribution from the manufacturing sector to the district gross value addition. This indicates that other types of industries have also developed in Ernakulam district.

Table 8.1: Contribution of Manufacturing Sector to District Domestic Product and Total Manufacturing Districts in Kerala, 2016-17

Districts	*Contribution of manufacturing sector to district domestic product (per cent)*	*Contribution to total manufacturing sector in the state (per cent)*
Thiruvananthapuram	6.1	6.5
Kollam	10.5	10.0
Pathanamthitta	4.0	1.3
Alappuzha	15.8	12.3
Kottayam	8.9	5.9
Idukki	5.7	2.3
Ernakulam	11.1	14.5
Thrissur	10.1	10.7
Palakkad	10.0	7.9
Malappuram	6.3	6.4
Kozhikode	10.4	9.2
Wayanad	2.4	0.5
Kannur	11.8	8.6
Kasaragod	11.0	3.9

Data Source: Government of Kerala, 2018.

MINERAL BASED INDUSTRIES

By 2016, 11.28 km^2 of the area was leased out for mining activities in Kerala. This covered China clay, Silica sand, Bauxite, Lime shell/Seashell, Limestone, Mineral sand, Quartz, Graphite, Iron ore, and Laterite. Minerals in Kerala are grouped under major and minor minerals. Major minerals cover Ilmenite, Rutile, Zircon, Sillimanite, Leucoxine, Monazite, Lime shell/Seashell, Limestone, and Graphite. They are mostly mined in Kollam, Alappuzha, Kottayam, Ernakulam and Palakkad. The total production of major minerals was 5.75 lakh tons by 2016-17 (Government of Kerala, 2018). Limestone and Ilmenite are the main products. Some of the industries developed based on major minerals are Indian Rare Earths Ltd., Kerala Minerals and Metals ltd., Malabar cement, Travancore Cements Ltd., Kundara Ceramics, English India Clays Ltd., Excel Glass Factory, and Kerala Clays and Ceramics Ltd.

Minor minerals cover Granite, Laterite, Bauxite (cement), River sand, Silica sand, China clay, Brick clay, etc. The total production of minor minerals was 531.57 lakh tons. Minor minerals are extracted from all 14 districts in Kerala. Granite and ordinary earth are the main minor minerals. Major industries developed based on minor minerals are stone crushers, production units of manufactured sands, and brick and tiles factories. The total revenue earned from minerals was Rs. 138.7 crores in 2016-17. Ernakulam district topped the list by contributing 24.5 per cent of total revenue collection from minerals followed by Palakkad (9.6%), Thiruvananthapuram (8.9%), Kollam (8.0%), and Malappuram (7.6%).

PUBLIC SECTOR UNDERTAKINGS

Public sector undertakings (PSUs) play an important role in industrialization, especially in the manufacturing sector of Kerala. It consists of state government companies and statutory corporations established by the government to carry out commercial activities. Kerala has 130 PSUs, of which 15 are non-working (Government of Kerala, 2019). Major central sector units include Hindustan Latex, Hindustan Organic Chemicals, Travancore Titanium Ltd. (FACT), Indian Rare Earths Ltd. (IRE), Hindustan Machine Tools (HMT), and Cochin Shipyards Ltd. Some of these major industries have been established before independence. Among the state-level public sector enterprises the significant establishments are Malabar Cements Ltd., Transformers and Electricals Kerala Ltd., Kerala Minerals and Metals Ltd. (KMML), Kerala State Electronics Development Corporation (KELTRON), etc.

The industries department control 40 PSUs. They cover the chemical sector (7 numbers), electrical sector (4), engineering sector (6), electronic sector (3), textile sector (8), ceramics sector (2), traditional sector (6), wood-based sector (1) and development sector (3). Among these 40 PSUs, 14 PSUs are profit-making and the rest 26 are incurring losses. Chemical sector units account for 48 per cent of total turnover. Kerala Minerals and Metals Ltd., Travancore Cochin Chemicals Ltd., Travancore Titanium Products, and Kerala State Industrial Development Corporation are profit-making units as reported during 2017-18. There is a need to evaluate working of all PSUs for their functions, production, resource base, and managerial efficiency. The Government of Kerala has set up Public Sector Restructuring and Internal Audit Board under the industries department to undertake reforms of PSUs.

INDUSTRIAL INFRASTRUCTURE DEVELOPMENT

Kerala State Industrial Development Corporation (KSIDC) formed in 1961 aims to promote large and medium-scale industries. It has promoted many pioneers in the corporate sector including KELTRON, Hindustan Latex Ltd., KMML, Malabar Cements, etc. KSIDC is a single point contact for the investments in Kerala. It has promoted full-fledged industrial growth centres in Alappuzha, Kozhikode, and Kannur districts. Besides, it is involved in developing a mega food park at Cherthala, an electronic hardware park at Kochi, a life science park, the second phase of light engineering park at Palakkad, infrastructure development at industrial growth centres, and medical service parks. By 2017-18, KSIDC has completed 8 projects and has 39 projects under implementation. Besides, financing several industrial units and implementing various mega projects KSIDC has undertaken various other activities to promote industries in Kerala.

The Government of Kerala had set up "Kerala Industrial Infrastructure Development Corporation" (KINFRA) in 1993 for accelerating the industrial development of the state by providing infrastructural facilities required for various industries. KINFRA has completed infrastructure development in 24 industrial parks of which 9 are catering to small and medium enterprises. By 2017-18, there were 757 units functioning in various industrial parks with a total investment of Rs. 1816 crore and directly employing 20,590 persons. There are 15 ongoing projects under KINFRA covering a wide range of products. So far KINFRA acquired 1639.7 acres of land of which 1186.7 acres of land are used for industrial development. KINFRA acts as the nodal agency for the Central Ministry of Food Processing industries, Kannur International Airport, National Institute of Fishing Technology (NIFT), Kannur, and for infrastructure development of export and allied activities.

There are several initiatives in Kerala to promote the development of industrial infrastructure. These include industrial co-operative societies, setting up industrial development areas, mini industrial estates, entrepreneur support schemes, industrial cluster development, etc. As of March 2018, there were 394 industrial co-operative societies functioning in Kerala. Out of these 394 industrial co-operative societies, 90 societies were registered by women. Rural employment generation through setting up micro-enterprises in the non-farm sector is one of the priority initiatives. There are also capacity building programmes for micro, small and medium enterprises.

MICRO, SMALL AND MEDIUM ENTERPRISES (MSMEs)

The Micro, Small, and Medium Enterprises (MSMEs) have emerged as important sectors of the economy both at the national and state level. The classification of micro, small, and major is based on the amount of money invested to develop the enterprise. When the investment is Rs. 20 lakh for industry or Rs. 10 lakh for service enterprise the unit is called micro-enterprise. When the investment is up to Rs. 5 crore, the unit is designated as small and when investment is between Rs. 5 to Rs. 10 crores the unit is classified as medium. This investment does not include the cost of man power. MSME is playing a crucial role in providing large employment opportunities with comparatively low capital cost and developing forward and backward linkages with other sectors. They serve as complementary units to large-scale industries and contribute significantly to socio-economic development. Given Kerala's excellent connectivity, availability of human resources, and relatively good infrastructure facilities MSME is highly suitable. It promotes industrialization and provides employment to rural areas. Directorate of Industries and Commerce (DIC) acts as a facilitator for MSME. According to DIC Kerala has 1,27,725 MSME units (Government of Kerala, 2019).

The district-wise distribution indicates that Malappuram district tops the list by housing 17.4 per cent of total enterprises followed by Ernakulam (13%) and Thrissur (12%). Wayanad district ranks the lowest by having only 2.5 per cent of total enterprises (Fig. 8.2). MSME has developed mostly around the industrial belt and major urban centres. Wide regional disparity in the distribution of the MSMEs is well evident.

During 2017, there was a scheme of enrolment of MSME under Udyog Aadhar Memorandums (UAM). A total of 34,518 MSME covering both the manufacturing and service sectors was enrolled this year. Ernakulam district with 4390 units led in the manufacturing sector and Thiruvananthapuram district with 3203 units led in the service sector (Table 8.2). The districts of Idukki, Wayanad, and Kasaragod together accounted for 5 per cent of total UAM MSME units in

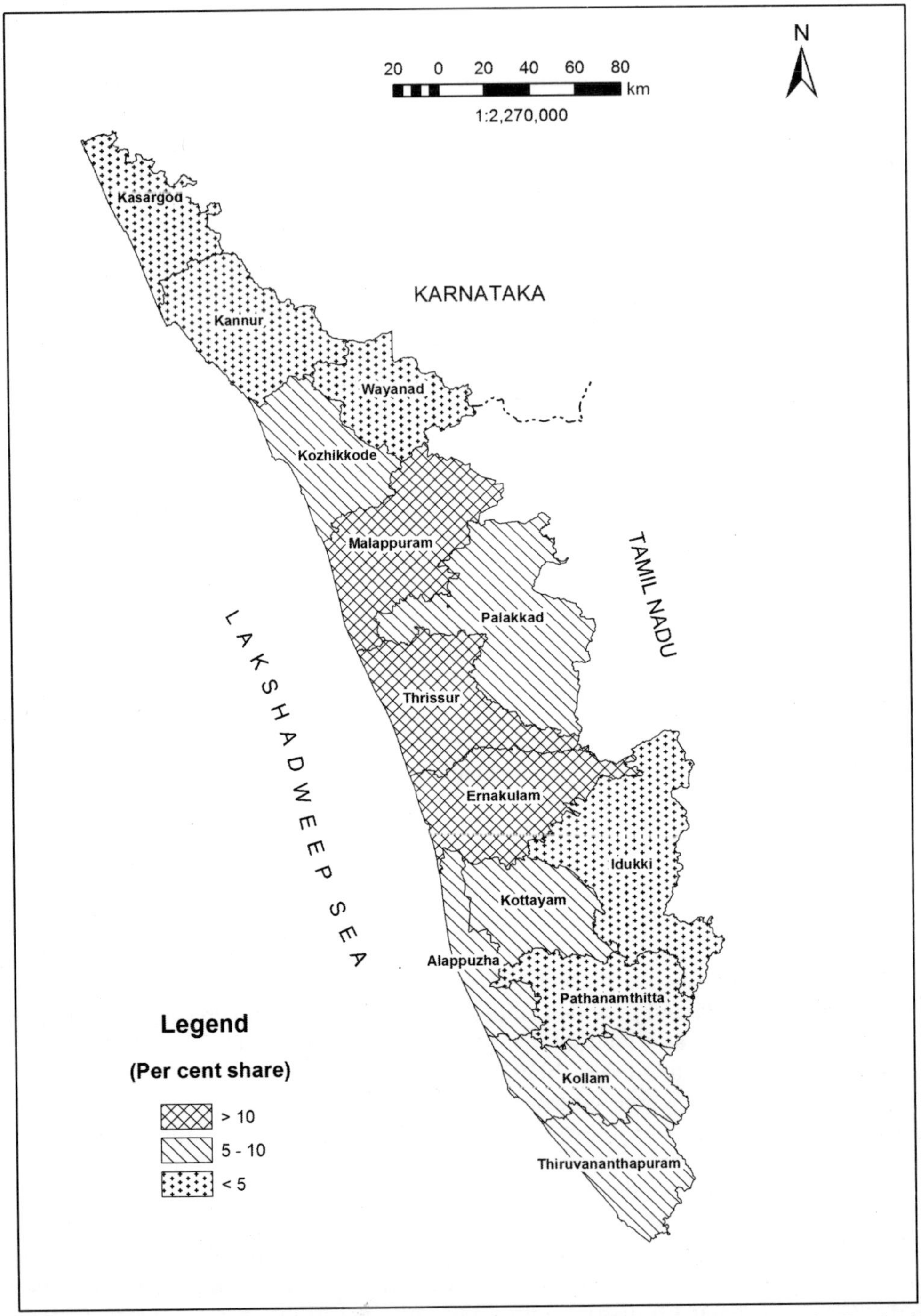

Fig. 8.2: Distribution of Micro, Small and Medium Enterprise (MSME) Units by Districts in Kerala, 2017.

Data Source: Government of Kerala, 2018.

Kerala. In the year 2016-17, there were 15535 new MSME units that began their operation with an investment of Rs.1387 crore and generating employment for 57,445 persons. The highest number of new units was in Ernakulam followed by Palakkad. The district of Idukki, Wayanad, and Kasaragod together accounted for 4.5 per cent of total newly registered units. This indicates that there is regional inequality in distribution of MSME units. The coefficient of variation of district-wise distribution of the manufacturing sector (68.5%) is lower than that of the service sector (122.5%). This signifies that inequality is higher in the case of the distribution of service sector enterprises. Only two districts, Thiruvananthapuram and Ernakulam together house 53.5 per cent of all UAM MSME units under the service sector.

Table 8.2: Distribution of MSME Units Registered under UAM by Districts in Kerala, 2017

District	*Manufacturing*				*Services*				*G. Total*
	Micro	*Small*	*Medium*	*Total*	*Micro*	*Small*	*Medium*	*Total*	
Tvm	3358	257	10	3625	2702	491	9	3202	6827
Kol	1888	231	8	2127	492	123	3	618	2745
Pat	734	64	6	804	231	38	0	269	1073
Ala	2299	222	21	2542	466	77	9	552	3094
Kot	1226	140	10	1376	434	102	2	538	1914
Idu	516	44	3	563	154	37	1	192	755
Ern	3398	951	41	4390	1766	551	18	2335	6725
Tri	2060	279	13	2352	783	178	7	968	3320
Pal	1113	206	15	1334	393	70	2	465	1799
Mal	1218	160	3	1381	244	55	1	300	1681
Koz	1383	178	5	1566	302	109	3	414	1980
Way	317	38	0	355	57	23	0	80	435
Kan	1097	175	6	1278	238	58	4	300	1578
Kas	404	69	0	473	79	40	0	119	592
Std. Dev	992	226	10.7	1183.1	780.5	167.3	5	953	2095.9
CV (per cent)	66	105	106	68.5	124.6	120	120	122.5	82.1

Data Source: Government of Kerala, 2018.

FOOD PROCESSING

The food processing industry has emerged as an important industrial sector with high growth potential. This industry is fast growing with a change in lifestyle and food habits, and economic growth. It accounts for 32 per cent of the country's total food market at all-India level. Kerala has recognized the food processing industry as a priority sector due to its potential for future growth and generating substantial employment. By 2013-14, Kerala had 1460 registered units for the food processing. Kerala Industrial Infrastructure Development Corporation (KINFRA) is actively engaged in promoting food processing industry in different parts of the state. One of the initiatives, KINFRA Mega food park, Palakkad proposes to establish a centralized processing centre with four primary processing centres at Ernakulam, Thrissur, Malappuram, and Wayanad. Food safety is an important issue with the growth of this industry.

TRADITIONAL INDUSTRIES

Traditional industries dominate the industrial scenario of Kerala. Coir, cashew, handloom, tiles, and handicrafts belong to this sector. *Beedi* is also one of the traditional industries functioning in Kerala, particularly in the Kannur district. These traditional industries provide employment to a large section of people next only to agriculture. We briefly discuss some of the industries here.

COIR INDUSTRY

Coir is perhaps one of the oldest traditional agro-based rural industries. India ranks first among the coir producing countries in the world, although, in the matter of coconut production, India's position is after Indonesia and Phillippines. Brazil, Sri Lanka, and Thailand are other major coconut producing countries in the World. Among the states in India, Kerala is the main coir producing state by accounting for more than 70 per cent of the country's total production. Tamil Nadu and Karnataka are the other two important coir producing states. Coir industry is the largest among the traditional industries in Kerala and it is mostly operating in the coastal plain. It provides subsistence to around 2 lakh families in the coastal belt. Coir sector employs 3.75 lakh workers in co-operative and private sectors and 80 per cent of employees are women (Government of Kerala, 2019). Manufacturing of coir ropes is perhaps the oldest industry in Kerala flourished with sea trade in the early years of the Christian Era (CE). It is difficult to exactly date the beginning of production of coir yarn and coir ropes, however, there are indications that the coir industry received a boost during the 16th Century with rising of the European maritime power (Rammohan, 1999). Coir and coir products have significant local markets. It is mainly a household industry nurtured by family people employing some outside labours. The coir industry continues to be a source of income for the family apart from meeting household demand. Coir is an important expert item from Kerala. It has multiple uses as a natural geotextile in various sectors including environmental management.

In 1961-62, the total quantity of coir and coir goods exported from Kerala was a little over 76 thousand tonnes, which declined to 24.7 thousand tonnes in 1985-86. Export started increasing in subsequent years and by 2015-16 it was 124.3 thousand tonnes. Some items like handloom mat, tufted mat, coir pith, power loom mat, rubberized coir, and coir fibre have registered an increase in export quantity, whereas there are some other items like curled coir, handloom matting, power loom matting, and coir geo-textiles which have recorded a decline in export quantity. Coir and coir products from India are exported to USA, Germany, UK, France, the Netherlands, and Italy. New exportable items cover coir pith and coir geo-textiles. Handloom products have high demand compared to power loom products in foreign export. Coir export increased by 250 per cent in value terms from 2012-13 to 2016-17.

The coir co-operative societies are engaged in the manufacturing of coir fibre, coir yarns, and other coir products. COIRFED the apex federation of primary co-operative societies manages part of the marketing. Societies themselves also market their own product. The purchase of husk, fibre, and production of yarn are also undertaken by coir co-operation societies. There are ten project offices to manage the co-operative societies in Kerala. Alappuzha district has the highest number of co-operative societies. Several organizations are functioning for the promotion of the coir industry.

As of 31.03.2017, Kerala had 1001 registered coir-cooperation societies, of which only 564 societies were working. There were 436 working societies in the yarn sector and the rest were in the product sector. These societies are grouped as primary co-operative societies (Yarn sector), manufacturing societies (Mat and Matting), small scale producers' co-operative societies, husk procurement and distribution societies, and fibre societies (defibring mill societies). Besides, there is a co-operative coir marketing federation, and other societies functioning in the state. Despite several initiatives, the coir industry continues to remain as sick traditional industry.

There are five activities linked with the coir industry:

i) Collection of raw materials–coconut husks
ii) De-fibring
iii) Spinning of coir yarn and making of ropes
iv) Weaving and
v) Marketing

This traditional industry faces problems in all five areas. It has been estimated that coir factories in Kerala require 216,000 tonnes of fibre, of which 60 per cent are available within the state and the rest 40 per cent are brought from outside the state (KITCO, 2003). One of the problems of the coir industry in Kerala is the shortage of husks. Around one-third of the units are operating at less than 50 per cent capacity utilization. In spite of the growth of the area under coconut plantation, the demand for husk cannot be met within Kerala. Root wilt diseases have affected coconut production in large parts of the state. Coconut trees in the Alappuzha district are mostly low yielding or non-yielding due to diseases and used up natural fertility of sandy soil in the strip of land bounded by the Lakshadweep Sea and the Vembanad lagoon. Traditional methods are still being used to ret the husk and extract fibre.

Lack of mechanization is another serious issue. Around 68 per cent of factories are in the category of either not mechanized or partially mechanized (*ibid.*). Problems of marketing and competition from synthetic fibre are other impediments to the development of the coir industry. Due to various problems, employment in this industry has come down from 4.3 lakh in 1989 to 3.75 lakh in 2018. All the workers engaged in the coir industry do not find employment throughout the year. Only 17 per cent of workers find employment more than 250 days in a year. Two-thirds of the workers get employment between 150 and 250 days and the rest falls below the 150 days category. Kerala State Co-operative Coir Marketing Federation (COIRFED) is the apex body of primary coir co-operative societies engaged in the procurement and marketing of products from the member societies. The other important agencies involved in the promotion of the coir industry include Kerala State Coir Corporation, Foam Mattings (India) Ltd., National Coir Research and Management Institute, Kerala State Coir Machinery Manufacturing Company, and Kerala State Coir Workers Welfare Fund Board. The Government of India also has several schemes in the coir sector.

The Government of Kerala has taken various initiatives to help the coir industry including organizing of international events with the participation of buyers from other countries. The coir sector in Kerala is facing competition from other states. The working environment of coir workers is not so hygienic and requires improvement. Strengthening of the coir sector warrants consolidation of the raw material base, modernization of factories, diversification of production, improved marketing strategy, and social security net for the workers.

HANDLOOM INDUSTRY

The handloom industry is the second most important traditional industry after coir in Kerala. The co-operative sector dominates the handloom industry by controlling 98 per cent of total looms as of 2017-18. In 1994, there were 83 per cent looms under the co-operative sector. The co-operative sector consists of two types of societies: factory type and cottage type. By 2017-18, there were 611 registered primary Handloom Weavers Co-operative Societies in the state of which 172 were factory type and 439 were cottage type societies. However, working societies consisted of only 98 factory type and 308 cottage type.

The HANTEX (Kerala State Handloom Weaver's Co-operative Societies) and HANVEEV (Kerala State Handloom Development Corporation) are two organisations established to promote the handloom industry in Kerala. The HANTEX is the apex body of handloom co-operatives overseeing the distribution of required inputs to primary co-operative societies, procurement, processing, and marketing of goods produced by the member societies. The HANVEEV established in 1968 was to uplift traditional handloom weavers in the unorganized sector in the handloom industry. Both these organisations are incurring financial losses.

The handloom industry is mainly concentrated in Thiruvananthapuram and Kannur districts. They are also found in Kollam, Ernakulam, Thrissur, Palakkad, Kozhikode, and Kasaragod districts. Kerala has 17,070 looms, of which 15,971 looms are in the co-operative sector. Production of handoom cloth was 38.97 million metre in 2017-18. The handloom sector generated 55.46 lakh man-days of work during 2017-18. There is a continuous fall in work generation. Even in 2013-14, employment generated was 92.19 lakh man-days. There were 874 powerlooms in Kerala during 2017-18 producing 56.66 lakh meters of cloth. Out of 874, as many as 860 powerlooms functioned under the co-operative sector. The productivity of powerlooms under the co-operative sector was 6,589 meter/loom (Government of Kerala, 2019)

The handloom sector faces several challenges. Lack of modernization, shortage of workers due to low wages, low productivity, competition from other states, and lack of product diversification in tune with changing trends are some of the problems affecting this sector. Appropriate policy intervention is necessary to address these problems. There should be continuous efforts to bring in improvement in product design and marketing.

CASHEW INDUSTRY

The cashew industry is another important traditional industry providing employment and generating foreign exchange. It is highly labour intensive. Nearly 1.5 lakh workers were engaged in this sector by 2016-17 and >90 per cent of workers were women. Kerala accounted for 11 per cent of cashew production and 35 per cent of all cashew processing units in India (Government of Kerala, 2018). Production is not sufficient to cater to the need of the factories. As a result, workers could not find year round employment. The Kerala State Cashew Development Corporation (KSCDC) and Kerala State Cashew Workers Apex Co-operative Society (CAPEX) are two agencies in the state engaged in the cashew processing sector. KSCDC could provide an average of 75 man-days of employment in 2016-17 whereas CAPEX could provide 120 days of employment. The major problem faced by this industry is non-availability of adequate quantities of raw materials. The domestic production of raw nuts is not enough to provide employment to workers even for three months.

Cashew grows throughout Kerala; however, the bulk of the production is from Kannur and Kasaragod districts followed by Kollam. Due to the seasonal nature of employment, the workers engaged in the cashew factory find it difficult to maintain their livelihood. The cashew industry is facing a serious problem. Many of the private factories are closed down rendering a large number of workers jobless. The availability of raw material is a major hurdle. The area under cashew is reducing over the years. It is important to enlarge the production base. Besides, due to the high wage rate, processing cost is also high in Kerala compared to other States. International competition, particularly from Vietnam is another factor affecting the cashew industry. Coordinated action at a different level is necessary both from the Government of India and the Government of Kerala to overcome the problems.

BAMBOO INDUSTRY

Kerala is rich in bamboo resources. It accounts for 19 per cent of the total area under bamboo in the country. There are 28 species of bamboo found in the state. Little over 59 km^2 area is under the plantations of cane, reed, and bamboo in the forest area (Government of Kerala, 2008). It is important to note that 67.3 per cent of bamboo comes from home gardens. The bamboo industry is a predominantly household based enterprise providing a livelihood for numerous households. It is estimated that nearly one lakh people in the state depend on bamboo for livelihood (Government of Kerala, 2019). Canes, reeds, and bamboos are useful for the production of mats and baskets, widely used for drying and storing of food grains, and construction of houses in the rural areas. Reed-basket and mat-weaving used to be traditionally undertaken by Scheduled Castes and Scheduled Tribe people as leisure time and off-season employment. About 90 per cent of production is by the household sector (Nair and Muraleedharan, 1983).

The Kerala State Bamboo corporation established in 1971 initially to function as a marketing agent and subsequently to oversee the collection of reeds and supply of reed to traditional weaves, co-operative societies, and other bona fide consumers. The main buyers are Central Warehousing Corporation and large sugar mills. The bamboo corporation is serving 10,000 bamboo workers. The Kerala State Bamboo Mission was constituted in 2003 to adopt a focused approach to revitalize the bamboo sector. It has taken several initiatives including skill development.

The bamboo industry in Kerala is mostly concentrated in the Angamaly-Kalady area in Ernakulam district and Nedumangad-Aryanad area in Thiruvananthapuram district. Easy availability of raw materials, proximity to reed forest, and transportation facilities favoured these two areas for the development of the bamboo industry. Products from the Angamaly-Kalady area cater to the external market such as Food Corporation of India (FCI) and Central Warehouse Corporation (CWC) whereas Nedumangad-Aryanad areas supply materials within the state. Pathanapura, Punalur, and Parakode are some of the Centres where mat production is done in the unorganized sector (Jayasankar, 2000). Baskets are produced in many areas, having access to raw material and concentration of traditional weavers. The recent initiatives of the corporation are (i) Flattened Bamboo Boards with support from the National Mission on Bamboo Application, and (ii) Bamboo Flooring Tiles project with financial support from Central and State Governments.

Availability of raw material, proper supply chain, mechanisation, the introduction of efficient and professional management co-operative, marketing, and protection of unorganized weavers are important challenges warranting urgent attention of the Government.

PROBLEMS OF MANUFACTURING SECTOR

Problems in the manufacturing sector in Kerala have been discussed from time to time (Subramanian, 1994). Here we list some of the salient issues that singularly and in combination beset industrial development in Kerala:

1. The structure of the manufacturing industry in Kerala continued to be made up of industries based on local raw materials and in terms of employment the composition is still dominated by traditional industries. Low levels of technology, low productivity, and low income characterise traditional industries in Kerala. There are apprehensions about labour displacement in the face of mechanization and technological innovations to upgrade the industries. The wage rate in Kerala is higher than the neighbouring states. Therefore,

the cost of production is high here. At the same time, a large number of employees engaged in traditional industries like cashew and coir do not get employment throughout the year due to a number of reasons including a shortage of raw materials.

2. Development of modern demand-based (foot loose) industries in Kerala is inadequate. The number of workers employed in the manufacturing activities connected with metal, machinery and transport equipment is significantly less.
3. Kerala's industrial base is lopsided with the dominance of local resource-based industries. Demand-based industries and capital goods related industries do not have a fair share (Subramanian and Pillai, 1986). Although a number of industries have sprung up during the last five decades, the industrial structures have neither ensured inter-sectoral linkage nor stimulated technological progress for the rapid growth of the manufacturing industry in Kerala.
4. Small sector has been playing a relatively larger role in the industrial structure in Kerala as compared to the all-India level. Conversely, the share of the large-scale sector is proportionately smaller in Kerala than that of the all-India average. The concentration of more than 35 per cent of all medium and large-scale units in one district (Ernakulam) indicates regional imbalance even within the industrially backward state. In fact, over the years concentration of medium and large scale industries in Ernakulam had grown from 25 per cent of all such units in the state in 1990 to 35 per cent in 2005 (Government of Kerala, 2006b). More significantly, in terms of types of industries, the investment is concentrated in few resource-based activities, and thereby industrial structure has remained rather lopsided.
5. In the matter of production cost, Kerala has a disadvantage as compared to the all-India average with respect to (a) emoluments and welfare payment to employees, (b) other inputs, (c) depreciation, and (d) interest. The highly organised labour in Kerala, though desirable in social democracy, has its implications on the private cost calculations of the enterprises. The cost of other inputs includes the cost of non-industrial services received from others, the cost of materials consumed for repair, and the maintenance cost of contract and commissioned work done by others. The higher component of depreciation reflected upon the age of machinery and equipment and hence the outdated nature of production technology. The high share of interest in the total cost could be seen in the context of the high capital gearing in financial management and the inefficient working capital and inventory management typical of enterprises both in private and public sectors in Kerala (State Planning Board, 1990).
6. General efficiency of manufacturing industries in Kerala has decreased over the years. The inefficiency of state public sector enterprises is manifested in the losses being accumulated by most of the units. Among 40 state PSUs under the industry department, only 10 made profit the rest 30 incurred losses as noted in 2017. A study in 1990 indicated that the performance of the companies in the private sector was not better than their counterparts in the public sector (Padmanabhan, 1990). Overall efficiency in the manufacturing industry appears to be decreasing over the years.
7. Public sector investment in the manufacturing industries in Kerala from the central sector is well below the state's share in the country's population. By March 2015-16 the central sector investment was 2.1 per cent of total investment in India. It is declining over the years. From 3 per cent during the 1970s it is now hovering around 2 per cent. The lowest share of central investment was noted in 1995 when it was only 1.23 per

cent. Kerala's rank is 11th among the states in India in the matter of investment through central Public Sector Undertakings (PSUs). Maharashtra tops the list with an 8.8 per cent investment. Kerala has 2.55 per cent of total employment in the central sectors in the country.

8. Most of the PSUs are small and aggregate investment is not substantial. In fact, industrial investment did not receive adequate attention in the past. One of the problems identified was a mismatch of organizational structure and the demand for the production process (Pillai, 1994). Another problem identified in the case of state sector enterprises is that Kerala has diffused its investment in all sectors. Instead of consolidating in few sectors, that had already been initiated, the trend was to spread its resources into several sectors. This has resulted in underinvestment in some sectors which required heavy investment like the engineering sector and weak linkage of some industries with the local economy in terms of purchase of inputs.

INFORMATION TECHNOLOGY (IT)

IT is a knowledge-based enabling technology. It is one of the fast-growing sectors in the World today. India along with China emerged as major contributors in the IT industry at the global level. The IT sector is contributing significantly to Indian economic growth since 1991. It is significant that production under this industry is largely in the private sector while Government intervention has been to provide an enabling environment (Mani, 2006). The growth of the IT industry in India is mostly concentrated in a couple of metropolitan cities located in western and southern India. Kerala did not figure in the IT industry in the initial phase of its development in the early nineties. With a subsequent emphasis on tier II and tier III cities as IT destinations, Kerala emerged as a serious contender and started attracting IT industries. By 2009, there were 14 IT organisations working in Kerala. There are four Special Economic Zones (SEZ) to house IT industries in the state. These are Technopark, Thiruvananthapuram, Infopark, Kochi, Cyberpark, Kozhikode, and Cyberpark, Kasaragod.

The Technopark at Thiruvananthapuram established by the Government of Kerala in 1990 is India's first technology park and one of the largest IT parks in India. By 2017-18, there were 400 companies functioning in 5 campuses spread over 760 acres of land. Technopark provided direct employment to 56,000 persons and offer indirect employment to 1,50,000 persons. Total export from Technopark in 2017-18 was worth of Rs. 6452 crore which was worth of Rs.1800 crore in 2009. The export value recorded an increase of 258 per cent in just 9 years. Infopark at Kochi distributed in three campuses, namely, Kochi, Cherthala, and Koratty in Thrissur, houses 396 IT companies and provides direct employment to more than 37,000 IT professionals.Cyberpark was established based on a hub and spoke model. It has three campuses in Kozhikode, Kannur, and Kasaragod. Thirteen companies are functioning with an employment of 266 persons. Major IT companies like TCS, Infosys, Wipro, HCL, along with mid-tier IT companies like UST Global are operating in Kerala.

The tier-II cities like Thiruvananthapuram and Kochi are now emerging as preferred destinations for IT professionals due to quality environment, health service, and educational facilities. The high quality of life and a calm environment are added attractions. The tier-III cities in Kerala are also potent for IT growth. The role of the IT industry in Kerala is very significant for the future industrial development of the state. The educated unemployed youth with proper skill up-gradation can be accommodated in this industry.

FUTURE CHALLENGES OF INDUSTRIAL DEVELOPMENT

Kerala has passed through a stagnant phase of industrial development. In recent years, there is considerable emphasis on re-industrialisation. To catch up with neighbouring states in industrial development, Kerala needs to achieve 10 per cent growth (Planning Commission of Inda, 2008a). The industrial policy of Kerala, 2018 proposed to set up industrial parks through Public-Private Participation (PPP) with an objective to achieve this growth rate (Government of Kerala, 2019). It is well accepted that availability of a large tract of land for setting up industries in Kerala is a limitation. There are no appreciable mineral deposits. At the same time the state has the distinct advantage of educated man power, connectivity, power supply, and other infrastructural facilities including a clean environment. The thrust may be on advanced technology centred development. Promotion of specific segments of chemical, medical, electronic and engineering industries, and enterprises based on biotechnology, nanotechnology, and life sciences are important components of future industrial development of Kerala. There is further scope for developing information technology and information technology-based services.

One of the top priority activities is the modernization of traditional industries of coir, cashew, and handlooms. There should be product diversification, value addition, and effective marketing strategies. In the mineral sector, particularly, rare earth, there is scope for value addition. Some PSUs in Kerala once made a significant mark in industrial development but over the years, functioning of these organisations was not up to the appropriate level. Moreover, several PSUs are small. There is considerable scope to reenergize PSUs. There are opportunities to develop post-harvest technologies, agro-based products, marine products, and forest-based products including medicines. The food processing industry has a large scope in internal and foreign markets. The strong local self-government, as present in Kerala now, can play an important role in facilitating small and medium enterprises. Self-help groups, particularly, women self-help groups also have a large role. Already, they are active stakeholders in certain sectors, which can be enlarged. Future industrialization process in Kerala may be oriented towards small, environment-friendly, knowledge-based and energy, efficient enterprises. The Government policy should facilitate such endeavour.

CHAPTER

9

Transport and Tourism

TRANSPORT AND DEVELOPMENT

Transport by definition is concerned with the movement of people and goods. It is essentially demand-driven and is required not for its own sake but because it is useful to satisfy some other needs. As an infrastructure, transport is a factor of production and plays important role in economic growth and development. It facilitates interaction and exchange, controls movements of people and commodities, and generates growth impulses by providing services or means to distribute surpluses and reaching goods to the place having demand. Ullman (1973) identified three main principles governing interactions involving transportation. These are complementarity, intervening opportunity, and transferability. The complementarity is a demand-supply relationship; intervening opportunity refers to alternative sources or substitutes and transferability is related to distance and time cost. The last item, transferability, has overriding importance in the matter of transport.

The interrelationship between transport and development is complex and multidimensional. Whether development follows transport or transport follows development is an egg and hen nature of the problem. A good transport system can help the resource-poor region to obtain a supply of resources from other areas and prosper. A proper distribution system can be effective only when it is supported by an adequate transport system. Intense economic activities in many of the transport junctions/nodes are manifestations of transport's role in economic development.

There are four main modes of transport: roadways, railways, waterways, and airways. For transport to be operational under anyone of these modes, there are four essential elements, namely, route, vehicle, motive power/energy, and terminal (Bamford and Robinson, 1978). All these elements with the possible exception of the route under waterways require huge capital investment. The waterways also need investment for maintaining navigability and to construct terminals (Jetties). The role of government and political-economic decisions are important in this context. Therefore, the history of transport development in developing countries like India with the colonial past and the developed countries are different. Profit motives of the colonial rulers were the prime drivers of transport development in the subjugated countries. Siphoning of resources through international ports and maintenance of law and order by suppressing national uprisings to facilitate unhindered flow of resources from the hinterland to the port were the main determinants underpinning colonial thrust for the development of penetrating transport network in these countries. Expansion of railways and roadways in India during the 19th and 20th century were primarily to achieve these objectives. However, it is also important to note, as has been indicated by Taffee *et al.* (1973), that transport facilities, although developed with a colonial motive, start serving the needs of the local economy over time.

TRANSPORT DEVELOPMENT IN KERALA: A BRIEF HISTORY

Geographical factors like relief, terrain condition, distribution of rivers, lagoons and estuaries, and land use play important role in transport development in Kerala. Main rivers flowing in east-west direction directly connect resource base in the Western Ghats to the markets (ports) along the west coast. The lagoons and cross-channels connect these rivers in the coastal plain and facilitate the north-south movement. Subsequently, canals were excavated to further strengthen and extend the watercourses by linking coastal water bodies. Kerala established international trade relations since the early days and the waterways which played an important role in movement of tradable commodities from the hinterlands grew along with trade. The discovery of a new sea route in 1498 boosted foreign trade in Kerala significantly, particularly on the Malabar Coast. This gave rise to a series of port settlements along the coast. Forest and agricultural produce constituting the bulk of exportable materials were brought to the ports along coast through rivers and interconnecting channels. As these resources were distributed all through Kerala and rivers provided a direct connection between the hinterland and port, the entire operation was in a decentralized mode and dominated by waterways. There was a port perhaps at every opening to the sea. Dense vegetation cover and undulated nature of topography restricted settlement expansion in the eastern midlands and foothills. People concentrated mostly in the coastal plain and along the river valleys. Major transport development is limited mainly to the coastal plain and adjoining lowlands till today.

The trade development and resource movement influenced transport development in the early days. Subsequent transport development in Kerala clearly bears the imprint of colonial rule, trade expansion and exploitation of agricultural potential both by the British and native kings. Before reorganisation of the linguistic states in India in 1956 present day Kerala was under three different administrations: Malabar under Madras Presidency, and Cochin and Travancore under two princely kingdoms. The historical evolution of the transport systems in Malabar, Cochin, and Travancore were different (Ibrahim, 1978). The main purpose of British rulers to expand the railways network for that matter transport system in India was to siphon natural resources from the hinterlands, suppress the nationalist movement, and consolidate their power. Extension of railway line up to Beypore by 1861, construction of important canals in Malabar area between 1840 and 1860, and road construction largely coincided with the first phase of tenant uprisings in Malabar and Mopla revolt (*Ibid.*). By the end of the 19th Century, almost all construction activities in the British ruled Malabar came to a halt with the decline of local revolts. From 1861, it took another 43 years for completion of the Calicut-Mangalore railway line by 1904.

The growing importance of Cochin port in external trade underpinned by the introduction of steamship and prosperity of the British Empire in Europe expanded the port-hinterlands in India and brought large areas in Kerala under trade relations. It was necessary to increase the transport facilities to move tradable goods. Cochin due to well endowed natural waterways commended the lion share of the trade with the Portuguese and the Dutch on the Malabar coast for nearly three centuries (Menon, 1995). All important canals were constructed during the period of Diwans Sankara Variyar and Sankunni Menon spanning over 40 years from1840 to 1880 (*ibid*). By the end of the 19th century, present-day Kerala was covered with a good network of roads in excellent condition. The Shornur bridge constructed in 1864-74 strengthened the direct connection of Cochin with Malabar and Coimbatore. Railways connection came to Cochin from Shornur by 1902 covering a distance of around 105 km (*ibid*). A small part of this line (around 29 km) ran through the area under the jurisdiction of the Travancore kingdom.

Transport development in the Travancore region coincides with the expansion of plantation agriculture in the high ranges. The Public Works Department set up in 1860 under the initiative of Diwan T. Madhava Rao took an ambitious programme of repairing existing roads and constructing new roads throughout the length and breadth of the native state of Travancore (Ibrahim, 1978). The Main Central Road (MC Road) connecting Thiruvananthapuram and the northern boundary of the Travancore kingdom was constructed in 1878. It connected all the major trading centres along the foothills from where feeder roads were laid both towards high ranges in the east and coast in west, thereby connecting the provenance of resources with the ports and markets. The expansion of road network provided an impetus to trade and agriculture. By 1900, the Travancore had more than 4800 km of roads. Waterways also developed along the coast. Varkala canal completed in 1880 provided an uninterrupted waterways from Thiruvananthapuram to Beypore in Malabar (*Ibid*). Railways entered Travancore by 1904 when Kollam was connected to Tirunelveli. The railway line connecting Kollam and Thiruvananthapuram was operational in 1918, and Thampanur figured in the railway map of India only by 1926.

The historical development of the transport system in Kerala indicates that there are variations in the spatial expansion of all three networks—waterways, roadways, and railways. Malabar, which received a head start in the second half of the 19th century experienced a stagnating situation in the 20th century and lagged far behind Cochin and Travancore regions. Still today, the Malabar region is considered to be backward in transport development.

TRANSPORT SYSTEM IN KERALA: PRESENT SCENARIO

Kerala is one of the few states in India which is served by all four transport systems: roadways, railways, waterways, and airways. We provide a brief account of all of these means of transport.

ROAD NETWORK

Kerala can boast of a well-developed road network connecting all parts of the state (Fig. 9.1). Hardly there is a place in Kerala without any road connection. The total road length is 2,29,349 km excluding un-metal roads under the local Self Government Department (Government of Kerala, 2019). The road density in Kerala is 5.90 km/km^2 against the national average of 1.70 km/km^2. The corresponding figures were 3.3 km/km^2 and 0.42 km/km^2 in 1990-91. Kerala has road density of 0.52 km/km^2 against the national average of 0.23/km^2 even in 1960s. Accessibility in Kerala was always better than that of the all-India average. Kerala records 687 km road per one lakh population, whereas the national average is 441 km per lakh population. The state accounts for 3.9 per cent of total road length in the country. Although the quality of the road is not good in many stretches, it has wide penetration. According to the study of the Confederation of Indian Industries (CII), 2003, with the rate of infrastructure penetration of 2.5 Kerala ranks fourth among the states in India in the matter of infrastructure penetration. Maharashtra tops with 2.8 points (Government of Kerala, 2004). Transport was one of the important variables considered for this rating.

Road network in Kerala consists of National Highways (NH), State Highways (SH), Major

District Roads (MDR), other PWD roads, roads maintained by various other organisations, and Village/Panchayat roads (Table 9.1).

Table 9.1: Types of Roads and Their Length in Kerala, 2017-18

Type of road	*Road length in Km*	*Per cent*
National Highway (PWD)	1,782	0.80
State Highway (PWD)	4,342	1.90
Major District roads and other roads (PWD)	27,470	11.95
Roads under Municipalities and Corporations	25,055	10.90
Forest Department Roads	4,576	2.00
Other agencies (Irrigation, KSEB, Railway etc.)	2,940	1.30
Sub total	66,166	28.85
Panchayat roads	1,63,184	71.15
Grand total	229349	100.00

Data Source: Government of Kerala, 2019.

Public Works Department (Roads Wing), Government of Kerala manages 31,812 km of roads under the categories of SH, and major district roads, and other district roads. The national Highway Wing of PWD and National Highway Authority of India are responsible for managing the national highways. There are 11 National Highways in Kerala (Table 9.2). The total length of these 11 NHs is 1782 km or 1.36 per cent of total road length under National Highways in the country.

Table 9.2: National Highways in Kerala, 2017-18

Sl. No.	*New No.*	*Old No.*	*From*	*To*	*Length in Km*
1.	NH 66	NH 17	Thalappadi	Edappally	421
		NH 47	Edappally	Kaliyikkavila	249
2.	NH 544	NH 47	Walayar	Edappally	168
3.	NH 85	NH 49	Bodimettu	Kundanoor	168
4.	NH 744	NH 208	Kollam	Kazhuthuruthy	81
5.	NH 766	NH 212	Kozhikode	Muthanaga Kerala-Karnataka border	118
6.	NH 966	NH 213	Kozhikode	Palakkad	125
7.	NH 966A	NH 47C	Vallarpadam	Kalamassery	17
8.	NH 966B	NH 47A	Wellington island	Kundanoor	6
9.	NH 183	NH 220	Kollam	Theni-Tamil Nadu border	216
10.	NH 183 A	-	Bharanikkavu	Pathanamthitta via Vandiperiyar	117
11.	NH 185	-	Adimaly	Painavu via Kumilli	96

Data Source: Government of Kerala, 2019.

National Highways are reorganized and newly numbered. The NH 17 and parts of NH 47 together form the main axis along the coastal plain running from Thalappady in Kasaragod to Kalyikkavila in Thiruvananthapuram through Edapally in Ernakulam district, where both the NHs meet. These two National Highways (newly numbered as NH 66) connect all district headquarters en route. The part of NH 47 from Edapally (Ernakulam) to Walayar in Palakkad, renamed as NH 544, and NH 213 (new no. NH 966) from Kozhikode converge in Palakkad. The Palghat gap situated almost in the central part of the Western Ghats is the zone of convergence for the NHs and railway lines connecting northern and southern tips of Kerala with the Tamil Nadu plain. National highways connect Kerala with Tamil Nadu and Karnataka and also link district headquarters and interior parts of Kerala. The transport network closely follows the physiographic trend. The main roads are in a north-south direction along the coastal plain and midlands. East-west roads connect the hilly hinterlands.

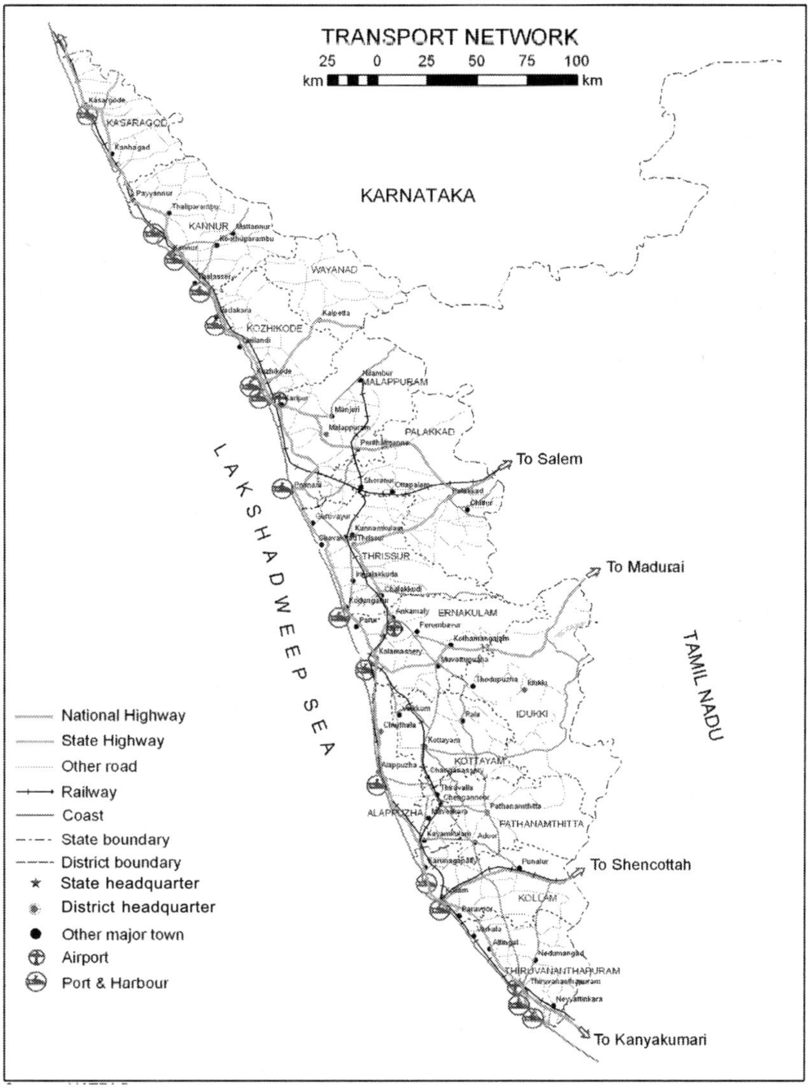

Fig. 9.1: Transport Network in Kerala.
Source: NATPAC.

State Highways and Other Roads

State Highways are arterial routes that link district headquarters with other important towns and connect them with National Highways. Kerala has 76 State Highways passing through all-important big and small towns across the state. The MC road (SH 1) is the main trunk road in the midlands of south Kerala which connects Thiruvananthapuram, Kottarakara, Thiruvalla,

Chenganacherry, Kottayam, Muvattupuzha, Kalady, Angamaly, and a host of small towns located between these towns that have sprung up all along the foothills. The MC road except Kottayam-Muvattupuzha stretch is passing through a densely populated area. Similar trunk roads have been developed in north Kerala and also further east like Punalur-Muvattupuzha. NH and SH constitute the basic frame of the road network by connecting all major nodes like district headquarters and important second-order towns. District roads are mostly feeder roads that link these nodes with interior parts. Panchayat roads at the lowest level of road hierarchy connect rural settlements/hinterlands with the nodes of the next higher order. Lengths of all categories of roads have increased over the years and the maximum increase is recorded in the case of State Highways and for the roads managed by the Municipalities and Corporations. Roads under Municipalities and Corporations are all link roads to connect various parts of the cities/towns with the main artery. Forest Department and KSEB have constructed roads to connect forest out-posts and various hydro-power stations respectively.

Distribution of PWD Roads by Districts

District-wise distribution of PWD roads in 2018 indicated wide spatial variations. The length of PWD roads varied from 48 km/100 km^2 in Wayanad district to 157 km/100 km^2 in Kottayam district with a state average of 82 km/100 km^2. Road development is significantly low in the districts of Wayanad, and Idukki, located in the Western Ghats. Palakkad district also lags well behind the state average in road development. Except for Kozhikode all other districts in Malabar fall below the state average (Fig. 9.2). Most of the north Kerala districts (Malabar districts) continue to lag behind in the matter of road development. The length of PWD roads has increased in all districts over time although the rate of increase is not uniform. Distribution of PWD road length bears a positive relationship with urbanization and population density. The coastal plain, lowlands, and adjoining midlands, particularly below 100 m altitude, have a well-developed road network. Road accessibility reduces towards eastern highlands.

MOTOR VEHICLES

Kerala has witnessed tremendous growth in the number of motor vehicles in the last few decades. The number of motor vehicles was 1.20 lakh in 1975, which increased to 19.10 lakh by 2000 and 120.4 lakh by 2017-18. Pressure on the road has increased accordingly (Table 9.3).

About 65 per cent of the registered motor vehicles are scooters/motorcycles distantly followed by four-wheelers (22 %) and auto rickshaws (6 %). There were 30,986 vehicles per 100 km^2 and 5251 vehicles for every 100 km of road length by March 2018. In terms of population, there were 361 vehicles per 1000 persons in Kerala against an all-India average of 18 vehicles/1000 persons. District-wise distribution of vehicles varies from 1.76 lakh in Wayanad to 17.97 lakh in Ernakulam (Table 9.4). Ernakulam district leads in two-wheelers, four-wheelers, and goods vehicles. It also leads in number of vehicles per thousand population (548).

The number of vehicles grew by 147 per cent in Kerala during the last 10 years and in last five years (2013-14 to 2017-18) the growth was 41 per cent. The average annual growth rate is a little over 8 per cent. It was over 10 per cent between 2010-11 and 2013-14. At the district level, the highest growth was observed in Kasaragod district (81.14 %) between 2013-14 and 2017-18. All districts in Malabar recorded a high growth rate (Fig. 9.3). The highest growth

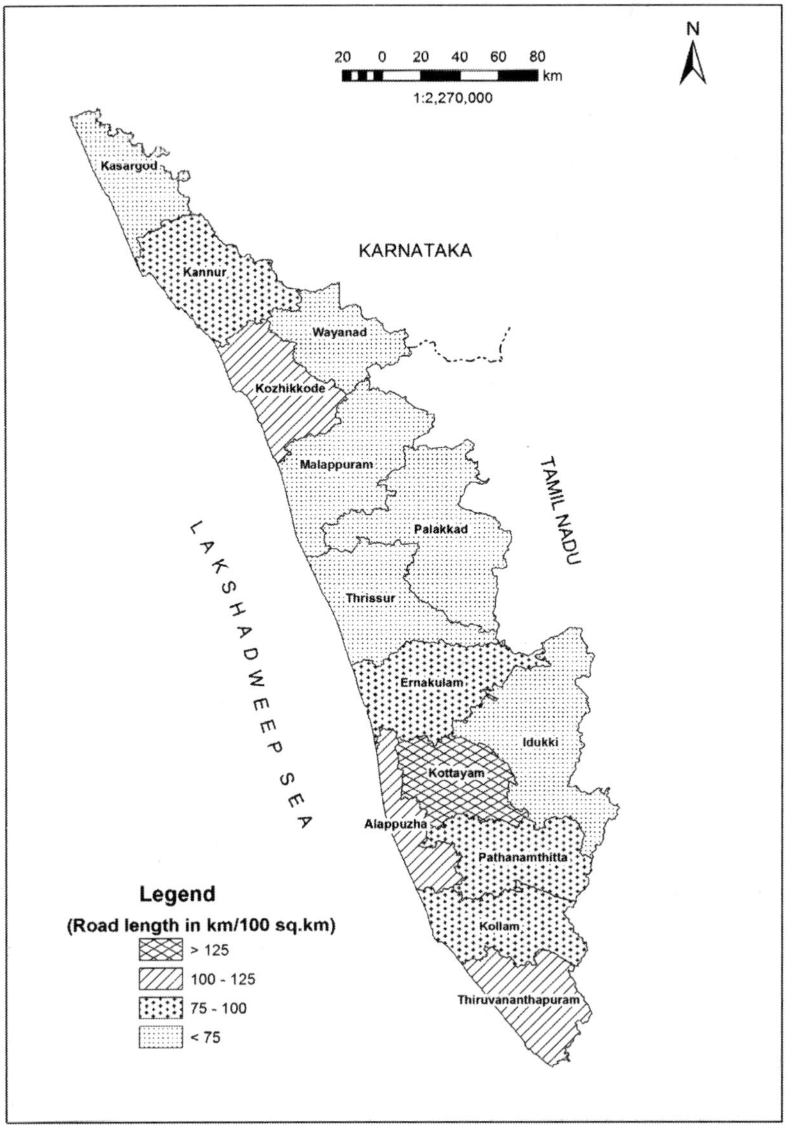

Fig. 9.2: Density Distribution of PWD Roads by Districts in Kerala, 2018.
Data Source: Government of Kerala, 2019.

was observed in the case of Alappuzha followed by Pathanamthitta in southern Kerala. Relatively backward districts are improving. It indicates, indirectly though, that Kerala's development is more equitably distributed.

Kerala does not have sufficient road space to accommodate all these vehicles; as a result, roads are overcrowded with heavy traffic particularly in the corporation cities. All the district headquarters appear overcrowded. Getting parking space is a problem in almost all cities in

Table 9.3: Growth of Road Network and Motor Vehicles in Kerala (1975 to 2018)

Item/Year	*1975*	*2000*	*2018*	*Growth (per cent)*		
				1975-2000	*2000-2018*	*1975-2018*
Road length in kms (NH & PWD)	14870	21731	33594	46.10	54.60	125.90
Panchayat/Rural road in kms	54812	81790	163184	49.20	99.50	197.7
No. of motor vehicles	119720	1910237	12042691	1496.00	530.4	995.90
No. of motor vehicles/km of PWD roads	8	88	358	1000.00	306.8	4375.0

Data Source: Government of Kerala, 2005, 2019.

Table 9.4: Distribution of Registered Motor Vehicles by Districts in Kerala (in Lakh), 2018

Dist.	*Total*	*Two Wheelers*	*Three Wheelers*	*Four Wheelers*	*Buses*	*Goods vehicles*	*Others*	*Vehicles/ 1000pop*
Tvm	15.23	9.98	0.8	3.41	0.26	0.53	0.25	461
Kol	8.68	5.72	0.53	1.81	0.06	0.41	0.15	329
Pat	5.02	3.04	0.26	1.34	0.03	0.22	0.13	419
Ala	8.31	5.89	0.3	1.54	0.06	0.37	0.15	391
Kot	7.45	4.18	0.43	2.14	0.08	0.41	0.21	377
Idu	2.42	1.3	0.23	0.61	0.03	0.16	0.09	218
Ern	17.97	11.65	0.61	4.19	0.15	0.92	0.45	548
Tri	12.48	8.41	0.63	2.52	0.13	0.57	0.22	400
Pal	8.19	5.78	0.47	1.31	0.07	0.40	0.16	291
Mal	10.97	6.86	0.8	2.25	0.08	0.73	0.25	267
Koz	11.14	7.78	0.55	2.05	0.08	0.53	0.15	361
Way	1.76	1.04	0.14	0.39	0.02	0.12	0.05	215
Kan	7.4	4.37	0.49	1.81	0.08	0.48	0.17	293
Kas	3.41	1.97	0.26	0.9	0.02	0.15	0.11	261
Total	120.43	77.97	6.5	26.27	1.15	6.00	2.54	361

Data Source: Government of Kerala, 2019.

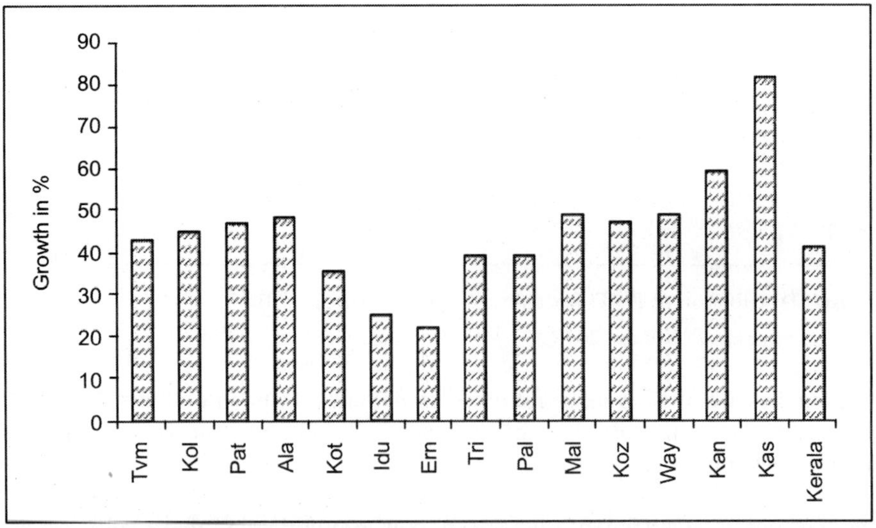

Fig. 9.3: Growth of Motor Vehicles by Districts in Kerala (2013-14 to 2017-18).

Data Source: Government of Kerala, 2015b; 2019.

Kerala. The segment of NH passing through the cities is narrow creating traffic congestion. There are NH bypasses in some cities; however, the traffic problem is getting compounded with an increase in the number of vehicles of all types. Due to overcrowding road accidents are frequent. There were as many as 40181 road accidents during the year 2018. It had increased by 13.4 per cent in the last 10 years. The accident rate in Kerala is higher than the all-India average. Quality of road is not up to the mark and this is perhaps one of the reasons for frequent road accidents. The Government of Kerala has taken up Kerala State Road Transport Project (KSTP) to upgrade the condition of roads, manage road safety and strengthen institutions. Several projects have been taken up under this initiative. Road Infrastructure Company Kerala (RICK) Limited was incorporated to implement state road improvement projects. Besides, there are research and development organisations in the road sector.

Bus Services

Kerala had more than 1.15 lakh buses of which 37.8 per cent were stage carriages, and the rest 62.2 per cent were contract carriages/omnibuses with valid registration by 2017-18. Stage carriages have increased by 54 per cent during the last five years (2012-13 to 2017-18), whereas the number of contract carriages/Omnibuses has decreased by 46 per cent during the same period. As a whole, the number of buses has decreased by 28.6 per cent but the number of cars and the number of two-wheelers have increased by 62.6 per cent and 47 per cent respectively during this 5 year period.

Buses form the backbone of the most urban transport system. Transport of passengers through roadways in Kerala is mostly vested with private operators. Kerala State Road Transport Corporation (KSRTC) holds only 13 per cent of stage carriages plying in the state. It had a fleet strength of 5635 buses by 2017-18 and operated 4688 schedules. The average distance run by a bus in a day was 343 km during 2017-18. Twenty per cent of total fleets are more than 10 years old. The total number of passengers commuted in 2017-18 through KSRTC buses was 10,593 lakh. Compared to 2008-09, there was a shortfall by 12.3 per cent in passenger volume carried by the KSRTC buses. Passengers using bus services are coming down. There is a growing tendency among people to use private vehicles for daily commuting. Thiruvananthapuram district held the maximum number of buses of 1395 and operating the maximum number of schedules of 1279. Thiruvanathapuram city being the state capital enjoys the maximum accessibility. Other districts are served mainly by the private sector.

Buses in the cities are overcrowded and have to negotiate congested roads. Besides, the number of buses is also not sufficient. The quality of public transport is deteriorating. This has resulted in a shift from public transport/bus travel towards individual motorized transport, and this has worsened the problem of road transport—slow down of speed and deterioration of traffic safety. In addition, there has been a serious impact on road transport services, the use of fuel, and overall environment. Full privatization of bus services is often suggested as a means to overcome the problem. However, with little scope for increasing road space and given the performance of existing private operators this alternative may not work. Increased efficiency of public transport may improve the situation. Kerala Urban Road Transport Corporation (KUTRC) was set up in 2014 to manage low floor buses procured with financial support under Jawaharlal Nehru Urban Renewal Mission, Government of India. The objective is to improve the quality of public bus services in the cities. It runs AC/non-AC buses. Low-floor buses have helped senior citizens to travel. By 2017-18, KUTRC operated an average of 709 buses per day, recorded 297 km run/day/bus, and carried 770 lakh passengers. This initiative has partly improved the transport service condition of daily commuters specially in urban areas.

RURAL ROADS

Kerala is well known for rural road development. It has the unique distinction of connecting all villages by all-weather roads. The total length of Panchayat roads grew from 54,812 km in 1975 to 104,257 km in 2005 and 163,184 km in 2018. Kerala's experience in rural transport development provides important lessons for other states (Ramaswamy, 1998). Panchayat roads, although have good penetration, are poor in quality as most of them are un-metalled/earthen roads. Another point to note is that although all Panchayats are connected by bus, due to the large size of the Panchayats and dispersed nature of settlement pattern, a sizable section of rural population is not able to enjoy the benefit of a cheap public transport system. There are also variations in accessibility across Kerala. The northern districts and highlands are poorly served. The coastal villages are relatively well connected and enjoy better transport facilities.

RAILWAYS

The railways in Kerala was introduced as early as 1861 when the Malabar area under the then Madras Presidency was connected with Madras. Railways entered Kerala through the Palghat gap, almost in the middle of the state, from where the railway lines diverged to the north and south of Kerala. Palakkad emerged as a major nodal point in railway development. There are branch lines connecting Pollachi to Palakkad. Another entry point was the Shencotta gap connecting Kollam with Madurai. The meter gauge line continued to operate till 2009. Thereafter, conversion to broad gauge line started. Now, all the railway lines passing through Kerala are broad gauge. At present, Kerala has a total railway route length of 1257 km. Thiruvananthapuram and Palakkad are two railway divisions jointly carrying out railway transport operations in Kerala. Palakkad division operates 76 express and 49 passenger trains and carries 2.16 lakh passengers daily. Thiruvananthapuram division operates 80 express trains and 60 passenger trains and caters to the needs of 2.6 lakh passengers daily. Railways in Kerala are basically passenger-oriented.

The railway line runs parallel to the NH in Kerala. In fact, it is the transport corridor in coastal plain where railways line, NH, and National Waterways run parallel to each other. Railway transport generally entails complimentary transport facilities by road at both ends, hence it is better justified where the traffic is to be moved over a long distance, or where the volume of traffic is too large to be moved by alternative modes. The expansion of the railway line is difficult in Kerala due to the hilly nature of the hinterlands. Movement can speed up with doubling of the existing single lines, construction, and strengthening of new terminals, and improving the road network connecting the railway stations with the hinterlands.

Kerala Rail Development Corporation Limited (KRDCL) is a joint venture of the Government of India and the Government of Kerala to undertake major railway infrastructure development project on a cost sharing basis. There are several projects being executed under this initiative. Kochi Metro Rail Project (KMRP) has been taken up to ease out transportation problem of Kochi city. It is already operational and covers an 18.3 km stretch from Aluva to Palarivattom (13.4 km) and Palarivattom to Maharaja's College (4.9 km). Works are continuing in various phases. Once completed, it is expected to cover a total stretch of 40 km. Metro rail has changed the daily commuter's movement in Kochi city.

INLAND WATERWAYS

The inland waterways played a significant role in the movements of goods and people in Kerala since early days. In fact, with the growth of foreign trade, in early years, the inland water transport also grew connecting hinterlands to the ports. The state has a total length of 1687 kms or 11.5 per cent of total inland waterways in the country consisting of navigable rivers, lagoons, estuaries and connecting canals. Thirty-four out of 41 west-flowing rivers are navigable in their lower reaches and navigable length varies from 1 km in the case of Ayroor and Mamom rivers to 74 kms in the case of the Pamba river. The navigable rivers constitute about 54 per cent of the waterways and the canals form the rest 44 per cent.

Historical records indicate that as early as 1766 a four-kilometer long canal, known as Sultan's canal, connecting Ramapuram to Taliparamba, was constructed by Ali Raza, who was ruling parts of Malabar area on behalf of Hyder Ali. A 55 kms long waterways could be made operational by constructing this 4 km long canal in the Hosdurg-Azhikkal section. Several canals were constructed to connect rivers and lagoons during 1844-1848 when Mr. Connolly was the Collector of Malabar under the Madras Presidency. Stretches of these canals like the one passing through Kozhikode town are known as Connolly's Canal. Similarly, canals and tunnels were constructed in the Travancore region to provide uninterrupted movement through the watercourses from Trivandrum to Kollam and further north.

Therefore, Kerala has a long history to interlink navigable rivers and lagoons by artificial canals as a means of water transport development. The continuous water course running for a distance of 560 kms from Trivandrum (Poovar) to Hosdurg is the main canal system connecting all major and minor ports and most of the industrial establishments in the state and is known as West Coast Canal. This canal is comprised of three main segments: Hosdrug-Kottappuram (324 kms), Kottapuram-Kollam (168 kms), and Kollam-Kovalam (68 kms) (SPB, 2006). The Hosdurg-Kottappuram section can be further sub-divided as Hosdurg-Azhikkal (55 kms), Azhikkal-Badagara (48 kms), Badagara-Beypore (74 kms), and Beypore-Kottapuram (147 kms). The waterways in the Azhikkal-Badagara sub-section are not continuous. There is a proposal to link Mahe, Anjarakandi, Tellicherry, and Valapattanam rivers by constructing three-link canals having a total length of 34.2 km. The Kottappuram-Kollam segment is the most important segment of the West Coast Canal system. The Kovalam-Kollam section has a bottleneck near Varkala, where there are two tunnels of 721 metre and 282 metre length, 5.08 metre height, and has the maximum width of 5.03 metre. These tunnels constructed more than a hundred years back are horse-shoe shaped with a depth of water just 30 cms, which is suitable only for navigating small crafts.

Kottappuram-Kollam Canal (National Inland Waterways No. III)

The Inland Waterways Authority of India (IWAI) identified three waterways as national waterways (NW). These are: (i) NW-I on Ganga-Bhagirathi-Hoogly from Allahabad to Haldia having a length of 1629 kms, (ii) NW-II on the Bramhaputra river from Sadiya to Dhubri having a length of 891 kms and (iii) NW-III on the West Coast Canal from Kottapuram to Kollam (168 kms) including Chambakara (14 kms) and Udyagmandalam (23 kms) having a total length of 205 kms.

The Kottappuram (Kodangunloor)-Kollam part of the west coast canal is the most important segment of the inland waterways in Kerala. Kochi, the main port in Kerala is located in this route at the northern end of the Vembanad lake (Fig. 9.4). With a length of 84 kms and an area of 205 km^2 the Vembanad lake constitutes the main activity centre in this part of the waterways.

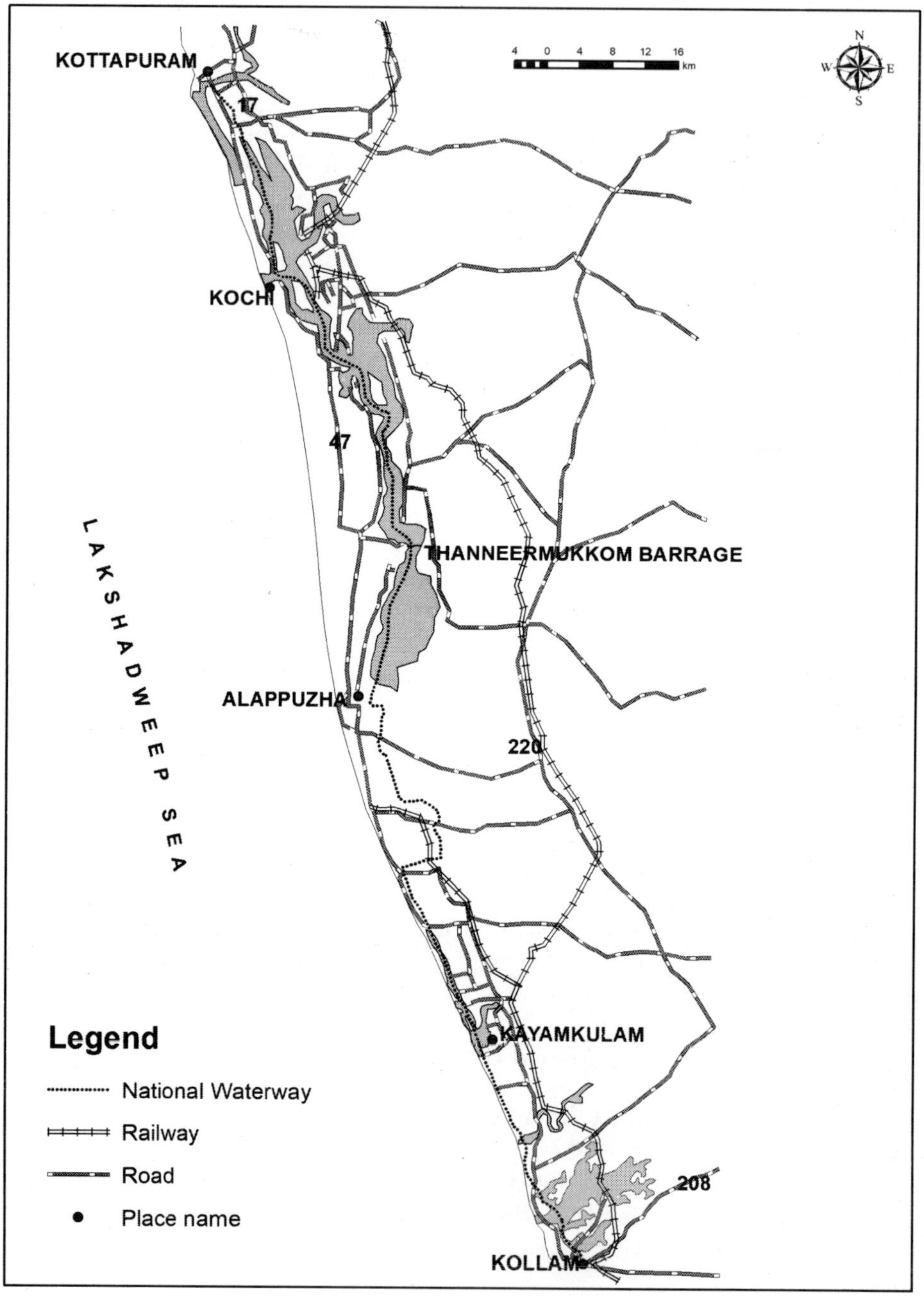

Fig. 9.4: Inland Waterways in Kerala.

Source: KSWTD, Government of Kerala, 2016.

It carries 60 per cent of the total tonnage of cargo transported by inland waterways in Kerala. Recognition of this part of the west coast canal as NW-III perhaps underlines the importance of this segment of waterways in the Kerala economy. The Kayamkulam lagoon and Asthamudi estuary are other two significant water bodies forming part of the NW-III. Kollam is located on the southern bank of the Asthamudi estuary. A stretch of 93 km covering the Kochi-Alappuzha segment has been improved under the programme of the Kerala State Transport Project (KSTP). The channel has been deepened by two metres by dredging to provide adequate draught. The Government of India accepted the proposal to extend NW-III from Kollam to Kovalam in the south and Kottapuram (Kodangalur) to Kasaragod in the north. With this, the waterways network will cover a length of 650 km in Kerala.

Feeder/Link Canals

The feeder/link canals connect interior parts with the West Coast Canal system either through the main river or criss-cross channels. NATPAC (1989) listed 175 important cross canals and link canals. The combined length of these feeder canals in Kerala is about 1000 kms (NATPAC, 1989). Important feeder canals are waterways connecting Kochi with surrounding industrialised areas like Ambalamugal and Udyogamandal. Several link canals connect the coastal settlements and Kuttanad area, stretching between Alappuzha-Puracaud in the West and Kottayam-Chenganacherry in the east. This area mainly depends on waterways for transportation.

Water Transport Services

Kerala State Water Transport Department (KSWTD) is the main agency to operate water transport services in the districts of Alappuzha, Kollam, Ernakulam, Kannur, and Kasaragod. There are fourteen operating stations in Kerala. Alappuzha district has seven operating stations followed by Kottayam (three stations). Other operating stations are in Kollam, Ernakulam, Kannur, and Kasaragod. By 2017-18, fifty-one passenger boats and five rescue boats operated generating more than 3 lakh trip schedules and carrying 135.43 lakh passengers. The number of passengers travelled in KSWTD boats was 160.27 lakh in 2008-09. Improvement of road connectivity in the areas dominated by waterways has impacted passenger movement through water transport. People prefer road transport for faster movement. Kerala Shipping and Inland Navigation Corporation Ltd. handled little over 5 lakh tonnes of cargo in 2008-09 (Government of Kerala, 2010).

Alappuzha-Kuttanad-Kottayam Segment

Alappuzha, the headquarter of the State Water Transport Department (SWTD) is the main centre where most of the water transport services originate. It principally caters to the need of the waterlogged Kuttanad area. There are 54 services from Alappuzha connecting 17 stations. Nedumudi receives 12 services from Alappuzha in a day followed by Krishnapuram (11 services per day) and Kottayam (five services per day). The first service from Alappuzha starts at 5.05 hr. and the last service is at 21.30 hr.

Future Prospects

There has not been much development in water transport facilities in Kerala in the post-

Independence period. In fact, water transport has not received the required attention perhaps due to the improvement of road transport. It is difficult to increase the route length of waterways due to physiographic settings; however, it is possible to enhance gross distance by operating more number of services and improving the navigability of the existing channels. An earlier assessment indicated that water transport handled only six per cent of cargo traffic and two per cent of passenger traffic (Mahesh Chand, 2002). There is ample scope to increase water transport's share of passengers as well as goods traffic in Kerala. The inherent advantage of Inland Water Transport to move things in bulk at a comparatively cheaper price and least demand on energy should be properly exploited. The employment generation potential of the sector is quite high. There is a scope for both direct and indirect employment generation. It has been estimated by the National Transport Policy Committee that with an investment of Rs. one lakh in Inland Water Transport (IWT) can generate employment to the tune of 33.59 man-years compared to 16.93 in the case of road transport and 4.3 in case of rail transport (NATPAC, 1989). On the energy front the estimate indicates that one horsepower can move 150 kg on road, 500 kg on rail, and 4000 kg on water.

Several initiatives have been taken to develop inland waterways in Kerala. The development proposals have been categorized as Waterways components and Tourism and Conservation components. The Water Resource Department will handle the waterways component and the tourism component will be handled by the Department of Tourism in association with various agencies. The waterways in Kerala has high tourism potential due to scenic beauty and for the fact that several heritage sites are located along with these water courses. The canal network in Alappuzha is being considered for declaration as a heritage site. It is important to pay attention to the successful implementation of various projects, so that inland water transport regains its rightful place in the development of Kerala and helps ease out part of the pressure on road transport.

AIRWAYS

Kerala is well known for out-migration, particularly to gulf countries. There is hardly a Panchayat from where people have not migrated abroad. People travelled by air to the outside countries were 6 times more than domestic air traffic. Three international airports, one each at Thiruvananthapuram, Kochi (Nedumbassery), and Kozhikode figure among the top 15 busiest airports in India. They handle more than 8 lakh domestic passengers and 50 lakh international passengers every year. Another international airport started operation from Kannur. Now Kerala has four international airports. The Thiruvananthapuram international airport is an important airport in south India and connects nine international air routes and eight domestic air routes. It also provides approach control service to eight major and minor aerodromes. The Kochi international airport (CIAL) set up in Public-Private Participation (PPP) mode is emerging as an important airport, experimenting with innovative models of operation. It ranked as the third best airport in the 50 to 150 lakh passenger category in 2016. Kochi airport is fully powered by solar energy. Kozhikode also figures among the busiest airports in India.

TRANSPORT DEVELOPMENT IN KERALA: SOME ISSUES

Geography plays an important role in the transport development of any area. Topography, ruggedness, placement of rivers, and overall configuration of land and population distribution provide a basic matrix over which transport systems develop. Road density is inversely proportional to the ruggedness/hilly condition and directly proportional to population density. Most of the roads and railways development are in plains.

Kerala, by virtue of its geographical configuration and topographic alignment show corridor type of transport development. Roadways, railways, and waterways run parallel to each other in the coastal plain. A 30 km strip from the coastline to the east will cover the majority of the transport lines and population in the state. Kerala's situation in the matter of transport network and the population distribution is often compared with that of greater Mumbai and therefore, Mumbai mode of transport infrastructure development can be followed in Kerala (Prasad, 2003).

Taking advantage of the topography and already existing transport corridors in the coastal plain the north-south corridor should be strengthened by improving the quality of highways, railways, and waterways. This main artery can be linked with feeder roads. A fast-moving coastal and midland highway should be developed. Although there are roads in every Panchayat, the overall quality of roads is far from satisfactory. Highways, urban roads, and rural roads warrant quality improvement. Hill roads cutting foot slope and passing through forests often cause environmental problems like landslides and encroachment. This issue might have to be factored in the plan for road infrastructure development. Urban roads are narrow and overburdened. Traffic congestion, delay in movement, and accidents are regularly reported. Kerala which is looking forward to the IT industry and tourism for its development impetus needs to improve its transport infrastructure to reduce travel time. Several initiatives are taken up under Kerala State Transport Project (KSTP), however, the pace of work needs acceleration to cope with the faster development trajectory of other sectors.

TOURISM

Tourism is considered a smokeless industry. Of late, it has been an important component of Kerala's economic development. Total revenue generated from tourism through direct and indirect sources amounted to Rs. 33.4 thousand crore as reported by the Department of Tourism, Government of Kerala (Government of Kerala, 2019). It grew by 154 per cent during the last 10 years (2008-2017). Tourism is not simply a visit to a particular place or site; rather it is a means to promote people to people contact, cultural understanding, interaction, mutual dependence, cooperation, and appreciation. Tourism brings people from different cultures and origins together and nurtures the inherent human quality of appreciation, harmony, and peace. Religious tourism in India is known for centuries. People from one part of the country used to move to other parts for visiting sacred places and worship. Modern-day tourism goes beyond religious places and includes social and cultural artifacts, historical places, sites of natural beauty, place of adventures, and eco-exploration. Adventure tourism attracts people, particularly young people willing to explore the wilderness. The development of tourism depends upon several factors from the quality of destination, comfort to safety, and security. According to the World Travel and Tourism Council (WTTC, 2019), globally, the travel and tourism sector contributes about 10.4 per cent of GDP and provides 10 per cent of total global employment.

Kerala is well known as a tourist destination across the world. Every part of the state has tourism potential. Kerala offers a wide variety of tourism from the seafront to high mountains. Diverse landscapes, high altitude, dense tropical rainforest, rich biodiversity, and cultural heritage offer tourist attractions to suit different tastes and interests. Ayurveda raises the interest of both domestic and foreign tourists. According to the Department of Tourism, Government of Kerala, domestic tourists grew from 75.9 lakh in 2008 to 146.7 lakh in 2018 and international tourists increased from 5.98 lakh in 2008 to 10.9 lakh in 2018 (Fig. 9.5). There was a growth of 5.15 per cent in foreign tourists and 11.39 per cent in domestic tourist arrival during 2017 compared to 2016. Kerala received around 11 lakh foreign tourists, which was 10.73 per cent of total foreign tourists visited India. The majority of foreign tourists (58 %) arrived during a period from November to March as evident from data available for 2017. The United Kingdom with a 16.57 per cent share led the list of the countries generating foreign tourists who visited Kerala during 2017. The other countries include the USA, France, Germany, Saudi Arabia, Russia, Australia, Canada, Malaysia, and Switzerland. District-wise distribution of arrival of foreign tourists indicates a regional bias. South and central Kerala attracted around 80 per cent of total foreign tourists through the gateways of Thiruvananthapuram and Ernakulam. Ernakulam district received 41.6 per cent of foreign tourists followed by Thiruvananthapuram district with 38.3 per cent during 2017. North Kerala received only 4 per cent of foreign tourists.

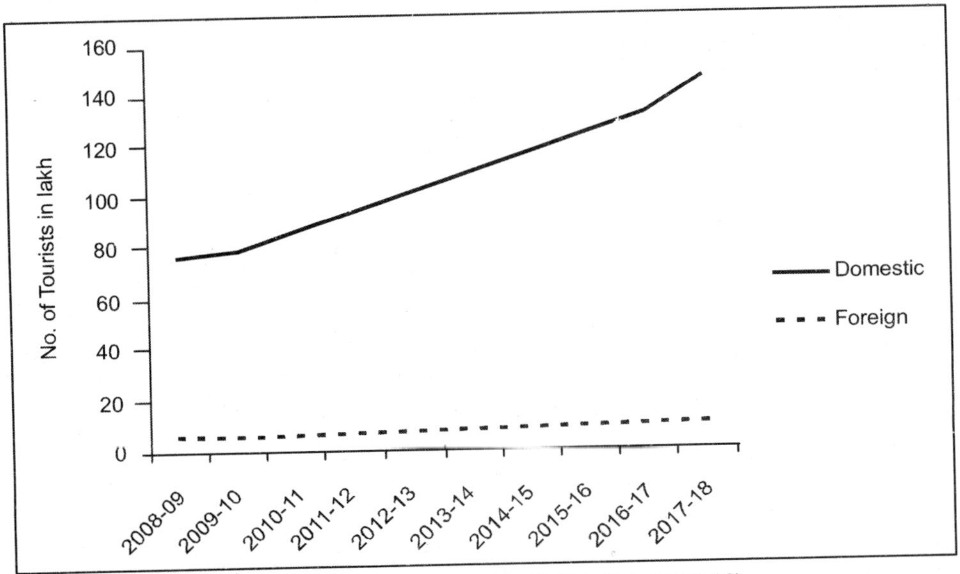

Fig. 9.5: Domestic and Foreign Tourists Arrival in Kerala (2008-09 to 2017-18).
Data Source: Government of Kerala, 2019.

Domestic tourists visit Kerala all through the year, however, the months of November and December witness a higher number of tourist arrivals than other months. Around 63 per cent of domestic tourists originated from Kerala itself. Among the rest 35 per cent of domestic tourists, the maximum tourists came from Tamil Nadu followed by Karnataka. District-wise distribution of domestic tourist arrivals varies widely. Pathanamthitta district received the least number of 1.6 lakh, whereas Ernakulam district topped the list of districts by receiving 32.8 lakh tourists (Table 9.5). Ernakulam, Thrissur, and Thiruvananthapuram district together received 57 per cent of all domestic tourists in Kerala. As domestic tourists include intrastate movement, it may be inferred that most of the people are visiting these three districts. Thiruvananthapuram and Ernakulam districts also attract people to avail of administrative, legal, medical, and educational services. The potentiality of other districts as tourist destinations should be explored to enlarge the tourism base of Kerala.

Table 9.5: Arrival of Domestic and Foreign Tourists by Districts in Kerala, 2018

District	*Total Domestic*	*Per cent to Total*	*Total Foreign*	*Per cent to Total*
Tvm	2505333	17.1	420719	38.54
Kol	381829	2.6	6227	0.57
Pat	164494	1.12	2003	0.18
Ala	433456	2.95	75037	6.87
Kot	468593	3.2	32350	2.96
Idu	1090086	7.43	42285	3.87
Ern	3285088	22.39	453973	41.59
Tri	2642546	18	10775	0.98
Pal	474180	3.23	1711	0.16
Mal	520832	3.55	18451	1.69
Koz	932345	6.35	13106	1.2
Way	815624	5.56	8995	0.82
Kan	695655	4.74	5123	0.47
Kas	263459	1.8	1115	0.1
Kerala	14673520	100	1091870	100.00

Data Source: Government of Kerala, 2019.

There are several initiatives from the government to promote tourism including addressing key issues of the tourism industry like waste management. Tourism development and quality of the environment are intimately related. The scope of environmental management has to be enlarged to include landscape management and preservation of the garden character of Kerala. The tourism potential of the state is yet to be fully explored. Due to connectivity and administrative thrust, tourism in Kerala, particularly foreign tourism, is limited to southern and central parts of the state. North Kerala deserves special attention. This part is equally rich in tourism potential. The opening of Kannur international airport may help promoting tourism in this part. In the case of domestic tourism also, there is wide spatial differentiation. This warrants proper attention while framing tourism policy for Kerala.

CHAPTER

10

Settlement, Population and Migration

SETTLEMENT SYSTEM

Physiographic set up has its imprint on settlement systems in Kerala. The basic structure of settlement pattern is linear and dispersed. A major part of Kerala is characterized by undulated topography with lateritic ridges and slopes intervened by alluvial valleys. Ridge runnel topography dominates the coastal plain. Plateau landscape covers parts of the Western Ghats. Settlements have developed along the ridges in linear fashion. Landform condition does not encourage compact or nucleated settlement like other states in India. The areas with compact settlements are adjacent to Tamil Nadu and are characterized by low to gentle sloping relief. Some compact settlements are found in plateaus. Coastal settlements are linear and dispersed along the coast and backwaters to enjoy equal fishing opportunities. The initial spread of settlements along the ridges and not on the valleys was primarily to avoid flood damage and to maintain food production. Due to the availability of water throughout Kerala and for the most part of the year, there is practically little or no restriction for settlement development in the state except some pockets in deep forest, swamps, and hard crust mesa surfaces. This physical advantage has facilitated the development of the dispersed nature of settlements. Physiographic control on the settlement system is well manifested.

The settlement pattern emerged in Kerala corresponding to landform assemblages are: (i) Coastal plain with runnel and ridge topography exhibits linear settlements along the sandy ridge, (ii) The vast alluvial patches in Alappuzha, Kottayam, Ernakulam and Thrissur districts have linear and dispersed settlements in the river terrace and levees, (iii) Alternate laterite ridges with intervening valleys host linear settlements along the ridges and upper slopes parallel to the valleys, (iv) Low rolling terrain of Palakkad type accommodate compact, nucleated settlement pattern, (v) Steep slopes and ridges show nucleated settlement in the valleys, and (vi) Plateau surfaces exhibit dispersed and nucleated settlements in the mounds and valleys (Chattopadhyay 1996).

Kerala's settlement system is changing. The rate of change has accelerated since globalization/economic liberalization. There are several socio-economic factors contributing to this change including global economic issues. Traditionally, settlements were horizontally spreading and each dwelling was surrounded by homestead gardens leaving wide space between houses. This distance or inter settlement space is reducing over the years, correspondingly, open space and homestead garden/biodiversity are declining. There is a tendency to build compact settlements along road junctions and newly constructed highway bypass/roads. The coastal plain and adjoining lowlands are being converted into a continuous built-up landscape. During the last couple of decades, there is a trend to grow vertically. Skyline of the main urban centres is undergoing changes with vertical growth of buildings. Multi-storied building complexes are being built throughout the state. As a whole, Kerala is witnessing a perceptible change in settlement structure and pattern. Considering the scarcity of land, high population pressure, and ease and economics of service delivery it is being argued in various quarters that Kerala's traditional settlement system may not be viable and there is a need to change the dispersed settlement system in favour of a more compact and nucleated settlements, particularly in the marginal/vulnerable areas, which are often exposed to life-threatening natural hazards like flood and landslides.

POPULATION COMPOSITION

Kerala's population represents an admixture of four racial strains, viz., the Negrito, Proto-Australiod, the Dravidian, and the Aryan (Menon, 1997). Archeological evidence indicates the presence of megalithic people in Kerala. Even from time immemorial, several foreign races and people contributed to the composite culture of Kerala through trade contacts. The Arabs, the Assyrians, the Babylonians, the Phoenicians, the Israelites, the Greeks, the Romans, and the Chinese had extensive trade contacts with Kerala. These early-day contacts had facilitated the introduction of world religions like Christianity, Judaism, and Islam in Kerala, which helped set in motion a process of cultural synthesis and demographic intermingling. The Census of 2011 reported that the Hindus accounted for 54.73 per cent of the total population followed by the Muslims (26.56 %), Christians (18.38 %), and others (0.33 %).

POPULATION CHARACTERISTICS

By 2011, Kerala housed 33.4 million people, which was 2.76 per cent of the total population in India. In 1951, Kerala with a population of 13.5 million accounted for 3.75 per cent of India's population. Population growth in Kerala has declined significantly. Kerala's achievement in the demographic sector is quoted as an example for the rest of India.

The annual growth rate of 0.49 per cent in 2001-2011 was the lowest experienced since the formation of modern State of Kerala. It was 0.91 per cent during the 1991-2001 Census periods. The decline in growth rate was attributed partly to a decline in fertility and partly to out-migration. Female literacy, the consciousness of general people about child health care, the substantial decline in infant mortality rate, and couples' preference for a small family in order to provide their children the best of livelihood opportunities have contributed to declining population growth.

In the case of population density, Kerala ranked first among the major states in India until 1981. West Bengal took lead in 1991 and 2001, while Bihar ranked first in 2011. The census record of 2011 indicated that the population density of Kerala was 859 persons/km^2, while all-India average was 382 persons/km^2. In 2001, Kerala's population density was 819 persons/km^2. As a high-density state, Kerala experiences considerable pressure on land and other natural resources.

DISTRICT LEVEL DISTRIBUTION

Distribution of population records spatial variations. According to the 2011 Census data, Malappuram was the most populous district with 4.1 million people followed by Thiruvananthapuram. Wayanad district housing only 0.82 million people stood at the bottom (Table 10.1).

Table 10.1: Distribution of Total Population, Population Density and Decadal Growth by Districts in Kerala, 2011

Districts	Total Population (in 000) 2011	Density of Population (persons per sq.km)				Growth (per cent)		
		1981	1991	2001	2011	1981-1991	1991-2001	2001-2011
Thiruvananthapuram	3301	1184	1344	1476	1509	13.5	9.8	2.25
Kollam	2635	873	967	1037	1056	10.7	7.2	1.72
Pathanamthitta	1197	426	450	467	453	15.6	3.8	-3.12
Alappuzha	2128	1319	1415	1489	1501	7.28	5.2	0.61
Kottayam	1975	771	830	884	896	7.71	6.5	1.32
Idukki	1109	193	215	252	254	11.2	1.72	-1.93
Ernakulam	3282	1053	1170	1050	1069	11.1	-10.3	5.60
Thrissur	3121	805	903	981	1026	12.2	8.6	4.58
Palakkad	2810	456	532	584	627	16.5	9.8	7.39
Malappuram	4113	677	872	1022	1158	28.9	17.1	13.39
Kozhikode	3086	958	1118	1228	1318	16.7	9.8	7.31
Wayanad	817	260	315	369	383	21.3	17.1	4.60
Kannur	2523	651	757	813	852	16.6	7.4	4.84
Kasaragod	1307	438	538	604	654	22.8	12.3	8.18
State total & averages	33404	655	749	819	859	14.3	9.35	4.88

Data Source: Census of India, Kerala, 2011.

Population size was small in the case of Idukki district also. All districts in Kerala recorded a steady decline in the growth rate of the population in varying proportions. Malappuram was the only district that had still recorded a double-digit population growth rate; nevertheless, the magnitude of population growth came down from 28.9 per cent in 1981-91 to 13.31 per cent in 2001-2011. During 2001-2011, the highest population growth rate was recorded in the case of Malappuram district followed by Kasaragod and Kozhikode. As many as 10 districts recorded a growth rate below the average growth rate of Kerala. Pathanamthitta and Idukki districts recorded negative growth rates during this period. Four districts, namely, Kasaragod, Kozhikode. Malappuram and Palakkad, which had recorded population growth rates above the state average in 2001-2011, belonged to the same category in the previous two decades also. The number of districts recording population density well above the average population density of Kerala is coming down. This indicates that population growth rates are falling throughout Kerala irrespective of the rural-urban composition of the district, and secondly, the relatively high population growth rate is getting confined to certain districts. Comparing population growth rate during the fifties and the last census (2001-2011), it emerges that there are certain spatial shifts of the areas having high growth of population from southern Kerala to northern and central Kerala. This regional shift in population concentration might have future implications in development as well as governance issues.

By 2011, population density at the district level varied from 254 persons/km^2 in the Idukki district to 1509 persons/km^2 in Thiruvananthapuram district (Fig. 10.1). Districts with high population

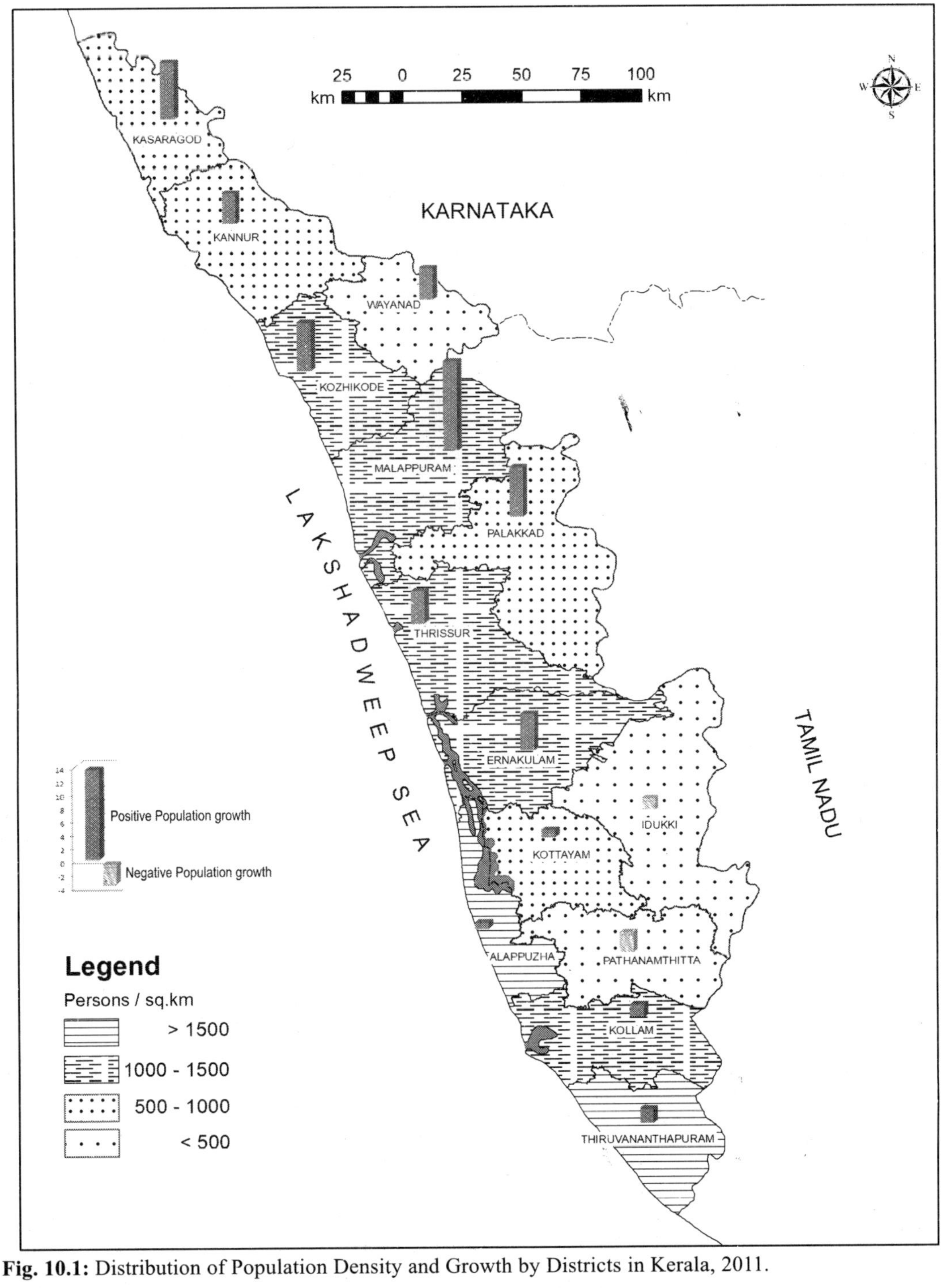

Fig. 10.1: Distribution of Population Density and Growth by Districts in Kerala, 2011.

Data Source: Census of India, Kerala, 2011.

density are distributed in three patches: Kozhikode-Malappuram, Ernakulam-Alappuzha, and Kollam-Trivandrum. This trend indicates that areas with high population density are emerging around the principal urban centres of Kozhikode, Ernakulam, and Thiruvananthapuram. Districts having high population density also contain a higher proportion of the urban population.

POPULATION DISTRIBUTION AND ALTITUDINAL ZONES

Population concentration decreases from the coast towards the east with an increase in altitude. As the majority of the districts (nine districts) extend from the coast to the Western Ghats, there is high density as well as low-density zones in each of the districts. Alappuzha district, which falls entirely in the lowland zone recorded the highest population density among all districts in Kerala till 2001 (Table 10.1). It recorded the second highest population density after Thiruvananthapuram by 2011. Population density is low in the districts of Wayanad and Idukki, located within the Western Ghats. The lowest population density is recorded in the Idukki district, which has high intra-district internal relief variation compared to the Wayanad district. The population density in northern districts like Kasaragod and Kannur is also low. There are extensive hard crust laterite surfaces which are not very hospitable for usual crop husbandry.

The coastal plain and adjoining lowland under the altitudinal category of lowland has the highest population density. All Panchayats and urban centres falling in this zone have a population density of > 1500 persons/km². Most of Kerala's population is concentrated in the coastal plain and adjoining lowlands. The local bodies in the Midland zone have a population density between 1000 persons/km² to 1500 persons/km². The highland region is characterized by the Panchayats with a population density of <500 persons/km². Population density decreases with an increase in altitude, which is evident from correlation analysis between population density and area under highlands by districts. The trend is inversely proportional with high negative correlation coefficient (r) value of (-) 0.90 (Fig. 10.2).

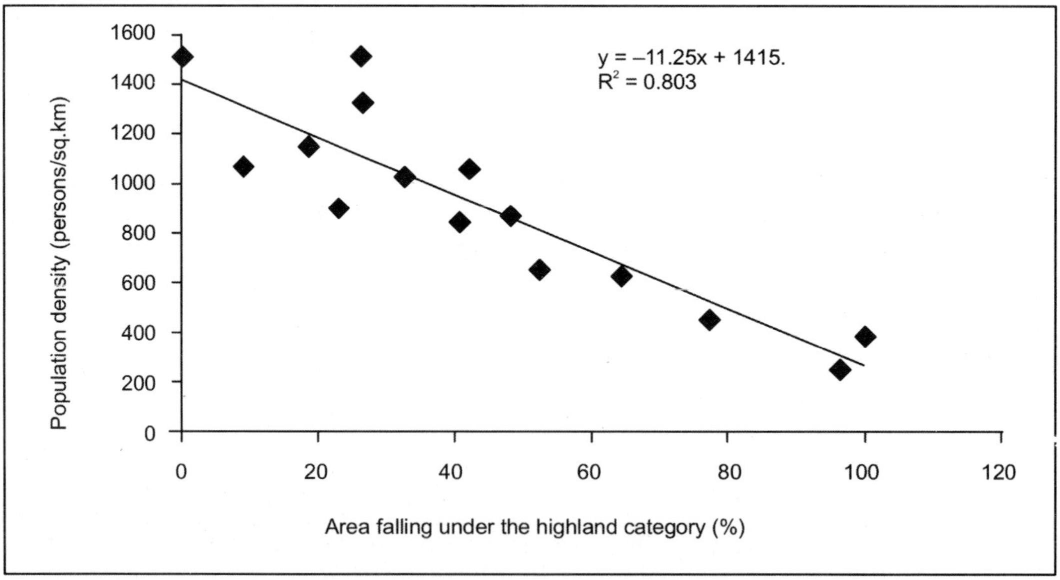

Fig. 10.2: Relation between Population Density and Area under Highlands in Kerala.

DEMOGRAPHIC TRANSITION

Demographic Transition (DT) refers to the transition from high birth and death rates to low birth and death rates. This change is linked with a country that progresses from pre-industrial to modern economic systems. The term DT was first used by an American demographer Frank W Notestien, although another American demographer Warran Thompson while examining historical demographic data tried to interpret the trends of birth and death rates and provided the foundation of the demographic transition model (Kirk, 1996). According to this model, there are four stages from pre-transition to post-transition. Recently, a fifth stage is also identified to denote a decline in population in some countries. A brief description of different demographic transitions is given here:

(i) Pre-transition characterized by high birth rate and high death rate resulting in small population growth,
(ii) Early transition marked by high birth rate but falling death rate. This phase witnesses rapid expansion of total population,
(iii) Late transition characterized by a falling birth rate but continued falling death rate. The phase is of low population growth and
(iv) Post-transition phase having a low birth rate and low death rate. Population growth is small and low fluctuating.

In general, the first three stages are considered and the fourth stage occurs with approaching zero population growth. Demographic transition is associated with economic growth. The first stage or pre-transition coincides with early economic conditions and early stages of development with poor diet, illiteracy, absence of health care, more dependence on human labour for family income, etc. It is also characterized by early marriage, high infant mortality, and high prevalence of the disease, lack of hygiene, famine, and epidemics. The second stage or early transition is linked with rising income level, enhanced productivity, improved diet, availability of food and health care, high literacy level particularly female literacy, improved mother and child care, fall in child mortality, and death rate. The third stage or late transition follows accelerated economic development, changes from agrarian to partially industrialized and tertiary sector development, increased per capita income and standard of living, greater participation of women in income earning activities outside the home, increasing urbanization, better all-round health care, and preference for small/nuclear family. The society attains very low population growth and it heads towards zero population growth.

Democratic transition in Kerala is widely discussed (Bhat and Irudaya Rajan, 1990, Zacharia and Irudaya Rajan, 1997, Planning Commission of India, 2008a). Kerala has achieved remarkably in demographic transition in a short span of time. The crude birth rate (CBR) in Kerala declined from 44 per 1000 population in 1951-61 to 14.3 per 1000 population in 2016 and crude death rate (CDR) came down from 20 per 1000 population in 1951-61 to 7.6 per 1000 population in 2016 (Government of Kerala, 2012a; Government of India, 2017a). Birth rate and death rate declined for India also, but the rate of decline in Kerala, particularly for birth rate, was much higher. The natural increase in population is only 7.9 per 1000 population in Kerala against the all-India average of 15.2 per 1000 population. The total fertility rate (TFR) in Kerala declined from 5.6 per woman in 1950s to 1.8 per woman in the 2015. In the case of all-India, TFR declined from 6.3 in the 1950s to 2.3 in 2015. Kerala's spectacular achievement in low TFR is primarily due to the high level of female education.

The decline in infant mortality rate played an important role in the decline in fertility. The infant mortality rate in Kerala drastically reduced from 120 per 1000 population in the 1950s to 10 per 1000 population in 2016. In the case of India, the reduction is from 139 to 34 during the same period (Government of India, 2017a). However, the marginal groups like Scheduled Caste and Scheduled Tribe population, fishing community, and economically weaker section still suffer from relatively high infant mortality rate. There is little difference between rural and urban areas in these aspects as access to social infrastructure in rural areas is comparable to that of the urban areas. The settlement pattern and improved communication facilities have contributed to the absence of rural-urban differences. With the low natural increase, low fertility rate, and low death rate, Kerala is in the late transition phase or third stage of demographic transition. It is projected that Kerala will attain Zero Population Growth within the next two to three decades, and subsequently, the population will start declining (Planning Commission of India, 2008a).

FACTORS CONTRIBUTING TO DEMOGRAPHIC TRANSITION

Several factors have contributed to the demographic transition in Kerala. Since 1980, when fertility started declining, the causative factors are being discussed. Historical developments as well as policy interventions are instrumental both in the reduction of fertility and also mortality. Kerala's cultural interactions with other countries since ancient times and the advent of Judaism, Christianity, and Islam from time to time have influenced her human behavioural environment. European influence brought a modern administrative system, modern education, and the public health system. The Christian Missionaries' concentrated effort for education and health ably supported by the local administration of Travancore-Cochin and indigenous tradition of health care have created a congenial environment to a wider reach of education and health care. It is important to note that by the end of the 19th century every village in Travancore and Cochin had a primary school.

Kerala has a long tradition of the Ayurvedic system of health care. Bio-Physical environment, particularly water availability and medicinal plants including herbs and spices like ginger, black pepper, turmeric, and cardamom that have medicinal values have endowed the state with a healthy environment. Through subsequent development of the Allopathic system of medicine, Kerala enjoyed better health care facilities than the rest of India. In response to the dispersed nature of the settlement, the health care system has developed throughout the state. People both in urban and rural areas enjoy almost equal access to basic health care facilities. Improvements in general educational status and health policies of the Government of Kerala have a positive contribution. The vibrant civil society also played its part in improving health services maintained by the state Government. As the health care offered by the state sector is good, the private health sector, recently developed, also offers good services at a reasonable cost. High morbidity reported from Kerala is due to the fact that, traditionally, people in Kerala are well aware of health, importance of reporting about health ailments and their faith on health service system instead of self-medication. An increase in lifestyle diseases because of rapidly changing standard of living have also contributed to high morbidity in recent years. In this matter also, there is hardly any difference between rural and urban areas.

Education, public awareness, agrarian reforms, land reform have contributed to various measures. Since the 1980s, the programmes like Universal Immunisation Programme, Total Literacy Campaign, World Bank population project, and increasing income due to gulf migration played important role in the demographic transition. Socio economic changes, particularly, emphasis on good quality of life, decent living, providing good quality education for children and upward social mobility cutting across socio-economic classes have influenced parent's choice for small families. With education as the prime means to enter into the trajectory of good quality of life through services, there had been a basic change in society by moving away from inherited status towards achieved status and from the interdependent community to individual competition and from traditional authority to modern bureaucracy (Planning Commission of India, 2008a). This sociological change has pervaded all social groups and it is commonly perceived that education and good health can bring the opportunity to improve their lives.

SEX RATIO

Of the total population of Kerala enumerated in the Census of 2011, 16.02 million were male and the rest 17.36 million were female. Kerala is well known for its favourable sex ratio towards females. The female population had been greater than the male population throughout the century. The 1901 census in Kerala recorded 1004 females for every 1000 males thrissurkerala.com/census-2011. This sex ratio increased to 1036 in 1991, to 1059 in 2001, and to 1084 in 2011. District-wise distribution showed that in 1901 all seven districts in Malabar except Wayanad recorded more female than male, the highest (1060) was in two districts of Kasaragod and Kannur. The rest of the districts had a female population less than males. Wayanad and Idukki districts with 805 and 839 females per 1000 males occupied the bottom two positions in 1901 (*ibid.*). These two districts continued to have less number of females than males throughout the last century (1901-2001). Sex ratio in favour of female population in Malabar districts and that in favour of male population in hilly districts of Wayanad and Idukki might be attributable to the fact that Malabar was ruled by Madras Presidency under British Government, who indentured labours for various activities like construction of railways, plantations, and clearing of jungles. While the out-migration of male labour from Malabar districts changed the sex ratio in favour of the female population, the influx of male labours in the hilly districts changed the sex ratio in favour of the male population. Besides, both Wayanad and Idukki districts have a sizable tribal population. By 2011, all districts in Kerala recorded more females than the male population with a variation from 1006 in Idukki to 1133 in Kannur (Table 10. 2).

CHANGING AGE STRUCTURE

The age composition of the population in Kerala is changing. Life expectancy at birth in Kerala is 72 years for the male population and 77.8 years for the female population (Govt. of India, 2017b). The number of elderly people (>60 years) is increasing over the years and at the same time number of children (<15 years) is decreasing. In 1961, elderly people were 6 per cent of the total population and the children were 43 per cent (Fig.10.3). By 2011, the proportion of elderly population and children were 13 per cent and 23 per cent respectively. In this 50 years' span, elderly people grew by 117 per cent, whereas children recorded negative growth of -47 per cent. The growth rate of the working population (20 to 64 years) is steadily coming down since 1981. As a whole, Kerala is gradually turning into a place of aged people, which may have serious implications for the development of the state. Population projection for the year 2031, when Kerala may attain zero population growth, indicates that the proportion of the elderly population will be one-fourth of the total population and that of children (<15 years) will be around one-

Table 10.2: Demographic Characteristics (selected) by Districts in Kerala, 2011

District/State	Sex ratio	Child population (<6 years) (per cent)	S C population (per cent)	S T population (per cent)	Fisherman population (Number)	Literacy (per cent)		
						Total	Male	Female
Thiruvananthapuram	1088	9.30	11.3	0.8	1,64,883	92.66	94.1	90.9
Kollam	1113	9.65	12.5	0.4	1,23,100	93.77	95.8	91.9
Pathanamthitta	1129	8.09	13.7	0.7	2,073	96.93	97.7	96.3
Alappuzha	1100	9.03	9.5	0.3	1,67,794	96.26	97.9	94.8
Kottayam	1040	8.84	7.8	1.1	24,420	96.40	97.2	95.7
Idukki	1006	9.53	13.1	5.0	691	92.20	94.8	89.6
Ernakulam	1028	9.27	8.2	0.5	1,33,387	95.68	97.1	94.3
Thrissur	1109	9.74	10.4	0.3	90,306	95.32	96.9	93.9
Palakkad	1067	10.76	14.4	1.7	2,534	88.49	92.3	85.0
Malappuram	1096	13.96	7.5	0.6	82,044	93.55	95.8	91.6
Kozhikode	1097	10.88	6.5	0.5	1,06,613	95.24	97.6	93.2
Wayanad	1035	11.29	4.0	18.5	230	89.32	92.8	85.9
Kannur	1133	10.87	3.3	1.6	60,208	95.41	97.5	93.6
Kasaragod	1079	11.92	4.1	3.7	43,342	89.85	93.9	86.1
Kerala	1084	10.40	9.1	1.5	10,01,625	93.91	96.1	92.1

Data Source: Government of Kerala, 2019.

sixth. By 2051, Kerala may have a 35.07 per cent population above 60 years and 14.71 per cent population below 15 years of age (Planning Commission, 2008a). The management and welfare of the elderly population are emerging as major challenges for Kerala society. As the number of young people is declining and there is a steady trend of out-migration, looking after elderly people for the day to day activities is an important societal task.

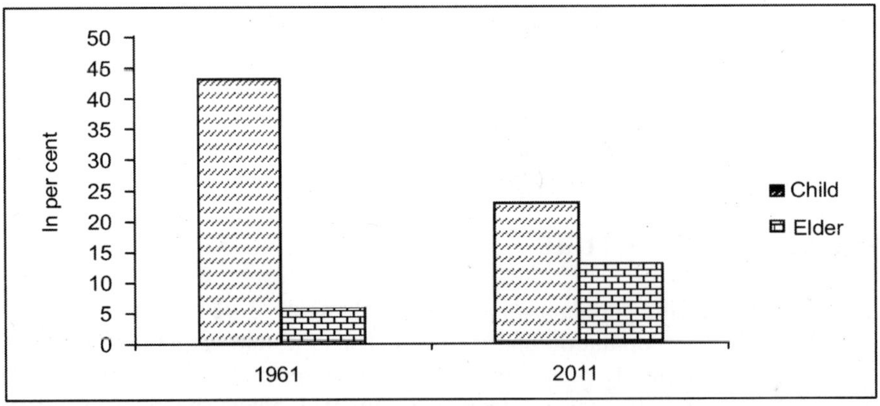

Fig. 10.3: Percentage Share of Child Population and Elder People to Total Population of Kerala, 1961, 2011.
Data Source: Census of India, Kerala, 1961, 2011.

The age-sex structure of Kerala for 1961 and 2011 shows the nature of the demographic shift in age groups (Figs. 10.4, 10.5). The pyramid shape characteristic of 1961 age-sex structure has given place to cylindrical shape (up to age group of 45-49) by 2011. The proportion of the child population has declined over the years. The female population dominates in most of the age groups except the child population (14 years). The present demographic structure of the higher female population compared to the male population in Kerala may change in the future.

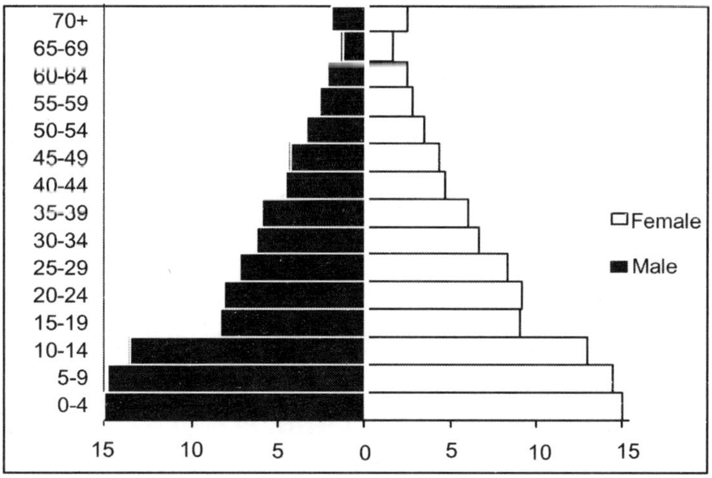

Fig. 10.4: Age Sex Structure of Kerala, 1961. X axis represent percentage of population and Y axis shows age groups.

Source: Irudaya Rajan, 2020 (personal communication).

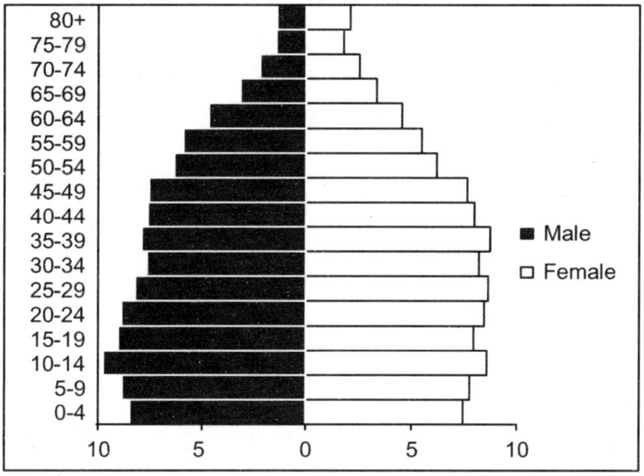

Fig. 10.5: Age Sex Structure of Kerala, 2011. X axis represents percentage of population and Y axis shows age groups.

Source: Irudaya Rajan, 2020 (personal communication).

CHILDREN UNDER 6 YEARS OF AGE

As people in Kerala are aging, it is important to discuss the child population (< 6 years). The 2011 census data indicated that Kerala's child population was around 3.3 million or 9.95 per cent of the total population of the state (Table 10.2). The proportion has fallen down from 11.91 per cent in 2001. District-wise distribution showed that the proportion of child population varied from 7.65 per cent in Pathanamthitta to 13.45 per cent in Malappuram. Considering the total child population in Kerala it is found that Malappuram district houses 16.6 per cent of all children (<6 years) in Kerala followed by Kozhikode (9.7 %). These two districts along with Palakkad, Thrissur, and Ernakulam districts account for little over 52 per cent of all child population in Kerala. Comparing district-wise data for 2001 and 2011, it emerged that the proportion of children declined the maximum (- 2.84 %) in Idukki district and the minimum reduction was in Kannur district (-1.12 %).

The male-female composition of the child population brings out some interesting points warranting the attention of policy-makers. Although Kerala has a more female population than the male population as a whole, in the case of the child population, the situation is reversed with the domination of the male child. The 2011 census indicated that there were 959 females for every 1000 male children. The number of female children was less than male children in all districts. This trend is contrary to the general trend of Kerala. If this trend persists, the demographic scenario may change in a couple of decades.

SCHEDULED CASTE AND SCHEDULED TRIBE POPULATION

Proportions of Scheduled Caste (SC) and Scheduled Tribe (ST) population to total population in Kerala are low compared to all-India figures. The 2011 census brought out that SC and ST population in Kerala comprises 9.1 per cent and 1.5 per cent of the total population respectively. At all-India level, these figures were 16.6 per cent for the SC population and 8.6 per cent for the ST population. The condition of the SC and ST population in Kerala is relatively better compared to other parts of India. While ST populations are largely concentrated in the hilly Western Ghats, SC populations are found to be distributed in all districts along with other communities.

The scheduled Caste population in Kerala was little a over 3.1 million or 9.81 per cent of the total population in 2001. Data for the last five decades indicated that the SC population increased till 1981 when it constituted 10 per cent of the total population and subsequently it started declining. This highlights that the growth rate of the SC population is lower than that of the general population. The condition of the SC population in Kerala is comparable to the other communities. At the district level, high concentrations were found in Palakkad and Thrissur districts in Central Kerala and Thiruvananthapuram and Kollam districts in south Kerala. These four districts together accounted for 47 per cent of total SC population in Kerala.

There were 0.48 million ST population or Adivasis reported from Kerala in 2011. The proportion of ST population was little above 1 per cent in all Census years. ST populations are broadly classified into four categories: hunters and gatherers, agricultural labourers, shifting cultivators, and settled cultivators (Planning Commission of India, 2008a). There are various clans and communities. Paniyans accounting for 21 per cent of the total ST population forms the single largest ST community. They are distributed in the districts of Wayanad, Kozhikode, and Malappuram. Originally these communities served as bonded labourers, however, after the abolition of bonded labour, they are now regarded as agricultural labourers. Five per cent of ST populations are still considered as a primitive tribe living in 398 settlements mostly concentrated in Kasaragod, Wayanad, Malappuram, and Palakkad districts. Wayanad district houses 31 per cent of the total ST population of Kerala followed by Idukki (11.5 %) and Palakkad (10 %).

The status of the SC and ST population in Kerala is relatively better compared to their counterparts in the rest of India. The general literacy rate of the SC population in Kerala was 82.66 per cent and that of the ST population was 64.35 per cent against the all-India averages of 54.32 and 47.08 respectively in 2011. Although the literacy rate is high for both the groups in all-India standards, it is significantly low for Kerala as the state has attained total literacy.

The incidence of poverty and deprivation are also high among these groups. Kerala registered remarkable success in poverty reduction. Incidence of poverty among the ST population in Kerala is 50 per cent compared to all-India average; however, rural poverty among the ST population in Kerala is an issue as it is more than two and half times than that of the rural population in Kerala. The districts like Wayanad and Idukki, which have a high concentration of ST population, also registered a high level of deprivation. Landlessness is perhaps the major factor for persisting poverty among the ST population.

FISHING POPULATION

The fishing community is another outlier group, mostly living in the coastal areas. They face multitudes of problems. Poverty among the fishing community is a serious issue not only in Kerala but in India and also across the world. A little over one million people in Kerala belonged to the fishing community. Among them, 7.7 lakh were marine fishermen and the rest 2.3 lakh were inland fishing people. Marine fishermen are found in those nine districts, which border the sea. Considering both marine and inland fishermen population, it was found that the Alappuzha district had the highest concentration of fishermen followed by Thiruvananthapuram (Table 10.2). If only marine fishing was considered, Thiruvananthapuram would rank first. The concentration of the marine fishermen population reduces towards the north with the relatively higher concentration in Kozhikode. Kasaragod district recorded the lowest number of marine fishermen, little over 42 thousand. The inland fishing population is distributed in all 14 districts. Ernakulam district housed the highest number of inland fishermen (62,432) followed by Alappuzha (60,590). These two districts account for 53.4 per cent of total inland fishermen in Kerala. This high concentration of inland fishermen population in these two districts is due to their location around the Vembanad lake, the largest coastal water body on the West Coast. Alappuzha also borders Kayamkulam lake. There are few fishermen in the districts of Idukki and Wayanad.

Fishermen belong to all three major religions—Hindu, Muslim and Christian. Fishing communities are relatively deprived compared to other social groups. Their accomplishment in several human development indicators like literacy, health care, housing, and living conditions are not comparable to those enjoyed by other communities. The livelihood status of the average member of the fishing community remained worse off than the average person of Kerala. Socio-ecological and political movements among the fishing community were intensified during the 1970s and 1980s. Fishermen are one of the most organsied communities in Kerala with considerable religious influence, particularly in south Kerala. Various movements from time to time emphasized on promotion of social security and livelihood schemes. Technological changes are also brought in. It is observed that fishing communities are trying to catch up with the rest of Kerala's population in recent years.

LITERACY

Kerala is one of the states in India that has shown tremendous progress in literacy rate. It was declared a totally literate state by 1991 when the effective literacy stood at 90 per cent. By 2011, Kerala's effective literacy rate was 93.9 per cent, with male and female literacy rates standing at 96.1 per cent and 92.1 per cent respectively. The literacy rate at the district level varied from 88.49 per cent in Palakkad to 96.93 per cent in Pathanamthitta. Besides Palakkad, there were other two districts of Kasaragod and Wayanad that had recorded less than 90 per cent literacy rate. Male literates accounted for more than 92 per cent in all districts and as many as seven districts recorded a male literacy rate of more than 97 per cent (Table 10.2). The distribution trend of female literacy followed that of male literacy. Palakkad district with a female literacy rate of 84.99 per cent remained at the bottom of the scale, whereas Pathanamthitta district with 96.26 per cent female literacy topped among the districts. Rural and urban variation in the matter of literacy rate is low (2 %).

Kerala progressed tremendously in the matter of literacy. In 1961, Kerala reported a 46.85 per cent literacy rate, which doubled to 93.9 per cent by 2011. The gap between the districts having the highest and the lowest literacy rates was around 28 per cent in 1961. By 2011, this gap was reduced to a little over 8 per cent. This achievement is attributable to various factors, partly historical and partly due to the initiatives of Government, civil societies, and people's science movements. The literacy rate is still low among the Scheduled Caste and Scheduled Tribe population and fishing people.

DISTRIBUTION OF RURAL AND URBAN POPULATION

Kerala is predominantly rural. Till 2001, the proportion of the total population living in rural areas was more in Kerala compared to all-India average. However, the situation has changed recently and according to the 2011 Census, Kerala's rural population component (52.3 %) was well below all-India average (69 %). The sudden growth of the urban population in Kerala reported in the 2011 census is due to the inclusion of categories like plantation workers under industry. The total rural population in Kerala was 17.5 million, with a decadal growth rate of – 25.90 per cent. The rural population density in Kerala (662 persons/km^2) is well above than that in the adjacent states of Tamil Nadu (273) and Karnataka (186). Rural-urban differences in the matter of demographic characteristics are low in Kerala (Table 10.3).

Table 10.3: Demographic Characteristics: Rural, Urban Components and Total Population in Kerala, 2011

Parameter	*Rural*	*Urban*	*Kerala*
Total population*	17,471,135 (52.3)	15,934,926 (47.7)	33,406,061 (100)
Male population*	8,408,054 (48.13)	7,619,358 (47.82)	16,027,412 (47.98)
Female population*	9,063,081 (51.87)	8,315,568 (52.18)	17,378,649 (52.02)
Population growth (2001-2011) (%)	(-) 25.89	92.76	4.8
Sex ratio	1078	1091	1084
Child population* (up to 6 years)	1,823,664 (10.44)	1,649,291 (10.35)	3,472,955 (10.40)
Literacy rate (%)	92.98	95.11	93.98
Male literacy rate (%)	95.35	96.95	95.80
Female literacy rate (%)	90.82	93.34	92.09

Data Source: Census of India, Kerala, 2011.
Note: * Figures in bracket are in per cent.

The district-wise distribution shows that the urban population outnumbered the rural population in six districts (Fig. 10.6). Wayanad, Idukki, and Pathanamthitta are predominantly rural. All three districts are located in the Western Ghats. Among the districts adjoining the coast, Kasaragod recorded the highest component of the rural population with 61.2 per cent of the total population in the district living in rural areas. Ernakulam district had only 32 per cent of people living in rural areas.

The regional trend indicates that the coastal plain and the lowlands are gradually turning into a continuous urbanized space with few exceptions like the Kuttanad arca, while the midlands and highlands continue to maintain rural character. The coexistence of rural and urban areas side by side is an important feature of the settlement system in Kerala. There is a rural-urban continuum and most of the urban services are available in the rural arcas also. This has impacted Kerala's internal migration trend till recently. Physiography plays an important role in rural-urban distribution.

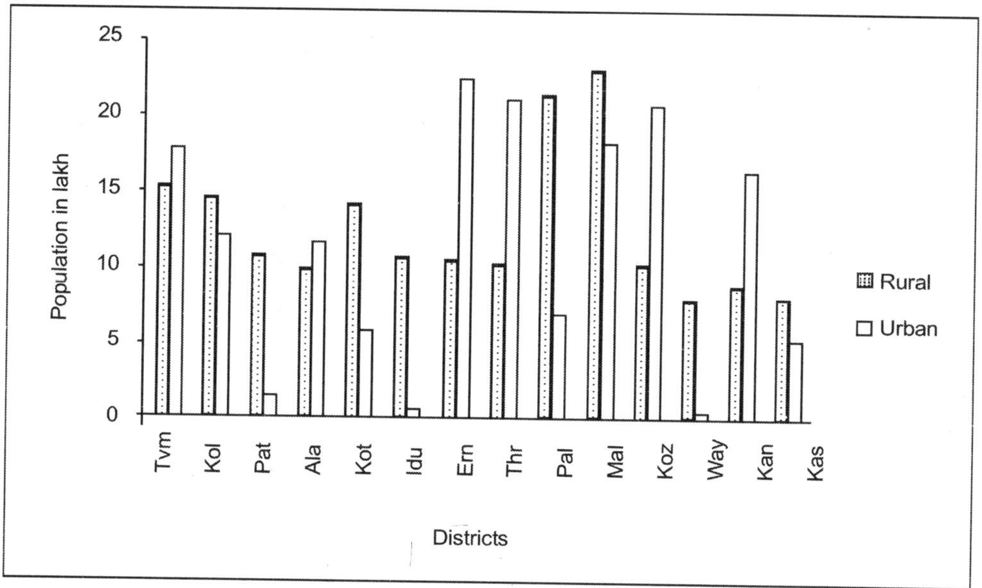

Fig. 10.6: Distributin of Rural Urban Population by Districts in Kerala, 2011.
Data Source: Census of India, Kerala, 2011.

OCCUPATIONAL STRUCTURE

The occupational structure provides an idea about the economic base of the area. A tenfold classification scheme followed by the Census department to consolidate employment data is normally used to discuss occupational structure under three sectors: primary, secondary, and tertiary. Workers engaged in the sectors of secondary and tertiary determine the urban non-urban character of an area. This scheme had changed since 2011. Total workers are now categorised as main workers and marginal workers. Each of these categories is further divided into four classes: (i) Cultivators, (ii) Agricultural workers, (iii) Workers engaged in the household industry, and (iv) Other workers. The discussion here covers the occupational structure of 2001 illustrating ten fold classification. Further, we also discuss district level occupational structure based on 2011 census data.

Occupational Structure, 2001

The 2001 census data brought out that 33.3 per cent of total workers were engaged in the primary sector comprised of cultivators, agricultural workers, and workers engaged in mining and quarrying, livestock, fishing, plantation, etc. The secondary sector accounted for 15.6

per cent of the total workers. This group included categories of household manufacturing and other than household manufacturing. The tertiary sector absorbed the largest share of workers (51.1 %). Workers in the primary sector decreased over the years, whereas the secondary sector had not witnessed much change. The significant increase was in the tertiary sector. Even in rural areas, the primary sector accommodated only 41 per cent of workers. The tertiary sector is an important category to provide employment to rural people also. Urban areas in Kerala also have some workers engaged in primary sectors. The 2001 Census data indicated that workers in the primary sector constituted 13 per cent of the total workers in urban areas. This scenario is due to the rural-urban continuum and juxtaposition of rural and urban areas.

Analysis of the tenfold classification of workers for the census periods of 1971 and 2001 revealed that there was an increase in the share of workers in the categories of construction, trade and commerce, transportation and communication, and other services (Table 10.4). The maximum increase was noted in the case of the construction sector.

Table 10.4: Occupational Structure of Kerala, 1971; 2001

Occupational Categories	*Workers in per cent to total workers*		
	1971	2001	Change
Cultivators	13.08	5.23	(–) 7.85
Agricultural labourers	23.87	9.52	(–) 14.35
Livestock, Fishing, Plantation etc.	5.88	11.33	(+) 5.45
Mining and Quarrying	0.50	0.77	(+) 0.27
Household manufacturing	4.85	3.31	(–) 1.54
Non-household manufacturing	13.58	13.31	(–) 0.27
Construction	1.88	9.89	(+) 8.01
Trade and Commerce	12.63	16.61	(+) 3.98
Transport and Communication	5.68	9.93	(+) 4.25
Other services	18.05	20.10	(+) 2.05

Data Source: Department of town and Country Planning, 2012.

In the primary sectors, the categories of agricultural workers and cultivators recorded a decline whereas livestock, fishing, plantations, etc. and mining and quarrying categories recorded an increased share of workers. The importance of agriculture is coming down, which is reflected through the negative growth experienced by the sectors of cultivators and agricultural workers. This trend manifests a change in the economic base and shift in land use practices in Kerala, which is emerging as a consumer state. Tertiary sectors, particularly construction, trade, commerce, transport, communication, and other services are important contributors to the economy. The decline in seasonal agriculture, increase in current fallow and substantial growth in the area under plantations, particularly rubber plantation are significant land use changes. This has been reflected in the change in occupational categories. Kerala is also witnessing a construction boom, especially after liberalization. The inflow of foreign remittance has fuelled this growth. Mining in Kerala is related to the construction sector. However, near-stagnation of the secondary sector is a matter of concern. There is little scope for increasing labour absorption capacity of the agriculture sector, therefore, the creation of non-agricultural activities particularly rural small-scale manufacturing enterprises is an important strategy for restructuring Kerala's economy.

Occupational Structure, 2011

According to 2011 Census data, Kerala had 34.3 per cent of the total population classified as workers with district-level variations from 25.8 per cent in Malappuram district to 46.5 per cent in Idukki district (Fig. 10.7). The high proportion of workers in Idukki, and Wayanad districts, both recording >40 per cent of the total population as workers indicated unique demographic characteristics of the hilly areas within the state. The occupational structure of 2011 indicated that 82.2 per cent of main workers and 73.6 per cent of marginal workers fall in the category of other workers (Table 10.5).

Table 10.5: Occupational Structure by Districts in Kerala, 2011

District	Main workers				Marginal workers			
	Culti. (per cent)	Agril. Lab (per cent)	House-hold Ind. (per cent)	Other workers (per cent)	Culti. (per cent)	Agril. Lab (per cent)	Household Ind. (per cent)	Other workers (per cent)
Thiruvananthapuram	2.9	6.1	2.1	88.9	3.9	12.5	2.9	80.7
Kollam	6.0	8.1	2.0	83.9	7.0	16.4	3.5	73.1
Pathanamthitta	10.7	12.0	2.6	74.7	9.9	20.3	3.5	66.3
Alappuzha	3.0	6.6	4.2	86.2	3.6	15.5	5.1	75.8
Kottayam	7.1	7.4	2.2	83.3	5.6	13.9	4.1	76.4
Idukki	20.6	27.0	1.2	51.2	15.1	33.2	1.9	49.8
Ernakulam	4.0	4.6	1.8	89.6	4.5	11.7	2.9	80.9
Thrissur	3.7	5.9	2.4	88.0	4.2	11.7	3.6	80.5
Palakkad	6.7	22.4	2.3	68.6	4.1	32.7	3.1	59.1
Malappuram	5.4	9.6	1.6	83.4	5.6	16.9	1.9	75.6
Kozhikode	2.6	5.0	1.2	91.2	3.4	13.7	2.1	75.8
Wayanad	17.6	26.2	1.1	55.1	8.3	42.3	2.2	47.2
Kannur	4.5	8.5	1.4	85.6	4.7	18.3	2.2	74.8
Kasaragod	5.5	7.4	4.8	82.3	3.5	12.2	7.2	77.1
Kerala	5.8	9.9	2.1	82.2	5.5	17.6	3.3	73.6

Data Source: Census Handbook, Kerala, 2011.

Share of cultivators among workers varied from 2.6 per cent in Kozhikode district to 20.6 per cent in Idukki district. Idukki district also recorded the highest proportion of agricultural workers (27 %). The primary sector continues to provide sizable employment in the hilly districts of Idukki, Wayanad, Palakkad, and Pathanamthitta. Only 2.1 per cent of workers were engaged in the household industry in Kerala. Two districts, namely, Kasaragod and Alappuzha recorded >4 per cent of workers in this sector. Traditional household industries like coir and cotton are important in both districts. So far as other workers were concerned, the distribution varied from 51.2 per cent in Idukki district to 91.2 per cent in Kozhikode district with the state average standing at 82.2 per cent. Although, other workers dominated in the category of marginal workers for the state as a whole, in the case of Idukki and Wayanad districts, the share of other workers was less than 50 per cent. Agricultural workers accounted for sizable marginal workers in all districts. Comparing aggregate data for all three time points, 1971, 2001, and 2011, it is found that the share of other workers is increasing over the years and all other categories are losing (Table 10.6).

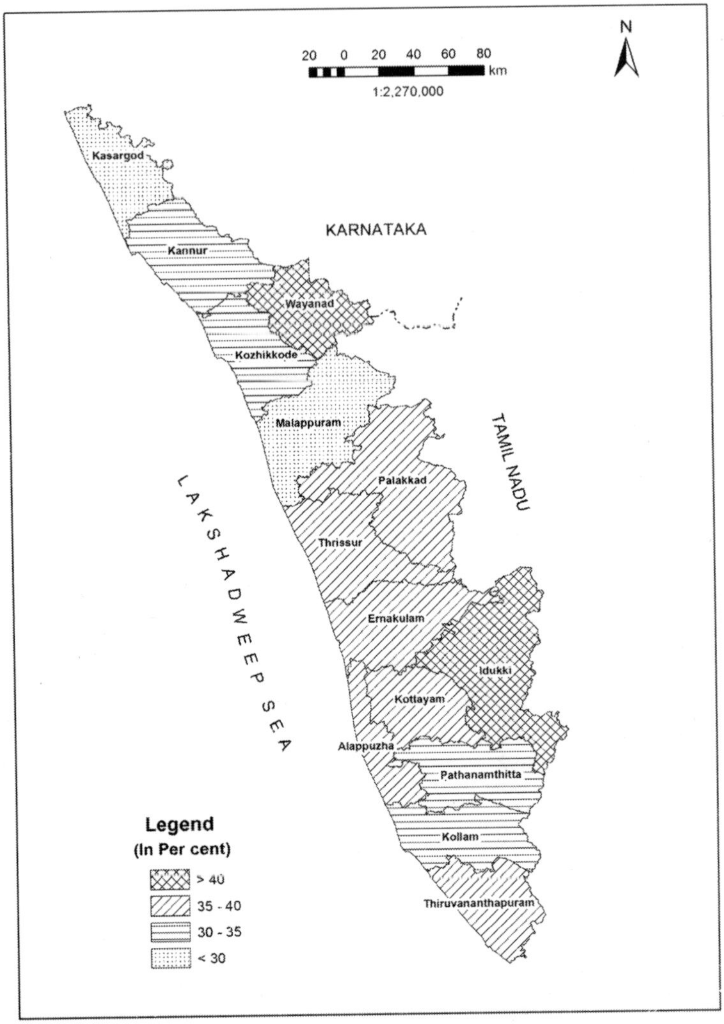

Fig. 10.7: Distribution of Total Workers by Districts in Kerala, 2011.

Data Source: Census of India, Kerala, 2011.

MIGRATION

People in Kerala are well known for out-migration. Due to the high population density and increasing pressure on land, there is a general tendency for people to move out of the state. As Kerala had interaction with other countries even before the Christian era, the mobility of people was a common feature. Kerala is experiencing out-migration since the 1930s. There are three types of migration noted in Kerala: migration to other countries, migration to other states within India, and internal migration or migration within Kerala. Besides, there is the migration of workers from other states to Kerala.

Table 10.6: Change in Occupational Structure in Kerala (1971 to 2011)

Categories	*Workers in per cent to total workers*			*Change (per cent)*		
	1971	*2001*	*2011**	*1971-2001*	*2001-2011*	*1991-2011*
Cultivator	13.08	5.23	5.80	– 7.85	+0.57	–7.28
Agricultural labourer	23.87	9.52	9.90	–14.35	+0.38	–13.97
Household industry	4.85	3.31	2.10	–1.54	–1.21	–2.75
Other workers	58.20	81.94	82.20	+23.74	+0.26	+24.00

Data Source: Census of India, Kerala 1971, 2001, 2011, *Only main workers considered.

Migration to Other Countries

External migration or international migration is an important socio-economic issue in Kerala. It brings remittances and as emigrants are distributed throughout the state, it impacts the local economy and also ecology. Land use change from food crop to a cash crop, flourishing of the construction sector and consumerism can be attributed partly to foreign remittances. According to the Kerala Migration Survey of Centre for Development Studies, the total number of Non-Resident Keralites (NRK) was estimated at 34.17 lakh in 2018 (Government of Kerala, 2019). This was 36.52 lakh in 2013. The number of emigrants was 21.2 lakh in 2018, which was 24 lakh in 2013. The number of return emigrants continued to increase from 7.39 lakh in 1998 to 12.94 lakh in 2018. Roughly there were 38.6 NRK and 29.3 emigrants per 100 households (Zachariah and Irudaya Rajan, 2018). The steady increase in the number of emigrants was noted till 2013, thereafter the number of emigrants started declining. Around 89 per cent of emigrants headed for Gulf countries of UAE, Saudi Arabia, Oman, Kuwait, Bahrain, and Qatar. UAE and Saudi Arabia together received around 62 per cent of total emigrants. The USA, UK, and Australia/New Zealand are major destinations outside Gulf countries. They accounted for 5.4 per cent of the total emigrants from Kerala. The number of migrants to the USA is increasing over the years.

Emigrants originate both from rural and urban areas. The 2001 census data indicated that the urban component of emigrants (31.8 %) is relatively higher when compared to the share of the urban population (25.5 %) to the total population. This is quite natural given the fact that opportunities and information available in the urban areas are relatively higher than those in rural areas. However, it is important to note that the rural people in Kerala are not far behind. The rural-urban difference is quite thin in Kerala, which is also evident in this sector. In the case of return migrants, there is an urban bias. District-wise distribution of emigrants showed the highest share of 19.1 per cent in Malappuram district followed by Kannur (11.8 %). Malappuram also recorded the highest number of return emigrants among all districts. The majority of external migrants (> 50 %) originates from the districts of Malappuram, Kannur, Thrissur, and Kollam. Idukki and Wayanad districts reported a small number of emigrants (Table 10.7). Over the years, northern districts have gained importance as a source of emigrants. There is a religious dimension of these migrants. Muslims constitute nearly 50 per cent of all migrants and the bulk of them (98 %) proceeds to the Gulf countries. Recent trend shows that the proportion of Muslim and Hindu migrants is increasing and there is a decrease in the proportion of Christian migrants.

Remittance from emigrants is an important issue in the Kerala economy. This inflow of money has impacted Kerala society in many ways—socio-economic, land use change, and environment. A neo-rich class has also emerged in course of time wielding economic and political power, which is not liked much by the traditional rich people. As a result, there are

Table 10.7: Distribution of Non-resident Keralites, Emigrants and Return Emigrants by Districts in Kerala, 2018

District	Non-resident Keralites		Emigrants		Return emigrants	
	No.	per cent	No.	per cent	No.	per cent
Thiruvananthapuram	230669	6.75	137007	6.46	93662	7.23
Kollam	406031	11.88	240527	11.33	165504	12.78
Pathanamthitta	164342	4.81	109836	5.18	54506	4.21
Alappuzha	236786	6.93	136857	6.45	99929	7.72
Kottayam	208400	6.10	166625	7.85	41775	3.23
Idukki	39030	1.14	32893	1.55	6137	0.47
Ernakulam	77347	2.26	53418	2.52	23929	1.85
Thrissur	335429	9.82	241150	11.37	94279	7.28
Palakkad	146729	4.30	89065	4.20	57664	4.45
Malappuram	715935	20.95	406054	19.13	309881	23.94
Kozhikode	312108	9.14	160691	7.57	151417	11.69
Wayanad	42696	1.25	30650	1.45	12046	0.93
Kannur	344291	10.08	249834	11.77	94457	7.30
Kasaragod	156891	4.59	67281	3.17	89610	6.92
Kerala	3416683	100	2121887	100	1294796	100

Data Source: Government of Kerala, 2019.

resentments and social tensions that may aggravate over the years due to demographic transition and change in population composition.

Out Migrants

The number of Kerala people living in other states in India was estimated to be 7 lakh by 2014, which was more than 11 lakh in 2003 (Zachariah and Irudaya Rajan, 2018). The return of out-migrants in 2014 was 3.89 lakh. This number is falling down indicating the preference of Kerala people to stay back outside or migrate to other countries from their place of living (*ibid*). Interstate migration (sum of out-migrants and return out-migrants) was 10.9 lakh in 2018, which was 14.4 lakh in 2011 (*ibid*). It was 21.09 lakh in 2003. Since then there was continuous decline. The trend of interstate migration from the districts changed significantly. Internal migration largely originates from Kannur, Pathanamthitta, Thiruvananthapuram, Kottayam, and Ernakulam districts. In the case of return out-migration, Kollam district topped among 14 districts with 21.4 per cent share followed by Thiruvananthapuram (16.5 %). There is a dominance of southern districts in the matter of return out-migration. The districts of Thiruvananthapuram, Kollam, Pathanamthitta, and Alappuzha together accounted for 57 per cent of total return out migrants. Considering interstate migration it was found that four districts of Thiruvananthapuram, Kollam, Pathanamthitta and Kannur accounted for 50 per cent of the total interstate migration in Kerala. Kannur district witnessed gradual increase in interstate migration over the years. It is the only district where both external and internal migrations are significant. Malappuram which had recorded the highest emigrants ranked low in the matter of out-migration within the country.

District-wise composition of out-migration and return out-migration indicated that out migration dominated in most of the districts. Return out-migration dominated only in two districts, namely Kollam and Alappuzha (Fig. 10.8). Around 90 per cent of the total interstate migration in Palakkad was composed of out-migrants. Idukki and Ernakulam districts closely followed Palakkad district in the matter of proportion of out-migration.

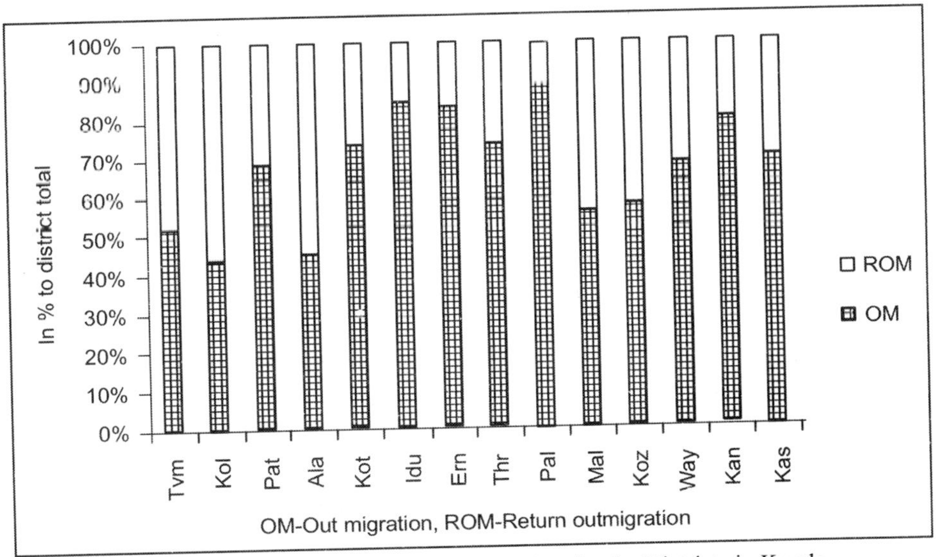

Fig. 10.8: Distribution of Out Migration and Return Out Migration by Districts in Kerala.
Data Source: Zachariah and Irudaya Rajan, 2018.

Karnataka, Maharashtra, and Tamil Nadu are the major destinations of out-migrants from Kerala. Together, these three states accounted for a little over 64 per cent of the total out-migrants in 2014, which was 73.7 per cent in 2011 (Zachariah and Irudaya Rajan, 2018). Karnataka's share went up but the share of Maharashtra and Tamil Nadu declined. Traditionally, Maharashtra was the major destination attracting the maximum number of people from Kerala; however, it is no more the number one choice for people of Kerala. Karnataka replaced Maharashtra by claiming 28.9 per cent of total out-migrants. Post-globalization growth of Bangalore is one of the pull factors for this changed scenario. Kerala out-migrants are now found moving to newer pastures like West Bengal, Rajasthan, and Assam (*Ibid*).

Intra-State Migration

Migration within Kerala, besides marriage migration, is also an important issue in the matter of demographic change and consequent land use change. There are four streams of intrastate migration: Rural to Rural, Rural to Urban, Urban to Urban, and Urban to Rural. Considering interdistrict and intradistrict migration within these streams it is found that rural to rural migration is dominant in both cases, however, the share is shrinking. During 1961, intradistrict rural to rural migration was little over 80 per cent and interdistrict rural to rural migration was 70 per cent, which had come down to 54 per cent and 58.5 per cent respectively by 2001 (Misra and Irudaya Rajan, 2018). Both rural to urban and urban to urban migration streams are growing. Intradistrict migration is dominating in all streams, which indicates that migration within the district is more than migration outside the district. It has been reported in an earlier study that 87 per cent of rural internal migrants had migrated to other rural areas within Kerala and about 68 per cent of migrants originating from urban areas had also moved to the rural areas while the rest 32 per cent of urban migrants staying outside Kerala moved from one urban area to another (Chattopadhyay, 1988).

Rural Kerala enjoys comparable facilities like that in urban areas. Villages are well served in the matter of amenities like electricity, transport facility, connectivity, health care facilities, fair price shop, educational facilities, post office, banking services, etc. The push-pull factors underpinning population movement within Kerala are not so effective due to low differences between rural and urban areas and easy accessibility. However, this scenario is changing in recent years. Urban to rural migration is shrinking and rural to rural migration is being replaced by urban to urban migration (Misra and Irudaya Rajan, 2018). Given the higher level of urbanization reported during the 2011 census migration from small towns to cities are bound to increase. According to 2001 migration data Kerala had 27.3 per cent of the population internally migrated with intradistrict migration of 19.9 per cent and interdistrict migration of 7.4 per cent (Table 10.8). Intradistrict migration varied from 11.7 per cent in Kasaragod to 32.2 per cent in Kannur in 2001. Thrissur and Ernakulam districts recorded high intradistrict migration. Thrissur district occupied the top position in intradistrict migration in 1991 also. Spatial distribution indicated that Central Kerala districts recorded high internal migration. Urbanisation and industrialisation play important role in intradistrict migration.

The trend in interdistrict migration is different. It was the highest in Idukki (18.1 %) followed by Wayanad (16.9 %) in 2001. Pathanamthitta also had sizable interdistrict migration. In the case of Idukki and Wayanad, the percentage shares of internal migration were elevated in 1991 but Pathnamthitta gained one percentage point during 2001.

Table 10.8: Internal Migration by Districts in Kerala (1991, 2001)

(Figures in %)

District	*Intra district migration*		*Inter district migration*	
	1991	*2001*	*1991*	*2001*
Thiruvananthapuram	11.2	14.0	3.4	3.6
Kollam	14.4	15.2	5.5	4.9
Pathanamthitta	18.3	20.3	12.4	13.2
Alappuzha	16.6	18.0	8.5	8.7
Kottayam	22.4	19.8	9.3	10.2
Idukki	15.5	18.3	22.3	18.1
Ernakulam	17.9	26.6	8.5	9.5
Thrissur	29.7	27.3	6.6	5.9
Palakkad	17.7	16.4	6.8	6.9
Malappuram	16.7	17.2	5.2	4.7
Kozhikode	23.4	19.7	6.4	5.2
Wayanad	15.8	12.4	19.9	16.9
Kannur	21.8	32.2	7.6	6.8
Kasaragod	15.4	11.7	7.3	6.5
Kerala	18.7	19.9	7.8	7.4

Data Source: Misra, Irudaya Rajan, 2018.

Wayanad and Idukki districts within the Western Ghats have a long history of receiving people from other parts of Kerala. The decades of 1921-31 and 1951-61 witnessed the major inflow in these two districts. The cheaper price of land in these hilly districts compared to the Travancore-Kochi area and lowland-midlands segment and higher productivity have acted as pull factors. Idukki received more people than Wayanad (Table 10.9). Kottayam is one of the districts from where the maximum number of people has moved to both places. Sixty-four per cent of immigrants in Wayanad are from Kottayam, Kozhikode, Ernakulam, and Malappuram. In the case of Idukki, the principal contributors are Kottayam and Ernakulam districts which together accounted for 81 per cent of immigrations. Kerala has a long history of internal migration from lowlands in the west to the high lands in the east and also from the Travancore area in the south to the Malabar area in north Kerala. Modern-day migration started with the introduction of plantation agriculture in the high altitudes and plateaus. It was followed by the initiatives of the "Grow More Food" campaign when forest lands and wetlands had been reclaimed to accommodate food crops. The nature of migration changed over the years. Nevertheless, it continued to be a force with long-term ecologic economic implications.

Table 10.9: Inter-District Migration in Wayanad and Idukki Districts of Kerala, 2001

From/To	*Wayanad*		*Idukki*	
	No. of Persons	*per cent*	*No. of persons*	*per cent*
Thiruvananthapuram	2889	2.1	6634	2.95
Kollam	2913	2.1	5515	2.45
Pathanamthitta	2777	2.0	11204	5.0
Alappuzha	2865	2.0	10367	4.6
Kottayam	28080	20.1	128661	57.4
Idukki	9083	6.5	-	-
Ernakulam	19865	14.2	53332	23.8
Thrissur	5471	3.9	2410	1.1
Palakkad	10403	7.4	1814	0.9
Malappuram	14339	10.3	827	0.4
Kozhikode	27075	19.4	916	0.4
Wayanad	-	-	1002	0.45
Kannur	13167	9.4	1031	0.45
Kasaragode	777	0.6	322	0.10
Total	139704	100.00	224035	100

Data Source: Census of India, Kerala, 2001 Migration Tables.

Migrant Workers

Kerala is experiencing in-migration from other parts of India, besides Tamil Nadu and Karnataka, from where traditionally people used to move to Kerala. Kerala state has turned into a destination for unskilled/semi-skilled and skilled labours from all-over India. There is no precise data on migrant labours. However, Gulati Institute of Finance and Taxation, Thiruvananthapuram estimated that by 2013 there were around 2.5 million migrant labours. With annual estimated arrival of 0.23 million, the current stock of estimated labour migrants in Kerala will be 3.4 million (Misra and Irudaya Rajan, 2018). According to the labour department, Kerala received 2.85 lakh labour from other states of India. They are found distributed in all districts varying from 2.45 per cent in Wayanad district to little over 21 per cent in Ernakulam district. Kozhikode, Thrissur, and Thiruvananthapuram are the other three districts attracting the bulk of migrant workers. These four districts together accounted for 50 per cent of all migrant workers in Kerala (Fig. 10.9).

The bulk of the recent in-migrants is from the state of West Bengal (41.4 %), followed by Assam (14.4 %), Orissa (10.7 %), Bihar (10.1 %), Tamil Nadu (9 %) Jharkhand (4.8 %) and UP (3.8 %). The rest 5.8 per cent comes from all other states including the union territory of Andaman and Nicobar islands. The majority of the migrants are engaged as labours in the construction sector, which has grown significantly in the post-liberalization period. Migrant labours are also engaged in brick kilns, clay factories, hotels, and other sectors requiring manual

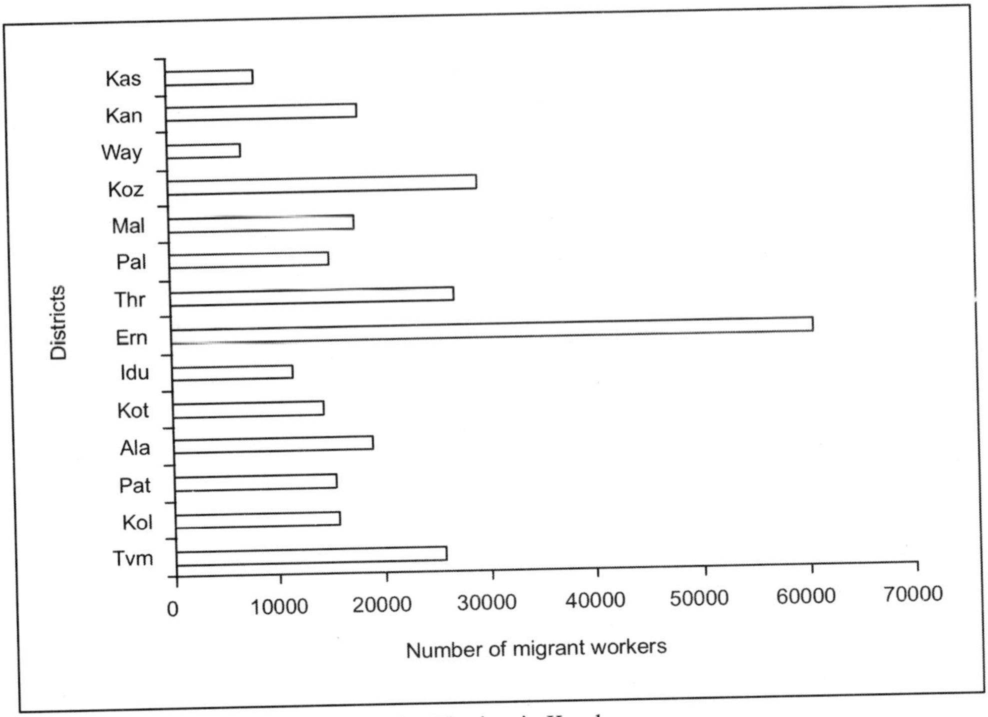

Fig. 10.9: Distribution of Migrant Workers by Districts in Kerala.
Data Source: Government of Kerala, 2019.

labours. Demographic transition, migration of skilled and semi-skilled labour to gulf countries, and consequent non-availability of local labours in Kerala have acted as pull factors for the in-migration of labours from other states. Apart from contractual labourers there are also migrations of small independent groups, who seek employment on daily basis on their own. The living conditions of these migrant workers are far from desirable. In recent years, labour migration has emerged as a serious socio-demographic issue in Kerala. While taking note of this issue, the Government of Kerala is trying to keep track of the migrant workers and introduces various welfare measures.

Migration in Kerala has various dimensions. People from Kerala move out and at the same time, people from other states in the country are moving in, which is of course a recent phenomenon. There is also a spatial trend in migration. The majority of external migrants (> 50 %) originates from the districts of Malappuram, Kannur, Thrissur, and Kollam whereas internal migration (migration to other states within India from Kerala) largely originates from Kannur, Pathanamthitta, Thiruvananthapuram, Kottayam, and Ernakulam districts. Kannur is the only district where both external and internal migrations are significant. Migrant workers are distributed in all districts, however high concentrations are noted in Ernakulam, Kozhikode, Thiruvananthapuram, and Thrissur districts. Idukki and Wayanad districts, contributing marginally to external migration and migration within India, are the main centres to receive migrants originating within the state. Management of the migrant population is an important component of Kerala's development.

CHAPTER

11

Urbanisation

DEFINITION OF URBANISATION

Urbanisation is a process of physical and social transformation encompassing all spheres of ecologic, economic, demographic, and cultural traits of human settlement. The Millennium Ecosystem Assessment (MEA) report considers the urban ecosystem as one of the ten ecosystems; the other nine are Marine, Coastal, Inland Water, Forest/Woodland, Dryland, Island, Mountain, Polar, and Cultivated land. The urban ecosystem covers 2.4 per cent of the terrestrial surface of the globe and the entire urban area is considered as a transformed ecosystem. However, the term urban is contested and there is little uniformity on its definition across the world. United Nations Organisation (UNO) defined urbanisation as the movement of people from rural to urban areas.

In India, an urban area is defined based on demographic parameters. According to the Census of India, an area may be classified as urban if the concerned area satisfies three demographic criteria: (i) population size of minimum 5000 persons, (ii) the density of population to be the minimum of $400/km^2$, and (iii) at least 75 per cent of the male working population engaged in the non-primary occupation. Besides, there are statuary administrative towns and notified areas. Since 1961, these criteria were being used to define urban areas. Some modifications were introduced in 2011 when plantation workers were classified as industrial workers. At the global level, India is considered as a low urbanized country with only 31 per cent of the total population living in urban areas compared to the World average figure of 51 per cent in 2011.

URBANISATION AS SPATIAL ORGANIZATION

The spatial organization of a society changes when a rural area turns into an urban system. In the process, a pre-industrial traditional society is transformed into an industrial/post-industrial/post-modern high energy oriented consumer society. Towns and cities in many countries, particularly those with a colonial past and a rich tradition of development like that in India, have experienced the transition from the traditional activities to post-liberalization/globalization related development activities. Traditionally, India's urbanisation pattern is an illustration of a highly polarised and spatially unbalanced urban system with sharp core-periphery differences. Rural hinterlands and city core seldom formed a continuum, and the relationship is more of an exploitative in nature where rural resources are either accumulated in urban areas or siphoned through it to higher-order urban centres with little scope of ploughing back in the hinterlands. However, recent trends show that urban industrial interaction fields are spreading through linking rural areas, and also small towns around the megacities and urban corridors are emerging along the transport routes (Dutt and Parai, 1999). Nature of labour participation and functional behavior of an area changes with urbanization. New institutions emerge and human-environment relations tend to change with the accumulation of wealth and man-made capital.

UNIQUENESS OF KERALA'S URBANISATION

Kerala's distinctiveness is also evident in the matter of its urbanisation pattern, which is characteristically different from the scenario observed in the rest of India. What makes Kerala's urbanisation different from the rest of India is the emergence of an intensive rural-urban interaction zone covering sizable parts of the state, where rural areas can be hardly distinguished from urban areas. There is an increasing co-existence between rural and urban functions, adjacent to the urban areas and also within the territorial limits of the urban centres. Kerala does not have a single primate city and three main corporation cities of Thiruvananthapuram, Kochi and Kozhikode lying along the coast with equal distance from each other offer complimentary functions, thereby undue concentration of people and urban functions in one place is largely avoided. Urban corridors are developing along the transport routes, roadways, railways, and waterways. The level of industrialisation is low. With high spatial diffusion of small cities, Kerala's case is somewhat unique and it calls for different perspectives to understand the spatial dynamics of the urbanisation process without undermining the role of capital formation and its manifestation on space.

In general, urbanisation model envisages the concentration of population in the cities as an inevitable outcome of economic growth. As the process of industrialisation advances, a growing number of people will shift their occupation from agriculture to industry and will also move from rural areas to the cities/towns, or new conurbations will emerge around the centres of industrialisation. The underlying assumption is that there is a clear cut distinction between urban and rural areas, which can be quantified and measured. This perceived distinction between urban and rural settlements has faced serious challenges. Mc Gee (1989, 1991) identified a distinct and spatial form of intensive rural-urban interaction zone adjacent to and between major urban centres in Asia. A number of cases include Jakarta-Bandung in Indonesia, Shanghai-Nanjing-Hangzhou in China's lower Yangtze, Taipei-Kaohsiung in Taiwan, and Seoul-Pusan in South Korea (Lin, 2001). A variant of this scenario could be marked in the wet zone of Sri Lanka and in the Kerala state of India (Casinader, 1987; Mc Gee, 1989).

The scenario in Kerala is comparable to that of Taiwan, where four major metropolitan city regions, namely, Taipei, Kaohsiung, Taichung, and Tainan dominate (William, 1999). Pannell's (1974) observation for Taiwan about the absence of primate nature of urban development in spite of having the conditions typically associated with urban primacy in underdeveloped countries, like, (1) small area and dense population, (ii) low per capita income, (iii) export or agriculturally based economics, (iv) high population growth, and (v) history of colonialism, fits well with the Kerala situation. However, in recent years, there had been a perceptible change in these conditions with the dominance of the tertiary sector, foreign remittance, and controlled population growth rate. Kerala is gradually yielding to the tendency of other large cities by having high rise buildings and continuous decline in inter settlement open space, especially in the three main corporation cities of Thiruvananthapuram, Kochi, and Kozhikode. Nevertheless, Kerala continues to present a distinct form of urbanization pattern.

TRENDS OF URBANISATION

According to the 2011 census, Kerala's urban content rose to 15,932,171 persons or 47.72 per cent of the total population against an all-India average of 31.16 per cent. The urban content of the World average is 50.6 per cent. Tamil Nadu, the neighbouring state, recorded an urban content of 48.5 per cent in the 2011 census. In 1951, the urban population in Kerala accounted for 13.48 per cent of the total population. It steadily grew over the years, although at a slow pace. The first phase of high growth was recorded between 1981 and 1991 and the next phase of high growth was between 2001 and 2011 (Fig. 11.1).

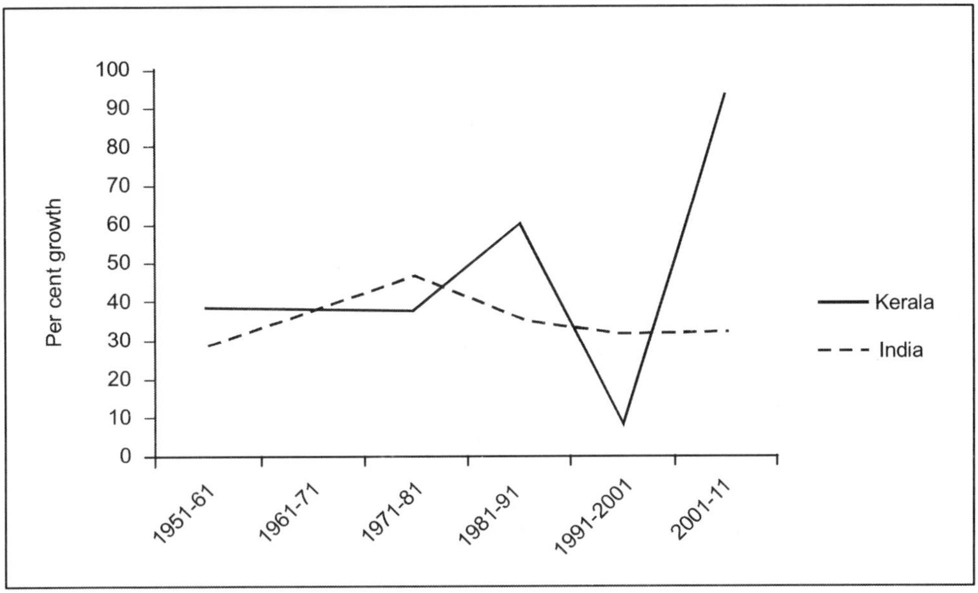

Fig. 11.1: Decadal Growth of Urban Population in Kerala and India.
Data Source: Department of Town and Country Planning, 2012.

Kerala's urban content was less than all-India average during 1961-71, 1971-81 and 1991-2001. The urban population in Kerala grew by 92 per cent between 2001 and 2011. This sudden high growth is partly attributable to a change in the definition of industrial workers for the census of 2011. The average pace of urbanization in India was 7.7 per cent during 50 years' period from 1961 to 2011, whereas Kerala recorded an average pace of 16.2 per cent during the same period (DTCP, 2012). It is significant to note that the urban area in Kerala has increased by 184 per cent in the period from 1961 to 2001 while the urban population has increased by 89 per cent during the same period. An increase in the density of the urban population was only 15 per cent during this period. This signifies that physical space is getting urbanized. Increased urbanization is not accompanied by a high concentration of population. This spatial spread over is possible due to a well-developed transport network, availability of social infrastructure, and other environmental facilities supporting human habitation in the peripheral areas.

DISTRIBUTION OF URBAN POPULATION BY DISTRICTS

According to the 2011 census report, the proportion of the urban population to the total population varied from 3.9 per cent in Wayanad district to 68.1 per cent in Ernakulam district (Fig. 11.2). There was no urban population enumerated in the districts of Kasaragod, Wayanad, and Pathanamthitta during the 1981 census when the state as a whole had less than 19 per cent of total population living in the urban sector (Table 11.1). Ernakulam district had as much as 40 per cent of its population in the urban areas in 1981.

Table 11.1: Urban Population and Decadal Change by Districts in Kerala, 1981-2011

Districts	*Urban population (per cent)*				*Change in proportion of urban population (per cent)*		
	1981	*1991*	*2001*	*2011*	*1981-91*	*1991-01*	*2001-11*
Thiruvananthapuram	25.3	34.0	33.8	53.8	8.7	–0.2	20.0
Kollam	13.2	18.6	18.0	45.1	5.4	–0.6	27.1
Pathanamthitta	—	13.1	10.0	11.0	13.1	–3.1	1.0
Alappuzha	15.9	30.6	29.5	54.1	14.7	–1.1	24.6
Kottayam	9.5	17.6	15.4	25.6	8.1	–2.2	13.2
Idukki	4.6	4.7	5.1	4.7	0.1	0.4	-0.4
Ernakulam	39.6	48.8	47.6	68.1	9.2	–1.2	20.5
Thrissur	21.1	26.3	28.2	67.2	5.2	1.9	39.0
Palakkad	10.1	15.8	13.6	24.1	5.7	–2.2	10.5
Malappuram	7.4	9.1	9.8	44.2	1.7	0.7	34.4
Kozhikode	27.2	38.4	38.3	67.2	11.2	–0.1	28.9
Wayanad	—	3.4	3.8	3.9	3.4	0.4	0.1
Kannur	23.4	51.0	50.4	65.1	27.6	–0.6	14.7
Kasaragod	—	16.5	19.4	38.3	16.5	2.8	18.9
Kerala	18.9	26.4	26.0	47.7	7.5	–0.4	21.7

Source: Census of India, Kerala, 1981, 1991, 2001, 2011.

Apart from Ernakulam, three other districts, namely, Thrissur, Kannur, and Kollam also registered a very high urban growth rate in the 1971-81 period. Kannur continued to register a high growth rate in 1981-1991 also. This district, by having 51 per cent of its population living in urban areas by the census year of 1991 emerged as the most urbanised district in Kerala, even ahead of Ernakulam. The same trend continued in 2001 also, although, the proportion of the urban population to the total population had marginally come down. As many as nine districts experienced negative growth in urban content between 1991 and 2001. The scenario changed in 2011. Kannur had 65 per cent of its population living in the urban sector by 2011. Besides Ernakulam and Kannur, the other districts with more than 50 per cent of the urban population in the 2011 census period were Thrissur, Kozhikode, Alappuzha, and Thiruvananthapuram. These six districts can be considered in the category of a high level of urbanization and together, they accounted for 68.8 per cent of the total urban population in Kerala in 2011, however, in 2001, this group of six districts housed around 77 per cent of the total urban population. This fall of urban content in these six districts together is attributed to a spatial change in the population growth pattern that Kerala is witnessing in recent years.

District level change in urban content during the 2001-2011 decade indicated a certain spatial trend. Malappuram, which had only 10 per cent urban population in 2001, enumerated 44 per cent in the urban sector by 2011, whereas another district with similar low urban

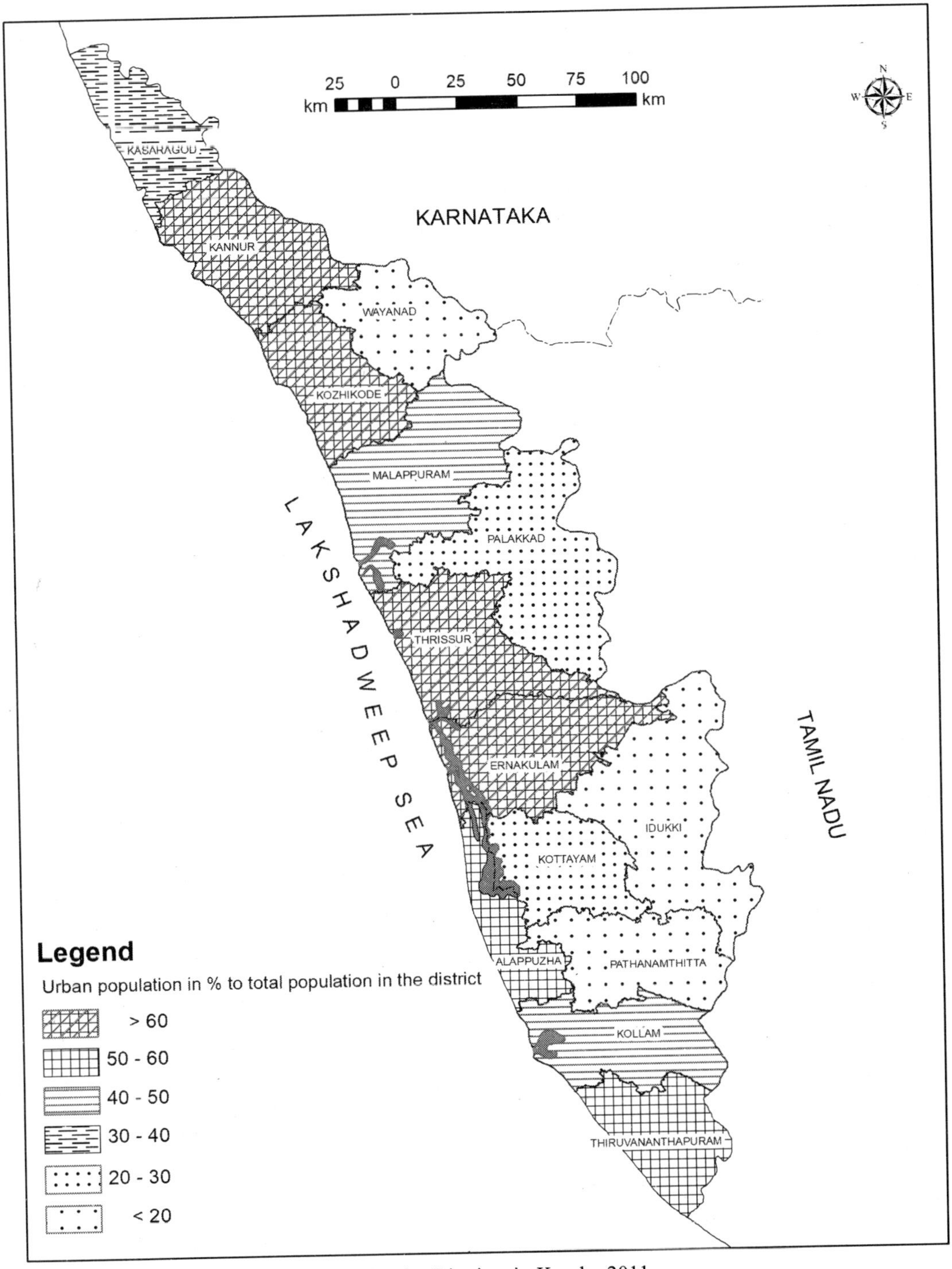

Fig. 11.2: Distribution of Urban Population by Districts in Kerala, 2011.

Data Source: Census of India, Kerala, 2011.

population content in 2001 like Pathanamthitta showed less than 1 per cent growth between 2001 and 2011. Wayanad also recorded very low change, of less than 1 per cent, and Idukki experienced negative growth. The census data of 1991 and 2001 indicated that three districts, Pathanamthitta, Kottayam, and Palakkad had numerically less population in urban areas in 2001 compared to 1991, and therefore, the growth rates were negative. Kannur district recorded the highest change (27.6 %) in urban content in the decade 1981-91 and Kasaragod recorded the highest change of 2.8 per cent point in urban content during 1991-2001. The decade of 1991-2001 was a decade of poor urbanization in Kerala.

Districts with high urban contents (districts having a proportion of urban population more than the state average) like Kannur, Kozhikode, Ernakulam, Alappuzha, and Thiruvananthapuram, each had recorded lower growth of urban population compared to total population growth in 1991-2001. However, during the 2001-2011 period, growth of the urban population was well above the growth of total population. Pathanamthitta and Idukki recorded negative growth even in the total population. The spatial trend of urbanization indicates that the concentration of the urban population in Kerala is shifting towards the central and northern parts. The relatively higher pace of urbanization marked in the southern part of Kerala during the 1960s and 1970s is now found in the northern part. Malappuram district recorded the highest decadal (2001-2011) increase in urban population followed by Kozhikode and Kannur districts. The second point is that the districts falling within the Western Ghats and having a sizable hilly area have low urban content. This pattern is perhaps related to the present trend of population growth, availability of habitable area, livelihood opportunities, and distribution of infrastructural facilities.

GROWTH OF URBAN CENTRES

During the first census operation in 1872, there were only seven towns in Kerala, and all of them were located in the then Malabar district of Madras Presidency. Altogether, 21 urban centres emerged in the Kerala region from Kannur in the north to Thiruvananthapuram in the south by 1901. As many as 10 towns out of these 21 towns were located along the coast and the other eight towns had developed either beside backwater or beside a navigable river (Chattopadhyay, 1988). Trade, commerce, pilgrimage, and administration, singularly and in combination, were the dominant functions performed by these urban centres, and almost all the urban centres were positioned in the coastal plain. In other words, the coastal plain was getting urbanised. Kerala being in the picture of world trade as far back as the eighth century B.C. (Ayyar, 1966; Menon, 1924) witnessed the emergence of a number of port-centric trade centres along the coast since historical past. The Kodungallur port, 30 km north of Kochi, is one such major trade centre referred to as Mousiris in Roman literature.

The number of urban centres was 94 in 1951, 92 in 1961, and 88 in 1971. This decadal decline in the number of urban centres was due to the declassification of some centres as the growth of non-agricultural workers could not keep pace with population growth. In the subsequent census years, there were growths in number of urban centres. In 1981, there were 106 urban centres and 5 urban out growths, which grew to 199 urban centres and 20 out growths in 1991. A change was brought in during the 1991 Census period in urban nomenclature. Till 1981, urban centres were classified as statutory urban areas and non-municipal urban areas. Statutory urban areas included Corporations, Municipalities, Cantonments, and Townships. Since 1991, census towns were introduced to denote non-municipal urban areas. Now, there are statutory urban areas and census towns. In 2001, there were 159 towns consisting of 60 statutory towns and 99 census towns. In addition, there were 33 out growths. Total number of towns grew to 520 in 2011 with 59 statutory towns and 461 census towns. The number of outgrowths came down to 16 in 2011. Out growths either became part of urban centres or recognised as a separate urban entity. With a total of 520 urban centres the average service area of an urban centre is around 50 km^2 area keeping aside the forest area. It can be further interpreted that average distance between geographical centres of two adjacent urban areas is only 8 km. Therefore, in most cases, there is an overlapping of fringe areas of adjoining urban centres resulting in formation of continuous urban mosaic.

District-wise distribution of statutory and census towns indicated that Ernakulam district had the highest number of statutory towns of 9 whereas Thrissur district houses the highest number of census towns of 128 (Table 11.2). More than 50 per cent of census towns are confined to the districts of Kannur, Thrissur, and Ernakulam. All three districts had urban population content of >65 per cent each. The number of Statutory towns came down to 59 due to the winding up of the Idukki township in 2011. There was no census town identified in Idukki and Wayanad districts. Most of the census towns are located in the coastal plain, lowlands, and midlands.

Table 11.2: Distribution of Statutory and Census Towns by Districts in Kerala, 2001, 2011

District/State	*Census 2001*			*Census 2011*		
	Statutory towns	*Census towns*	*Total towns*	*Statutory towns*	*Census towns*	*Total towns*
Thiruvananthapuram	5	-	5	5	26	31
Kollam	3	-	3	3	24	27
Pathanamthitta	3	-	3	3	1	4
Alappuzha	5	6	11	5	33	38
Kottayam	4	2	6	4	13	17
Idukki	2	-	2	1	-	1
Ernakulam	9	16	25	9	47	56
Thrissur	7	21	28	7	128	135
Palakkad	4	1	5	4	17	21
Malappuram	5	-	5	5	39	44
Kozhikode	3	10	13	3	48	51
Wayanad	1	-	1	1	-	1
Kannur	7	38	45	7	60	67
Kasaragod	2	5	7	2	25	27
Kerala	60	99	159	59	461	520

Source: Census of India, Kerala, 2011.

URBAN AGGLOMERATIONS

The Urban Agglomerations (UA) were identified for the first time during the 1981 Census year. This has become a consistent feature of regional urbanization in India (Chakraborty, 1988). Kerala had nine UAs in 1981 and by 2011 there were as many as 19 UAs, of which seven fell in the category of million-plus cities (Table 11.3). Five new UAs like Kayamkulam, Ottappalam, Changanassery, Chalakudy, and Kothamangalam emerged during the 2011 Census period. Some of the UAs formed due to coalescences of the peripheral zone of two urban centres. Kannur-Tellicherry in north Kerala is an example of this trend. Kannur and Tellicherry were two separate municipal towns till 1971. By 1981 each of them emerged as Urban Agglomerations, and in 1991 both these UAs merged together to form a single UA. Thrissur

UA presented a different picture where a cluster of 11 small satellite towns developed around Thrissur town in 1981. By 2001, there were three UAs in the Thrissur district and by 2011 all these towns were considered together to form a single Thrissur UA. All districts in Kerala except Wayanad, Idukki, and Pathanamthitta had one or more UAs in 2001 itself. Alappuzha, and Palakkad districts each had three UAs in 2011. The population in the UAs was growing steadily. There were 55 per cent of the total urban population concentrated in UAs during the census period of 1981, which had reached a little over 94 per cent in 2011.

Table 11.3: Urban Agglomerations, Population and Primacy Ratio in Kerala, 2001, 2011

Name of Urban Agglomeration	*District*	*Population, 2001*	*Population, 2011*	*Primacy ratio, 2011 #*
Kochi	Ernakulam	1,355,406	2,117,990	0.165
Kozhikode	Kozhikode	880,168	2,030,519	0.157
Thrissur	Thrissur	330,067	1,854,783	0.142
Malappuram	Malappuram	170,364	1,698,645	0.128
Thiruvananthapuram	Thiruvananthapuram	889,191	1,687,406	0.127
Kannur	Kannur	498,175	1,642,892	0.124
Kollam	Kollam	379,975	1,110,005	0.080
Cherthala	Alappuzha	141,512	455,408	0.031
Kayamkulam	Alappuzha	**	427,091	0.029
Kottayam	Kottayam	172,867	357,533	0.025
Palakkad	Palakkad	197,281	293,566	0.020
Alappuzha	Alappuzha	282,727	241,072	0.016
Ottappalam	Palakkad	**	238,238	0.016
Kanhangad	Kasaragod	129,364	229,706	0.016
Kasaragod	Kasaragod	76,011	192,761	0.013
Changanassery	Kottayam	**	127,971	0.009
Chalakudy	Thrissur	**	114,901	0.008
Kothamangalam	Ernakulam	**	114,574	0.008
Chittur-Thathamangalam	Palakkad	67,928	80,563	-

Source: Census, 2011, Department of Town and Country Panning, 2012. ** UAs emerged only in 2011; # Primacy ratio worked out for Class-I urban agglomerations (>1 lakh population).

Disaggregating the data at the district level it was found that Kannur and Ernakulam districts reported 100 per cent urban populations within the urban agglomerations. In other words, all urban centres in these two districts fell within the UAs. Among the rest nine districts, Kasaragod recorded 84 per cent of the urban population in the UAs and the remaining eight districts fell in the category between 84 per cent and 100 per cent of urban population living in UAs. Comparison of inter census data brought out that between 1991 and 2001 the proportion of population came down in as many as 11 UAs, and only 5 UAs recorded an increase. Between 2001 and 2011, only Alappuzha UA recorded a decrease in population, and all other UAs showed an increase.

DIMINISHING IMPORTANCE OF THREE MAJOR UAs

Three major urban agglomerations namely, Kozhikode, Kochi, and Thiruvananthapuram together accounted for 66.5 per cent of the total population living in all the urban agglomerations during 1981. This figure had come down to 53 per cent in 2001 and 39 per cent in 2011. While the proportion of the urban population residing in UAs is increasing over the years, the dominance of three major UAs in the matter of population size is declining. There are seven UAs with a million-plus population, and two UAs, namely, Kochi and Kozhikode each has crossed 2 million population marks. The primacy ratio worked out for UAs in Kerala indicates that Kochi UA with the highest urban population in Kerala has a value of only 0.165, which is well below the primacy ratio of 1.08 for Bangalore and 0.70 for Chennai (DTCP, 2012). Kerala does not have a primate city.

The growing concentration of urban population within the UAs indicates that urbanization is spilling over adjoining areas and peripheral zones are getting urbanized. Coalescence of the individual towns within an agglomeration and regional spread of agglomeration together represent the spatial spread effect of urbanization where small towns or peripheral towns are not depopulated due to migration, rather they grow on their own strength, and the intermediate space between two towns recognized as a zone of intense interaction gradually transforms into urbanized space.

There is a tendency toward the formation of continuous urban space in each district. This may lead to the emergence of unbroken urban space from one end of Kerala to the other end. This corroborates the argument that Kerala's urbanization is of diffused nature and characterized by a spread effect. Over the years, the urban population is spreading over larger space, and thereby Kerala's urbanization continues to be decentralized. It also signifies the spatial trend of concentration of non-agricultural activities in the coastal and adjoining lowland-midland segments.

SIZE CLASS DISTRIBUTION: DOMINATION OF SMALL TOWNS

The uniqueness of Kerala's urbanization is the domination of small towns. Considering all towns individually, out of urban agglomeration, it was found that out of 520 towns only 9 are Class I and 29 are Class II. Class I and Class II towns accounted for only 20.5 per cent and 11.5 per cent of the total urban population respectively during 2011 (Table 11.4, Fig. 11.3). Their shares were 44.7 per cent and 19.2 per cent of the total urban population in the same order during 2001. In fact, Class I cities recorded a fall of 11.6 per cent of their population in this decade. This was due to the classification of outgrowths as separate towns, inter size class movement of towns, and decline in population in class I towns. This point warrants elaboration. The Census of 2001 recorded 10 towns under the Class-I category, of which five towns namely, Thiruvananthapuram, Kollam, Kochi, Thrissur, and Kozhikode were a preexisting municipal corporation. The rest five Class I towns were municipal towns of Alappuzha, Cherthala, Kottayam, Palakkad, and Kanhangad. Among these 10 municipal towns, eight towns continued as Class I town during the 2011 Census also. Two towns namely Chethala and Kottayam moved down to the Class II category. In the case of Cherthala, there were two growths in 2001, which were classified as separate towns during 2011, thus there was a reduction of population. Kottayam recorded negative growth of population. The municipal town of Malapuram, which was in the Class II category in 2001 moved to class I category in 2011. This trend is in sharp contrast to the rest of India, where Class I towns dominate and continue to grow over the years. It is perhaps only in Kerala that Class I town slips down to the Class II category.

There were 21 statutory towns and 3 census towns in the category of Class II towns in 2001. Some of these towns in this Class II experienced changes in status. Eighteen statuary towns out of 21 statutory towns continued with their status in 2011 and three census towns also maintained their status. Out of three statutory towns that moved to other Classes, it is found that one Class II town, Malappuram moved upwards to Class I category and other two Class II towns, namely,

Table 11.4: Growth of Urban Population by Size Class of Towns in Kerala, 2001, 2011

Size class (population)	2001			2011			Change in urban population 2001-2011 (per cent)
	Number of towns	Population	per cent	Number of towns	Population	per cent	
I (>1,00,000)	10	3, 692,165	44.7	9	3,262,380	20.5	-11.6
II (>50,000)	24	1,587,908	19.2	29	1,888,254	11.9	18.9
III (> 20,000)	72	2,296,456	27.7	254	7,925,828	49.7	245.1
IV (>10,000)	37	566,635	6.9	159	2,352,637	14.8	315.2
V (>5000)	16	119,062	1.4	61	467,045	2.9	292.2
VI (<5000)	1	4,699	0.1	8	36,027	0.2	666.7
All classes	159	8,266,925	100	520	15,932,171	100	92.7

Source: Census of India, Kerala, 2011.

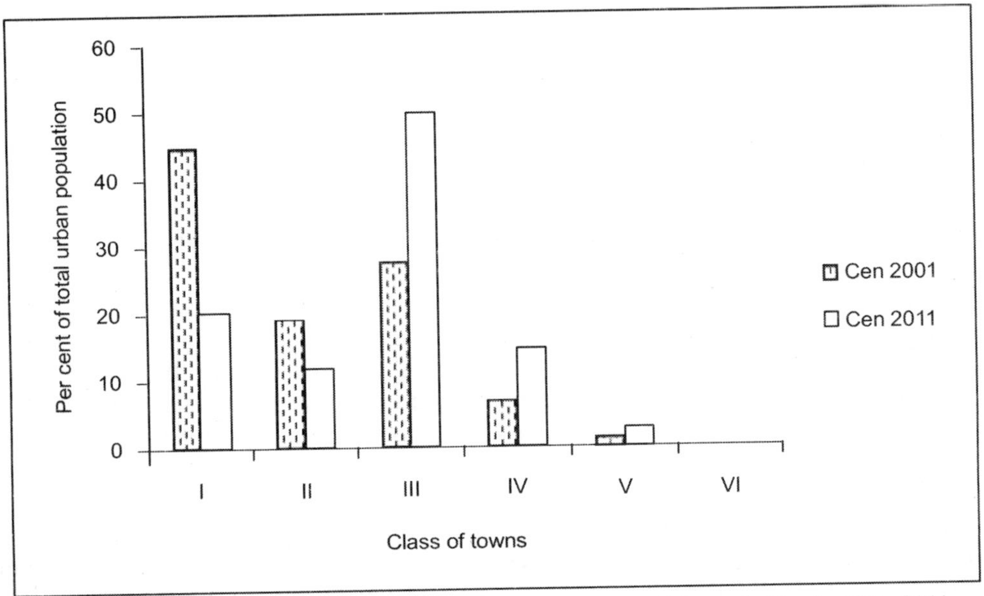

Fig. 11.3: Distribution of Urban Population According to Size Class of Towns in Kerala 2001, 2011.

Data Source: Census of India, Kerala, 2011.

Changanassery and Chittur-Thattamangalam moved down to Class III category. Kottayam moved from Class I to Class II and another two Class III towns, Ottappalam and Thodupuzha moved up to the Class-II category. Thus the number of statutory towns in the Class II category remained the same in 2011 as it was in 2001, although the composition of towns in the Class changed. There were five Census towns added to the Class II category in 2011. The total population living in Class II towns had increased by 18.9 per cent during the decade of 2001-2011.

The bulk of the urban population (49.7 %) was living in Class III towns in 2011. This was 27.7 per cent in 2001. The number of Class III towns was 12 in 1951, which increased to 64 in 1981. By 1991, the number came down to 46 as many of the Class III towns were part of Urban Agglomerations. The number increased to 72 in 2001. There was tremendous growth in the number of Class III towns between 2001 and 2011. People living in Class III towns grew by 245 per cent during this period. Still higher growths were observed for the towns down the scale (Table 11.4). The dominance of small towns was noted since 1981. The spatial spread of urbanization underpins this development.

DENSITY OF URBAN POPULATION

Spreading nature of urbanisation in Kerala against concentrated nature as evident in the rest of India is also manifested in the density distribution of the urban population. The state is well known for its high population density in the country. Kerala now stands third after Bihar and West Bengal in the matter of population density. However, urban population density as well as the growth of urban population density is relatively low. From 1961 to 2001, overall population density grew from 435 persons/km^2 to 819 persons/km^2 or by 88 per cent, while density of urban population increased from 2204 persons/km^2 to 2542 persons/km^2 or by 15 per cent. At the individual city level, it is found that Kollam UA ranks first in Kerala with the highest density of population of 6095 persons/km^2 in 2011, whereas Chennai and Bangalore, the first ranking UAs in Tamil Nadu and Karnataka respectively recorded population density of 24676 and 11574 in the same order. The population density of the UAs in Kerala varied from 1531 persons/km^2 in Malappuram to 6095 persons/km^2 in Kollam.

Three main Urban Agglomerations, namely Trivandrum, Kochi, and Kozhikode recorded population density of 5006, 2996, and 3746 respectively. UAs experienced an increase in population density at the all-India level, whereas, there was a declining trend in the case of Kerala. Low and declining population density of UAs indicate that (i) there is hardly any overconcentration of the population in the higher-order urban centers as migration from rural to urban areas is not taking place, (ii) natural growth of population has been substantially reduced, and (iii) urbanization in Kerala is spatially spreading out. There are core-periphery variations in the matter of population density distribution in the UAs. Core areas record a higher population density compared to the periphery. However, the growth of population density in the periphery is higher than that of the core area. This substantiates the fact that population density and the growth of population are inversely related in urban Kerala. It also indicates a very special feature of people in Kerala who prefer to live in a relatively rural setting having a house surrounded by a homestead garden. However, there are symptoms of change in this attitude. The emergence of multi-storey apartment buildings in major cities in Kerala is a remarkable recent phenomenon.

URBANISATION AND PHYSICAL SPACE

Understanding the relationship between physical space and urbanization is important to study Kerala's urbanization. Traditionally, Kerala is divided into three topographic zones: Low land (coastal plain), Midland, and Highland. The coastal plain is narrow and characterized by a chain of lagoons connected through channels/canals. Topographic undulation starts from the coast itself and increases from the west to east. The majority of the urban areas are concentrated in the coastal plain, lowlands, and western part of the midland (Fig. 11.4). The census data of 2011 indicated that 50 per cent of urban areas were in the coastal plain/lowlands, followed by 43 per cent in the midlands, and 7 per cent in the highlands (DTCP, 2012). Distribution of urban population followed this trend with 49 per cent urban population residing in the coastal plain/lowlands, 45 per cent in the midlands and 6 per cent in the highlands. Kerala is experiencing urbanization at a faster rate since 1981. The entire coastal plain and adjoining midlands record an increased pace of urbanization, which is reflected by a rapid increase in urban areas in this part of the state. The

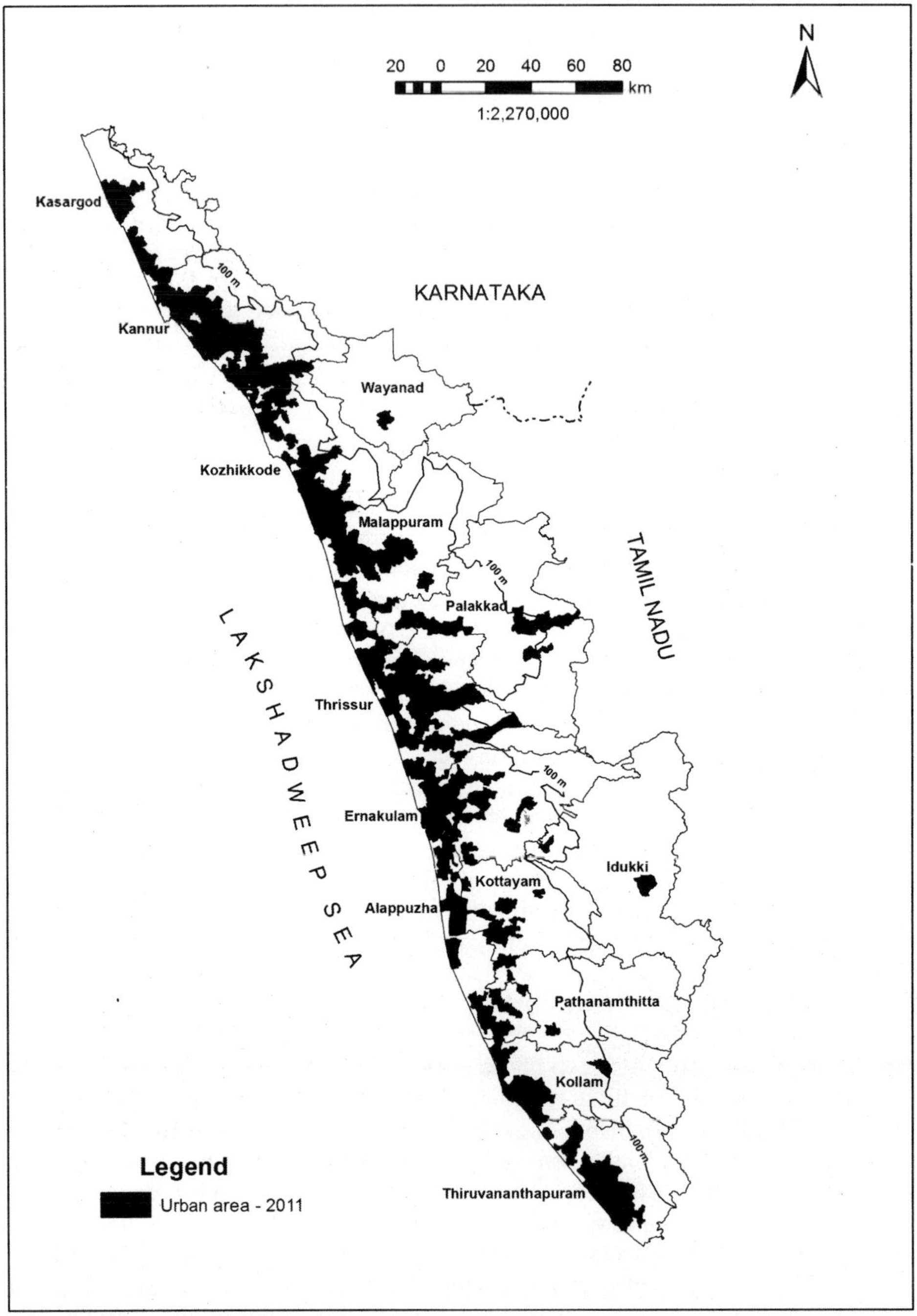

Fig. 11.4: Distribution of Urban Areas in Kerala, 2011. Note concentration of urban areas along the coastal plain and adjoining midlands.

Source: Department of Town and Country Planning, 2012.

coastal plain is emerging as the zone of contiguous settlements and urban centres are coalescing mostly along the longitudinal axis or topographic grain of the state. Almost all urban centres in the coastal plain are connected through roadways, railways, and waterways.

State highways and major district roads connect all urban centres in Kerala. Waterways of 1687 km length have further strengthened movement among the urban centres in the coastal plain. National Highways and railways link all Class-I towns in Kerala and provide connectivity to the neighbouring states of Tamil Nadu through the Palghat gap, Kollam-Shencotta gap, and Thiruvananthapuram-Nagercoil plain, and to Karnataka through Wayanad plateau and coastal plain in Kasaragod district. Ecologically sensitive areas like wetlands are increasingly consumed under rapid urbanization. Prime agricultural lands are also under pressure throughout Kerala and this is one of the most important land management challenges in the state so far as urban expansion is concerned.

Coverage of cluster of the built-up area indicated that most part of the coastal plain falls in the category of 75 per cent to 100 per cent cover (DTCP, 2012). However, this does not indicate that the entire stretch is full of built-up areas. A sample study by DTCP (2012) in Thiruvananthapuram Corporation brought out that clusters of 75 per cent to 100 per cent urban coverage has the built-up area of only 28 per cent. This is the unique feature of Kerala's urbanization where built-up space is encircled by natural space. Another peculiarity of Kerala's urbanization is the prevalence of primary activity in many of the census towns. In 2001, there were as many as 63 urban local bodies having agricultural potential. This number has rose to 179 in 2011. The occurrence of primary and non-primary activities side by side and mutual non-exclusiveness are important spatial dynamics in Kerala, which impacts urbanization patterns.

LOCATION AND RESOURCE BASE OF URBAN CENTRES

The emergence of urban centres depends upon (i) natural factors like topography and resource availability, and (ii) man-made factors like culture, administration, economic activities, market, communication, trade, and commerce. Locational determinants internalize both these factors. There are several instances where a single urban centre serves more than one function. Multi-functionality of the urban centres is increasing with improved communication facilities and the advancement of science and technology.

The historical records showed that a number of small ports had developed all along the Kerala coast to handle the trade of spices, coconut products, forest produces, and timber. Kerala was well known as a 'land of black pepper' in the historical past. By the end of the 19th century, besides Kannur, Thalassery, Kozhikode, and Kochi, as many as 33 minor and medium ports emerged in a stretch of 300 km to the north of Kochi. It was not a benevolent effort of the British to decentralise these trade but pure cost-effectiveness as hinterlands of these ports supported the same type of product, and, raw materials could be brought down to the coast straightway through rivers or roads. Almost all these trade centres were subsequently identified as urban centres (Chattopadhyay, 1995). At present, there are one major port at Kochi and 17 non-major ports distributed all along the coastline of the state. Foreign trade has been more or less centralized through the Kochi port. Kochi is also the major industrial hub of Kerala and forms the focal point of the emerging industrial corridor of Kochi-Thrissur-Palakkad-Coimbatore.

Many coastal towns are equipped with fishing harbour facilities and fish processing industrial units as a consequence of the development of fisheries and associated fishing industries. There are 14 completed fishing harbours and works are in progress in another 10 harbours (Government of Kerala, 2019). Internal trade and initial collection and processing continue to operate through a number of small and medium towns all along the coast. The coir industry, mostly tradition-bound, is an important activity in the coastal plain. Alappuzha district is well known for coir industries in Alappuzha town and surrounding small and medium-sized towns. An earlier study indicated that around 80 per cent of the urban centres manufactured one or more of the commodities related to plantation-based products such as tea, coffee, rubber, coir, *bidi*, coconut products, pepper, cashew nut, plywood, vegetable oil, paper, hardboard, tapioca stands, thread and fibre as sole or major important products (Chattopadhyay, 1988). A series of small towns have emerged in the foothills all through the state as initial procurement centres of local products. Hill products find their way to plains through these centres, which, in turn, supply consumer goods and other essential materials to the hills. These small centres are well connected with the large urban centres in the coastal plain mostly through roads, and in some cases through railways. Towns along backwaters and navigable parts of the main rivers enjoy connections through waterways also.

The tertiary sector is one of the main contributors to the growth of urbanization in recent years. It is observed that the growth of the urban population is related to the growth of workers in the tertiary sector and the growth of road density. The tertiary sector includes transport, storage and communication, trade, hotels and restaurants, banking, insurance and real estate, public administration, and other services including Information Technology. It contributed as much as 61.7 per cent of the Gross State Value Addition (GSVA) in Kerala during 2017-18 at constant prices (Government of Kerala, 2019b).

Resource base, settlement distribution pattern, and accessibility, although mutually dependent, have provided Kerala a basic frame of decentralization. Spatial dispersion of towns is mainly due to the decentralization of non-agricultural activities and nature of settlement type leading to the development of urbanised fringes or commuter villages which in turn results in non-distinction of rural-urban cultural patterns (Sreekumar, 1993). The urbanization process in the state is primarily governed by this basic structure. Rural-urban interaction and expansion of intense intermixing zones are taking place along transport corridors. As the size of the state is small, and accessibility is high, and amenities are well distributed there is not much attraction for people to migrate permanently to the cities. However, this trend is changing in recent years with people's increasing preference to live in urban centres.

CHALLENGES OF URBANISATION

Urbanisation in Kerala culminates in a unique spatial structure as exemplified in the foregoing discussion. The nexus between industrialization and urbanization, the emergence of primate cities, and well distinguished rural-urban divide are largely non-existent in Kerala. The intense rural-urban mix, the coexistence of agricultural and non-agricultural activities within urban areas, and specific landscape character facilitating the evolution of such a settlement system added to the uniqueness of Kerala's urbanization. The classical model of urban transition that describes the concentration of population in large cities as an inevitable spatial outcome of economic growth is not applicable to the case of Kerala, which is also quite evident. The physical growth and spatial trends of urbanisation indicate that the entire coastal plain and part of the midlands are getting urbanized with an unbroken chain of urban settlements of varying character and population density. Perhaps, this will be further consolidated through the transformation of interstitial areas among various urban centres. As the tertiary sector is the main driving force of urbanization, the small towns mostly in the peri-urban areas are evolving as multi-functional service centres surrounded by a landscape of the dominant primary sector. Urbanisation in Kerala throws several challenges, some of which are briefly discussed here.

1. Nature of economic growth expected with a high level of urbanization is missing in Kerala. Physically spread urbanization has brought potential agricultural areas including prime agricultural lands under the ambit of urban areas. While it dilutes the economic base of the urban areas, it also impacts usual rural land use practices through steep competition from non-agricultural use especially, from real estate sectors underpinning high appreciation of land prices. The challenge is to strengthen the economic base of the urban centres, bring compactness to urban settlements, and safeguard rural economic base.
2. Wetlands and similar natural areas both inter and intra urban spaces are reducing at a faster rate. The state as a whole has experienced a drastic decline in the area used for the cultivation of rice.With the spread of settlements all along the coastal plains and lowlands, there is a high probability of land conversion from agriculture to non-agricultural purposes throughout the state as a corollary to the spread effect of urbanization.
3. There is a need for complementary development of rural and urban areas. Given the recent thrust on organic products, and people's willingness to pay higher prices for organic substances there is ample scope for the rural hinterlands to cater to the need of the adjoining urban centres and also produce specialized products to match the emerging demand of urban consumers. Some sporadic initiatives are already noted in some parts of the state. It is necessary to consolidate and upscale these initiatives. Urban areas can serve as the market for rural products and thereby function as a growth engine.
4. Average service area and number of people to be served by urban amenities like housing, water supply, health, education, etc. have increased with the physical spread of urbanization. However, the growth of urban amenities could not keep pace to commensurate with this increase. As a result, existing urban infrastructure is under pressure and quality of service delivery tends to deteriorate over time. Besides, the cost of service delivery also escalates due to an increase in the service area. Improvement of quantity and quality of physical and social infrastructure in the urban areas is a big challenge. Containment of urban development is necessary to address this challenge. Efforts as well as expenses to provide urban services come down substantially in case of compact and concentrated settlements compared to diffused and linear settlements.
5. Kerala is gradually turning towards a consumer-oriented society. Solid waste disposal is emerging as an intriguing problem for urban management. According to the report of the Government of Kerala around 2500 tonnes of solid wastes are generated everyday. Around 50 per cent of these wastes are collected and the rests are left to decompose on road margins, drains, canals, water bodies, and open space. This provides an ideal breeding ground for pathogens and disease vectors. Leachates from these dumped untreated wastes can cause groundwater pollution. Sourcewise generation of solid wastes indicates that 44 per cent originate from market-related sectors followed by 28 per cent from household sectors. Volume in both these sectors will increase with the growing urbanisation trend and the entire coastal and lowland stretches will seriously be affected. It will be a real challenge for the Government to work out an effective waste management plan as part of urban development.

6. Although urbanization is related to regional economic development, its direct contribution to reduction of the poverty is marginal. The incidence of poverty is recorded in the pockets among tribal settlements in the highland segment and fisherman villages along the coast. The impact of urbanization on fisherman is limited to the market opportunities with little shift in occupation, as a result, fishing sector is overcrowded. It should also be noted here that relative poverty is more important in the case of Kerala than absolute poverty and in the urban sector challenge lies in addressing the issue of relative poverty.
7. Increase in built-up areas in the coastal plain and lowlands associated with urbanization leads to more and more addition of impermeable surfaced area resulting in high surface run-off compared to infiltration. Kerala's distinction from the rest of the country was notable in settlement distribution patterns in the urban areas also. Each household is surrounded by a courtyard with tree crops dominated by coconut. Kerala is famous for household biodiversity, be it in an urban area or rural area. Adequate inter settlement space covered by trees contributed to the greenness of the city, so unique compared to many other cities in India. However, in recent years, there is a reduction of this green space, and surfaced areas are increasing causing elevated land surface temperature, excess surface run-off, urban flooding, and various other environmental problems. Urban centres are turning into heat islands.

Several factors have contributed to Kerala's urbanization process. Historical legacy, settlement system, political-economic conditions, decentralized resource distribution, transport network, and landscape ecology played their part. Considering the high rate of urbanization and its physical spread there is a clear need for a strategic plan for urban development in Kerala, which should also provide guidelines for containment of urbanization, densification of settlements, consolidation of the urban core, and peri-urban development. Given landscape character and settlement distribution, there is a need for settlement policy in Kerala in conformity with land use policy, urban development policy, and strategic plan for urbanization.

CHAPTER

12

Social Sector

Kerala Model, Education, Health, Human Development and Poverty

KERALA MODEL

Education and health are part of the social capital that promotes human development, ensures empowerment and regular functioning of labour. Kerala is well ahead of many other states in India in the matter of basic education and human health. The state has achieved Millennium Development Goals (MDGs) under education and human health sectors well ahead of other states in the country (Government of India, 2017a). In fact, Kerala's accomplishments in these two sectors, since the late seventies and early eighties, in the face of low industrialisation, low urbanisation, and registering per capita income well below the national average, till recently, have been widely acclaimed as 'Kerala Model' and drew national and international attention. We provide here a brief discussion on the Kerala model deliberating on education and health.

By the end of the 1970s and early 1980s, the term 'Kerala Model' or 'Kerala mode of development' entered in development discourse. There were demographic, educational, and health data showing Kerala's edge over other Indian states. Given the average socio-economic conditions of Kerala prevailing in those days, compared to the national scenario, this trend invoked considerable interest among social scientists. The CDS-UN study on Kerala in 1975 underscored the importance of giving attention to 'educational and health services which help to build up human capital and make qualitative differences to the whole process of development' (CDS, 1975, 2000). This development experience of Kerala subsequently referred to as the 'Kerala Model' signifies that social development is possible through effective public action without the simultaneous rise in per capita income, growth of urbanisation, and industrialisation unlike developed countries (Dreze and Sen, 1989; Pillai, 2008). How did society, caste-ridden, a practitioner of untouchability and even where the lower caste peoples were kept away from the sighting range of certain upper caste people, can overcome these problems and turn into a front ranking-state in the country in spreading literacy and improving human health invoked considerable interest, nationally and internationally. The issue has been widely studied, debated and examined (Tornquist and Tharakan, 1995; Parayil, 1996; George, 1998; Parayil (ed), 2000; Franke and Chasin, 2000; Thomas Isaac and Franke, 2000; Veron, 2001; Heller, 2005; Ravi Raman (ed), 2010).

Germination of the 'Kerala Model' can be traced to the historical legacy of education and health development initiatives, geographical positioning of the state in terms of availability and decentralised natural resource base, affirmative public action, and proactive civil society. Since the formation of Kerala, whatever is the composition of the government, left or right, the social welfarist approach continued, and space for vibrant forward-looking civil society emerged empowering even common people to articulate their ideas and thoughts through discussion and discourse on various aspects of societal concern. It was well realized that human capacity building through education and health is the cornerstone of Kerala's development initiatives. Accordingly, there was a substantial budget allocation for the social sector including health and education from 1957 onwards and the growth rate of health expenditures outweighed total government expenditure in the initial decades (Ramankutty, 2000). The vibrancy of civil society is evident from the fact that Kerala Sasthra Sahitya Parishad (KSSP), a people's science movement organisation launched its own literacy mission in 1977, more than a decade before the National Literacy Mission was set up in 1988 (Chattopadhyay, 2017). Recently in 2018, the seamless teamwork to contain the Nipah virus, and relief and rescue operation undertaken by people from all walks of life in face of the devastating August 2018 flood again manifested strength and resolve of public action in Kerala.

Since formation, Kerala moved steadily over the years, undertook a series of unique initiatives, that helped overcome political geographic impediments and set footprints in land reforms, creation of physical and social infrastructure, improvement in health and education, control population growth, the spread of literacy, civil society movement, earning from foreign remittance, improving per capita income, and decentralised planning and governance. All these sectors are mutually reinforcing, and through synergy contribute to human development and improved quality of life.

Spatial Dimension of Kerala Model

It is often argued that Kerala's achievement in demography, health, and education sectors has put the state on firm footing to strive for sustainable development and therefore warrants closer examination (Parayil, 1996; Chattopadhyay and Franke, 2006). Sustainable Development as an alternative mode of development paradigm is inclusive and is expected to minimise spatial inequalities using the instrument of regional/local development initiatives (Schleicher-Tappeser, 1997). An overarching challenge is to examine the spatial dimension of development and to identify spatial inequality within a given territory so that future development initiatives can be properly planned and directed.

General discussions on the 'Kerala Model' are mostly based on aggregate average data whereas there are regional disparities in achievements of social and economic development. There are outlier communities lagging behind the central tendency and certain social groups seem to have been bypassed (Kurien, 1995; Kabir, 2010; Chakraborty *et al.*, 2010). To address this issue we stressed spatial variations in our deliberations on education and health.

EDUCATION

Kerala is the first state in India to achieve total literacy in 1991. While many other states in India are still striving to provide elementary education to all sections of society, control dropout rates, bridge the gender gap in literacy, bring down pupil-teacher ratio and train all teachers and para teachers under Sarba Siksha Avijan, Kerala has successfully tackled the first generation problems of eradication of illiteracy and near-total enrolment. Education in Kerala is inclusive and accessible to all sections of society. Over the years, Kerala's performance in the educational sector has been so impressive that it could receive the distinctive acclamation as the 'Kerala Model'; and some recommended 'Keralization' of the whole education system in India (Lewis, 1997; Tilak, 2001). This accomplishment of Kerala has a long history and the lessons are very important. We highlight some of the points here.

Spread of Modern Education

Educational facilities were quite widespread even during pre-British period (prior to 1793) (Nair, 1989), although access to these facilities was restricted as in those days educational opportunities were limited within the upper castes and directly linked with land ownership (Logan,1931). Brahmins and Nairs, upper castes in Kerala, had well developed educational institutions for their children, while a vast majority of people considered lower castes did not have any access to educational institutions. The scenario started changing, and at the beginning of the 19th century there were educational institutions functioning in many places including villages, however, most of these institutions were riddled with financial problems. The Travancore Kingdom responded positively and soon free and compulsory education in Travancore started under state control (Census of India, 1951). The village school scheme introduced in 1871 envisaged one school in every village, local people's participation in providing school building, and Government grants for teaching facilities commensurate to the number of pupils. However, children from backward communities used to find it extremely difficult to attend schools. Travancore Government took further initiatives to open schools exclusively for backward communities in 1895/96. Scholarships were instituted and teachers were also recruited from backward communities.

Christian missionaries played a pioneering role in introducing western education in Kerala. The First English school in Travancore was started in 1806 by a Prussian missionary. Primary schools were opened in all villages. The Government also encouraged missionaries to establish schools by providing government grants to their educational institutions (Tharakan, 1984). Several socio-religious groups and private agencies came forward to establish educational institutions over the years. Government intervention through educational reforms and programmes coincided with these initiatives. Various socio-political organisations and political agencies during pre-independence and post-independence period have contributed to the spread of literacy and school education, which received high priority in government policy. There were several innovative policies since the formation of Kerala, which had helped the expansion of school education. Some of the significant policy interventions in recent years cover "compulsory and free education, basic education, direct payment of salaries to private school teachers, reduction of regional and inter-community disparities in school education, promotion of female education, school-feeding programme, whole promotion policy and travel concessions to students" (Planning Commission of India, 2008a: 255).

Spatio-Temporal Growth of Literacy

According to the 1961 Census data, literacy rate of Kerala was 47 per cent, which grew to 94 per cent in 2011, while the corresponding figures for India were 28.3 per cent and 74.0 per cent respectively. The overall literacy in Kerala doubled in five decades. The highest growth of literacy rate was noted during the census decade of 1981 to 1991, when Kerala attained 20 per cent growth and was declared as the total literate state, first of its kind in India. Kerala's achievement inspired many other states in India and the total literacy campaign began in many parts of the country. The overall literacy rate remained more or less the same in Kerala for the census years of 1991 and 2001. It grew by 3 per cent between 2001 and 2011. Kottayam district ranked first among all districts in Kerala in the literacy rate till 2001. Pathanamthitta district with a literacy rate of 96.9 per cent overtook Kottayam in 2011. The literacy rates of Kasaragod, Wayanad, Malappuram, Palakkad, Idukki, and Thiruvananthapuram districts were below the state average in 1991, when Kerala achieved total literacy. All these districts continued to remain in the same category in 2011 also. However, Malappuram district moved upwards compared to Palakkad. The gap between the lowest and the highest-ranking districts narrowed down over the years, and by 2011, it stood at 8.5 per cent point (Table 12.1).

Table 12.1: Literacy, Gender Gap and Selected Educational Institutions by Districts in Kerala

Districts/State/CV	*Total Literacy (per cent)**	*Gender gap in literacy (per cent)**	*School density (Number/ $10km^2$)*	*Number of schools/lakh population*	*Number of higher secondary schools/lakh population*	*Number of colleges/lakh population*
Thiruvananthapuram	92.7	3.3	4.6	30	5.4	0.67
Kollam	93.8	3.9	3.8	36	5.2	0.57
Pathanamthitta	96.9	1.4	2.8	62	8.0	0.84
Alappuzha	96.3	3.1	5.5	36	5.9	0.61
Kottayam	96.4	1.5	4.2	47	6.8	1.17
Idukki	92.2	5.3	1.1	45	7.3	0.72
Ernakulam	95.7	2.9	3.3	31	6.4	0,76
Thrissur	95.3	3.1	3.4	33	6.5	0.71
Palakkad	88.5	7.3	2.3	36	5.4	0.53
Malappuram	93.6	4.2	4.4	38	6.0	0.51
Kozhikode	95.2	4.4	5.5	42	5.9	0.58
Wayanad	89.3	6.9	1.4	38	7.5	0.73
Kannur	95.4	4.0	4.4	52	6.3	0.52
Kasaragod	89.9	7.8	3.0	59	8.1	0.46
Kerala	93.5	4.0	3.3	39	6.2	0.65
Coefficient of variability (per cent)	3.0	48.6	37.6	24	14.7	26.9

Source: *Census of India, Kerala, 2011; Government of Kerala, 2018.

In general, districts in the Malabar part of Kerala were slow to pick up the momentum of the literacy movement. The literacy rate was below 90 per cent in Kasaragod, Wayanad, and Palakkad districts (Table 12.1). These three districts are dominated by SC, ST population. In 2011, literacy rates of SC and ST population were 88.7 per cent and 75.81 per cent respectively against all-India average figures of 66.07 per cent and 58.96 per cent in the same order.

Gender Gap in Literacy

The literacy rate among the female population is less than that of the male population throughout the country. It is a national quest to elevate the female literacy rate to the level of male literacy so that overall literacy progressed to the desired level of total literacy. The difference between the literacy rates of the male and female population is termed as the gender gap in literacy. It was only 4 per cent in Kerala as against 16.3 per cent for the Indian average in 2011. The male literacy in Kerala touched 96 per cent and female literacy reached 92 per cent by 2011. Palakkad district with 92.3 per cent of male literacy and 85 per cent of female literacy remained at the bottom among all the districts in Kerala in the matter of literacy. Alappuzha district topped among all districts with 97.9 per cent male literacy, and in the case of female literacy, the top position went to Pathanamthitta district with 96.3 per cent (Fig. 12.1). The gender gap narrowed down from 7.5 per cent point in 1991 to 4 per cent point in 2011. Scheduled Caste and Scheduled Tribe population recorded gender gaps of 7.6 per cent and 9.7 per cent respectively during 2011, whereas at all-India level these gaps were 19.7 per cent and 19.1 per cent in the same order. Kerala is successfully bridging the gender gap in literacy way ahead of all other states in India. The gender gap ranges from 1.4 per cent in Pathanamthitta district to 7.3 per cent in Palakkad district (Table 12.1). All districts made significant progress in male and female literacy, although there were four districts each having female literacy below 90 per cent. District level variation was the least for male literacy and relatively high for female literacy. Low literacy rates among Scheduled Caste and Scheduled Tribe population have impacted the overall literacy rates of the districts.

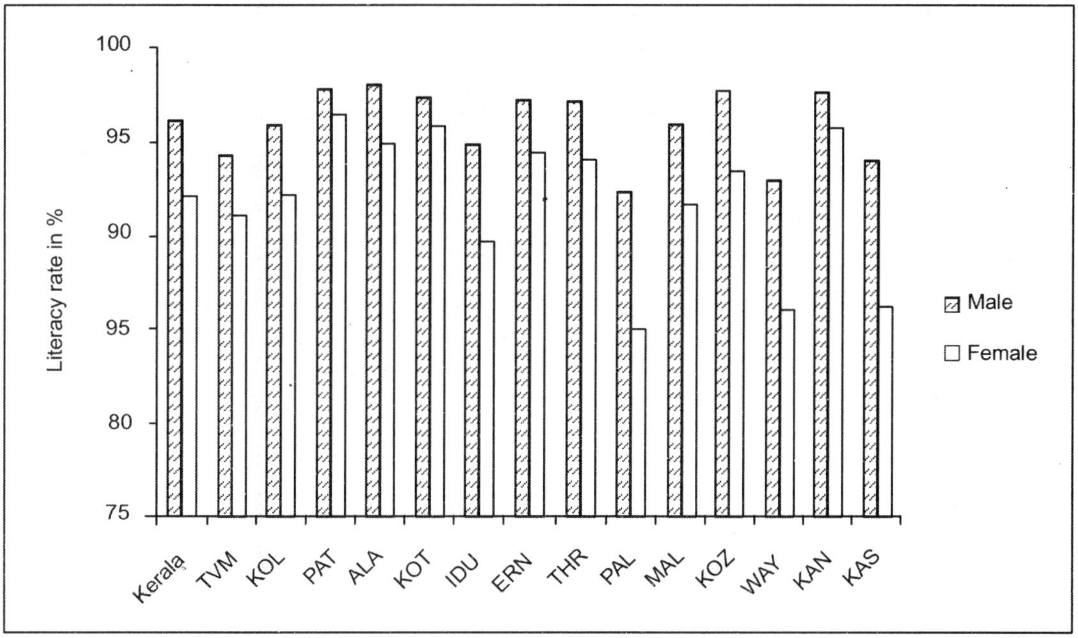

Fig. 12.1: Distribution of Male Female Literacy by Districts in Kerala, 2011.

Data Source: Census of India, Kerala, 2011.

SCHOOL EDUCATION

The school system in Kerala has been standardised into three stages: (i) Lower Primary stage from standard I to IV, (ii) Upper primary from standard V to standard VII, and (iii) Secondary stage from standard VIII to X. This 4+3+3 pattern differs from most other states, which follow the 5+3+2 pattern as recommended by the Kothari Commission (Planning Commission of

India, 2008a). There are three types of schools in Kerala, Government schools, Government aided private schools and Government unaided private schools.

Distribution of Schools

Kerala witnessed the steady increase in a number of schools. In 1947, there were only 3708 schools, of which 62 per cent schools were under government management and the rest 38 per cent were aided schools. By 2016-17, there were 12,981 schools offering state syllabus, and 1436 schools offered CBSE and ICSE syllabi and also included Kendriya Vidyalaya (KV) and Jawahar Navodaya Vidyalaya (JNV). Each district has one school under the JNV scheme. These JNV schools set the standards and nurture meritorious students. This is significant as school education is primarily administered by the government sector. Due to the equitable distribution of schools, all students in the state enjoy equal opportunities. Out of 12,981 schools following state syllabus, 6874 schools are in the category of lower primary, 2988 schools are upper primary and 3119 schools are high schools. Around 56 per cent of all these schools are government-aided private schools followed by 36 per cent government schools and the rest 8 per cent are private unaided schools. Malappuram district has the largest number of schools (1558), followed by Kannur (1308) and Kozhikode (1283) (Table 12.1). The opening of more number of schools in the Malabar areas perhaps manifests Kerala's attempt to universalise education and provide equal opportunity to schooling across the state.

The number of aided schools has increased over the years. Unaided private schools are also growing fast indicating the emergence of a market for private schools. School density (number of schools per 10 km^2) varies from 1.14 in Idukki to 5.47 in Kozhikode, the state average being 3.34. The number of schools per one lakh population varies from 30 in Thiruvananthapuram district to 62 in the Pathanamthitta district with the state average of 39 (Table 12.1). There are eight districts lying in the category of below state average. This group includes urbanised/industrialised district like Ernakulam. The position of districts alters slightly when schools offering other than state syllabuses are also considered. The distribution of schools in the districts indicates that there is hardly any difference between the core and periphery or developed and under-developed districts.

The number of government schools is more in the districts dominated by the rural population. Schools are well and uniformly distributed, thus they are accessible within short distances. One school serves an average area of 3 km^2. This area will substantially come down if only the settled areas are considered. Little over 94 per cent of the rural population is served by primary schools/sections within a distance of 1 km and 98 per cent of rural people is served within a distance zone of 2 km. Upper primary schools/sections are available within a distance of 3 km for 96.2 per cent of the rural population. Around 25 per cent of the rural population enjoy secondary education facilities within 2 km and for 97.8 per cent of the rural population secondary schools are available within 6-8 km (Government of Kerala, 2002; Government of Kerala, 2006c). Students enjoy easy movement as there are student concessions, and transport facilities are available adequately. Schooling facilities and associated infrastructure are more evenly and adequately distributed across geographical regions and among social groups in Kerala compared to many other states in India (Gasper, 2003). Provisioning of school facilities to the students in both rural and urban areas is perhaps the first step in universalising school education. Kerala has succeeded and the results are well evident.

Enrolment

School (I to X standard) enrolment of the total number of students showed a declining trend in Kerala since 1991-92. It was 32.7 lakh in 1961-62, which rose to the maximum of 59.07 lakh in 1991-92, and thereafter, the number of student enrolment steadily declined to 36.8 lakh in 2017-18. Present enrolment (2017-18) is around 12.3 per cent of the total population of >6 years of age in Kerala. Enrolment of school students per one lakh population of >6 years of age varies from 7,838 in Pathanamthitta district to 20,280 in Malappuram district with the state average of 12,296. School enrolment in all districts in the Malabar area is above the state average (Fig. 12.2). Available data indicate that, in 1960-61, the Malabar region accounted for 29 per cent of the total enrolment in Kerala. Share of Malabar region grew to 46.5 per cent in 1990-91 (Salim, 2002), and 55.2 per cent by 2017-18. The growth of enrolment in the Malabar region is due to various steps initiated by the respective governments in Kerala, since the formation of the state and total literacy movement. The present growth in enrolment is in tune with the population growth rate of the districts. Low enrolment in schools is directly related to the low birth rate.

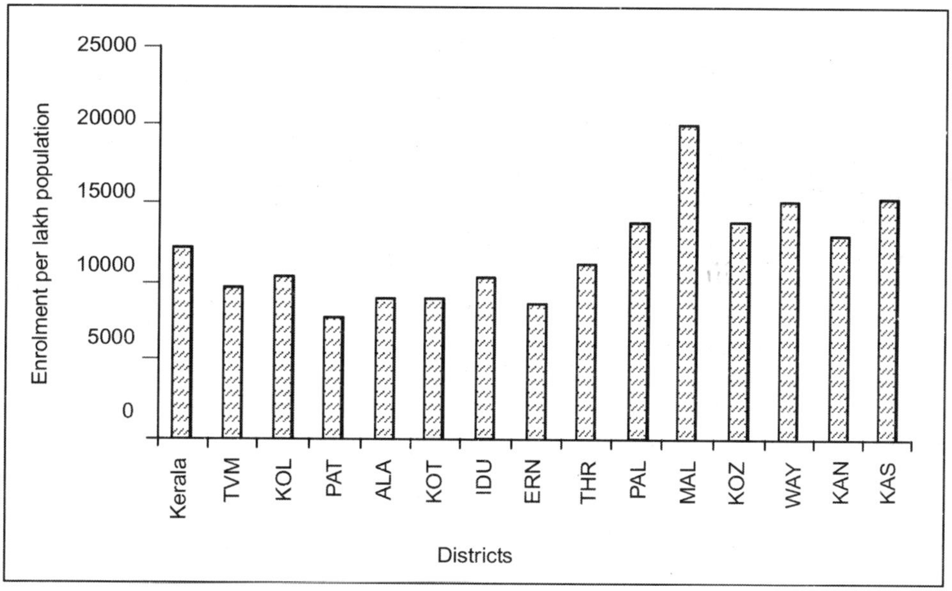

Fig. 12.2: School Student Enrolment by Districts in Kerala, 2017-18.

Data Source: Governmnet of Kerala, 2019b.

Since 1991-92, absolute enrolment in government schools and aided schools came down by 51 per cent and 38 per cent respectively, but it increased by 189 per cent in private schools. Increasing enrolment in private schools at the same time declining enrolments in government and aided schools are challenging issues and deserve policy intervention. There are several schools in the government and aided sector rendered as uneconomic due to lack of the required number of students. By 2017-18, the number of uneconomic schools rose to 5723, of which 55 per cent were under the aided category and the rest 45 per cent were government schools. Around 76 per cent of uneconomic schools were in the category of the lower primary. Population control and economic prosperity are the major causes underlying this change. A larger number of families opting for a single child and having economic means now prefer private schools. Education is being commercialised and has turned into a big business. The post globalisation era witnessed the fast growth of the private sector in the education system. There are efforts to improvc the conditions of government schools and attract students. This is expected to yield some results as evidenced by higher enrolment in government school during 2018-19.

Gender Gap in Enrolment

Kerala has successfully narrowed down the gender gap in enrolment. Girls accounted for 49 per cent of total school enrolment in 2017. The overall ratio between girls and boys stood at 96:100. In lower primary school (Classes I-IV), there were 98 girls enrolled for every 100 boys, in middle school (Classes V-VII), it was 97 and in high school (Classes VIII-X), the number was 94. All-India average showed, that, there were 93 girls for 100 boys in primary school (Classes I to V), 95 girls for 100 boys in middle school (Classes VI to VIII), and 90 girls for 100 boys in secondary school (Classes IX and X) during 2014-15. The average gender gap in Kerala is 2 per cent point. It varies from 1 per cent point in Ernakulam and Thrissur districts to 4.6 per cent point in Idukki. Kasaragod, Malappuram and Pathanamthitta are the other three districts with a relatively high gender gap.

Enrolment of SC and ST Students

Traditionally, Kerala emphasised spreading education to backward communities. The rulers of the Travancore kingdom opened exclusive schools for the backward community way back in the 19th century. The education system in Kerala is inclusive and deeply penetrative. Enrolments of SC and ST students in Kerala are also high. In 2017-18, SC and ST population accounted for 9.1 per cent and 1.45 per cent of the total population in the state, and their share of students stood at 10.6 per cent and 2.07 per cent respectively. It is evident from these figures that the spread of education among the SC and ST population is commensurate to that of the general category for the state as a whole. However, there are location-specific variations at the district level. The majority of Scheduled Caste and Scheduled Tribe students study in government schools. The concentration of government schools is relatively high in the districts dominated by SC and ST population like that of Wayanad district, which has >22 per cent of the population under SC and ST category and records 57 per cent of the total schools in the district under government sector.

School Enrolment and Socio-Economic Group

Kerala has succeeded to achieve educational equality among various socio-economic groups and between boys and girls. Analysing NCEAR data it has been noted that enrolment of children in the age group of 6-11 years is almost equal between the highest and lowest income groups in case of Kerala and Himachal Pradesh, whereas in all other states proportion of enrolment increases with increase in income (Tilak, 2001; Government of Kerala, 2005).

Dropout Rate

Dropout is a major problem in education management in India. Kerala registered the lowest dropout rate of 0.22 per cent among all the states in India during 2017-18. Scheduled Tribe students recorded the highest dropout rate of 2.27 per cent and in the case of Scheduled Caste students, the dropout rate was 0.26 per cent. Considering dropout rates across all three stages of schooling, it is found that the highest dropout rates prevail among the high school students, followed by lower primary students. Dropout rates are low among upper primary students. District-wise distribution of dropout rates indicated that Wayanad district had the highest dropout rate of 2.8 per cent in high schools, whereas in the lower primary level, Idukki district recorded the highest dropout rate of 0.55 per cent. Concentrations of Scheduled Tribe population are high in both these districts. Thiruvananthapuram district has the lowest dropout rates among high school students and Kollam recorded the lowest dropout rates among lower primary students.

Pupil Teacher Ratio

Kerala enjoyed a comfortable position in the Pupil-Teacher Ratio (PTR) since the formation of the state in 1956. In 1960-61, Kerala had an average of 3.4 teachers for 100 students against the all-India average of 2.96. The average number of teachers in Kerala rose to 3.6 for 100 students in 2000-01; however, for all-India, it continued to be less than 3. By 2017-18, there were a total of 162,608 teachers in LP, UP, and High schools teaching 36.8 lakh students in Kerala (Government of Kerala, 2019b). The number of teachers for every 100 students rose to 4.4. Declining enrolment as a consequence of demographic transition contributed to this scenario. The average PTR in Kerala is 23:1. It is 33:1 in LP school, 27:1 in UP school, and 15:1 in High school. At all-India level, PTR is 24:1 in elementary schools and 27:1 in secondary schools. The Right of Children to Free and Compulsory Education (RTE) Act, 2009 envisages that the PTR at LP level should be 30:1 and that at UP level should be 35:1. The Rashtriya Madhyamik Shiksha Abhiyan (RMSA) framework stipulated PTR at the secondary level at 30:1. Kerala is well placed in all these sectors. Women teachers constitute around 73 per cent of total teachers in Kerala. They account for 76 per cent of teachers in LP section, 71 per cent in UP section, and 72 per cent in High School against all-India average of 47 per cent, 45 per cent and 43 per cent respectively. The proportion of women teachers is the highest in private schools in Kerala with an 83 per cent share. Women teachers constitute 69 per cent and 73 per cent of total teachers in government schools and aided schools respectively.

Higher Secondary Education

Higher secondary schools covering 11th and 12th standard courses were introduced in Kerala in 1990-91 in accordance with the National Education Policy (NEP). By 2017, the number of higher secondary schools in Kerala was 2073, of which 854 schools were under the category of aided schools, 833 were government schools and 386 were unaided and technical schools. There is an average of 6.21 higher secondary schools per one lakh population in the state (Table 12.1). The availability of higher secondary schools is relatively high in Kasaragod, Wayanad, Idukki and Pathanamthitta districts. The total number of batches offered during 2017 was 7245 with a total enrolment of 382,051 students. Malappuram district has the largest number of higher secondary schools of 248 accommodating 14.5 per cent of batches and 14.9 per cent of enrolled students. The total number of vocational higher secondary schools in Kerala is 389, of which 261 schools are Government schools and the rest 128 schools are aided schools. Kollam and Thiruvananthapuram districts together have 93 schools or 24 per cent of all vocational higher secondary schools in the state.

HIGHER EDUCATION

The Europeans, particularly the Protestant missionaries settling in various parts of Travancore and Cochin states in the early 19th century laid the foundation for modern higher education in Kerala. The first college within the geographical boundary of modern Kerala was opened by Syrian Catholics in Kottayam by 1816. In 1834, Her Highness Swathi Thirunal, the then Maharani of Travancore, established an English school in Thiruvananthapuram, which was later upgraded as a college, the Maharaja's college, and subsequently, it was rechristened as University College of Thiruvananthapuram (UCT) in 1866. The UCT completed the 150th anniversary in the year 2016. The Government Victoria College in Palakkad started in 1883; the Zamorin's College in Kozhikode opened by 1888, and the Tellecherry Brennen College was established in 1890. Initially, all these colleges were set up as schools to impart English education and they were upgraded in subsequent years. Thus, college education began throughout Kerala by the end of the 19th Century. The Travancore University, which was subsequently renamed as Kerala University, was established in 1937 as the 16th University in India.

At present, there are one Central University and 13 state Universities, of which four Universities are of general nature, and the rest nine offer specialised and professional courses. Besides, four institutes of national importance, namely, Indian Institute of Science Education and Research (IISER), Thiruvananthapuram; Indian Institute of Space Technology (IIST), Thiruvananthapuram; Indian Institute of Management (IIM), Kozhikode, and Indian Institute of Technology (IIT), Palakkad are functioning in the state for conducting teaching and research in specialised areas. These Institutes have provided new impetus to higher education in Kerala. Indian Institute of Information Technology-Kerala (IIIT-K) is contributing to specialised education system particularly, in the Geomatics sector. According to the national survey by the Ministry of Human Resources Development (MHRD) (2013), there are 654 colleges under government and private (aided and unaided) sectors covering all streams. This number swells up to 1073 when all types of colleges including PG Centres, Off-Campus Centres and recognised Centres are considered together. Colleges in Kerala are of three types: Government, Aided Private, and Unaided Private. Out of 271 Arts and Science colleges under government and aided sector, 61 are government colleges and the rest 156 are aided colleges (SPB, 2018). Total student enrolment in BA, B.Sc., and B.Com, and MA, M. Sc and M.Com. courses were 2.96 lakh in 2017. Girl students constituted 68.7 per cent of all students.

Distribution of Colleges

Considering the district-wise distribution of Arts and Science Colleges under government and aided sectors, it is found that Ernakulam district has the highest number of colleges of 25 followed by Kottayam (23 colleges). Wayanad and Kasaragod districts have only 6 colleges each. One Government or Aided College caters to an average area of 143 km^2 and 1.23 lakh population in Kerala. In other words, for every one lakh population, there are 0.65 colleges in Kerala (Table 12.1). The number of colleges per one lakh population varies from 0.46 in Kasaragod to 1.17 in Kottayam. Both Wayanad and Idukki districts lying within the Western Ghats figure among the seven districts which are placed in the category of the above state average. This indicates that regional differences are relatively low in the matter of college education facilities in Kerala.

Higher education facilities in Kerala are comparable to all-India averages. In fact, except for the gender parity index, Kerala's condition is less favourable than that are prevailing in both Karnataka and Tamil Nadu. The gross enrolment ratio in Kerala is above the national average but below that reported from the neighbouring two states.

Table 12.2: Status of Higher Education in Kerala, Tamil Nadu, Karnataka and India

Sl. No.	*Parameters*	*Kerala*	*Tamil Nadu*	*Karnataka*	*India*
1.	People in the age group of 18-23 years (in per cent)	2.26	5.53	5.3	100.0
2.	No. of colleges/lakh population (18-23 years)	30.0	30.0	44.0	25.0
3.	Average enrolment per college	559	774	289	707
4.	Ratio between government and private college	1: 3.2	1:6.8	1:3.7	1:2.7
5.	Enrolment in Government colleges (per cent)	22.0	17.0	29.0	39.0
6.	Colleges offering general courses (per cent)	55.0	58.0	65.0	66.0
7.	Enrolment in higher education in per cent to total population in the age group (Gross enrolment ratio)	23.1	38.2	24.0	20.4
8.	Students enrolled in Under Graduate courses (per cent)	75.0	67.9	71.3	79.0
9.	Students enrolled in Post Graduate courses (per cent)	13.5	15.7	12.4	11.8
10.	Students enrolled for M.Phil. and Ph.D. (per cent)	0.57	0.85	0.50	0.41
11.	Gender parity Index	1.39	0.86	0.90	0.88

Source: Government of India, 2013a.

Data for the years of 1972-73 and 1986-87 indicated that the enrolment ratio in general higher education came down in Kerala like that in few other states. This is a matter of concern specially for Kerala as the state has achieved the enrolment ratio near 100 per cent in primary and upper primary education and the enrolment ratio at the secondary level is reasonably high (Tilak, 2001). The general presumption that higher education does not contribute to economic growth and therefore the thrust of the policymakers is more on primary and secondary education compared to higher education. This has perhaps impacted education policy even in Kerala. The problem of higher education in Kerala is linked to larger socio-political issues. The competing interest of different groups renders it difficult for governments to shape and enforce the regulation, hence most of the attempts focus on problems selectively. As a result, 'the casualty in the process is a policy environment that would provide incentives to deliver quality higher education while also ensuring regulations that prevent profiteering or the use of education to dispense patronage'(Kodath, undated).

CHALLENGES IN EDUCATION SECTOR

Kerala's achievement in the educational sector is commendable. She has met all the national targets, and in the matter of millennium development goals, its achievement is well ahead of time. Many other states in India are striving to emulate Kerala's success. However, there are several challenges that are the causes of major concern. While the coefficient of variation is the lowest in the case of total literacy (3.0 %), it is quite high (48.6 %) in the case of the gender gap in literacy (refer Table 12.1). Special attention is necessary to improve female literacy particularly in the districts of Kasaragod and Palakkad, which noted a wide-ranging gender gap in literacy. Improvement of school quality, particularly government and aided schools covering teaching, classroom, libraries, laboratories, playgrounds etc. is a big challenge and necessary provisions may be made (Chandrasekhar *et al.*, 2001).

Many of the schools are uneconomic due to low student strength as a consequence of low birth rate, besides there is a strong pull from private institutions. Inclusiveness, socio-economically non-discriminatory, and gender equality, strong points of Kerala's education system, are now facing serious challenges particularly after globalisation/economic liberalisation introduced during the 1990s. The gap between the poor and the rich is increasing. This impacts the education system, which is potent to seriously impair the very foundation of Kerala mode of development. Private schools are increasing and pre-primary playschools are emerging in a big way. The problems include increasing cost of education borne by the students, growth of self (student) financed institutions, strengthening of non-financial entry barriers, and inadequate attention to the problems of the disadvantaged groups (Ajitkumar and George, 2009). Growth of disposable income, increase in single-child family, the willingness of parents to provide the best education to their wards, decline in quality of education in Government schools are some of the reasons underlying the growth of the private sector. Commercialisation, social segregation, and class differentiation in education may lead to increasing spatial disparity as quality education may be more costly and inaccessible to the economically weaker sections. An inclusive society is now turning to a segregated society which in long term may affect people's participation in public action/civil society movement, one of the unique features of Kerala.

HEALTH

Kerala is a high-performing state in India in the matter of human health as reflected through life expectancy at birth, the infant mortality rate (IMR), the maternal mortality rate (MMR), birth rate, immunisation, institutional delivery and similar parameters. It has achieved most of the Millennium Development Goals (MDG) set for this sector. Kerala not only outshines other states in India but also compares favourably with that of the developed countries in the matter of several health indicators.

How does Kerala achieve this status is a matter of detailed scientific deliberations, which has been attempted by various scholars (Krishnan, 1985; Jeffrey 1992; Kannan *et al.*, 1991; Ramankutty and Panikar, 1995 among others). The development of health infrastructure facilities depends on socio-economic, cultural and political background. Kerala has a long history of well-developed traditional health care systems well before the European landed in India. One of the significant factors of improved health condition in Kerala is awareness among people to avail health services, availability of health care facilities, and their better utilisation. Over the years, a system developed among people to approach caregivers during sickness rather than to resort to self-medication (Ramankutty, 2000). There were significant state government initiatives to improve health services for the people in Kerala since its formation in 1956. Even before, there were noteworthy initiatives that had created a congenial environment to improve health services. Medicinal plants in Kerala are well known as part of indigenous medicine. The consciousness of people about health care was also evident in their food habits. A well-entrenched tradition evolved to drink boiled water mixed with medicinal herbs. This habit continues still today. People even boil tap water supplied by the Kerala Water Authority for daily consumption. The discussion here will cover some of the health indicators and health care system of Kerala.

Infant Mortality Rate (IMR)

Kerala recorded one of the lowest mortality rates among the Indian states in the 1950s itself (Planning Commission of India, 2008a). In several health indicators, Kerala moved further ahead and made considerable strides in improving its status in the next few decades. One of the main factors contributing to Kerala's success is that Kerala has succeeded to reduce the mortality rate among the very young population, particularly among the < 1-year group. The high survival rate of newborn babies is a unique achievement of Kerala. In 1971 itself, IMR in Kerala was significantly lower than the all-India average and the trend continues to date. By 2015-16, IMR in Kerala was 6 against an all-India average of 41 and rural Kerala recorded IMR of 5, whereas, for rural India, it was 46 (Government of India, 2017b). Time series analysis of data indicates that the fall in IMR can be noted in three distinct phases: 1956-1966; 1967-1976, and post-1976 (Planning Commission of India, 2008a). The first phase is most probably related to vaccination programmes against smallpox and other infectious diseases, and also the expansion of health care facilities to the traditionally backward northern districts (Krishnan, 1985).

The immunisation programme was launched in Kerala in the late 1970s, well before the launching of WHO sponsored universal immunisation programme in 1985. Expansion of immunisation programme through length and breadth of Kerala played important role in bringing down the infant mortality rate. Availability of medical attention during pregnancy and child birth has contributed significantly and this is evident from the fact that institutional delivery even in rural areas of Kerala had increased from 26 per cent in 1973 to 99 per cent in 2013-14. Higher institutional delivery is one of the factors to bring down neonatal mortality rate (NNMR) and infant mortality rate. Kerala and Goa, two high performing states with the lowest IMR reported institutional delivery at the rate of 99.9 per cent and 100 per cent respectively. Medical attention during pregnancy, overall health condition of expecting mother, and immunisation are important contributors in bringing down the IMR. Female literacy also contributes to declining IMR with a correlation coefficient of 0.51 and a level of significance at 95 per cent (Fig. 12.3).

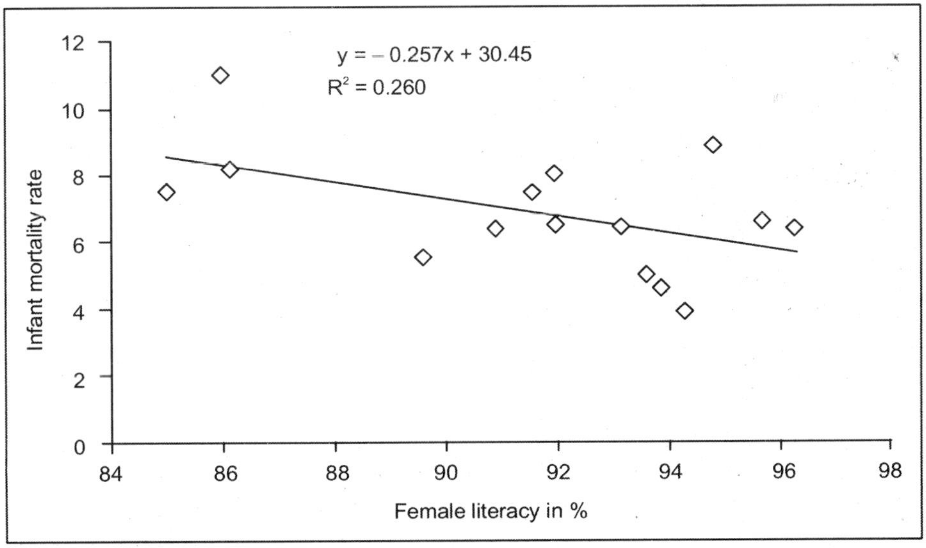

Fig. 12.3: Relation Between Infant Mortality Rate and Female Literacy in Kerala, 2013-14.

At district level, IMR varies from 3.9 in Ernakulam to 11 in Wayanad district. Wayanad is the only distrct with doule digit IMR. Districts with high level of urbanisation record low IMR. Spatial variability is relatively high for IMR with coefficient of variability of 27.2 per cent (Table 12.3). Distribution of Crude Birth Rate (CBR) shows a spatial trend. All districts in Malabar area except Palakkad recorded higher CBR compared to state average (Table 12.3).

Table 12.3: Status of Selected Health Indicators and Availability of Selected Health Care Facilities under Allopathic System of Medicine by Districts in Kerala

Districts/State/ CV	*IMR, 2013-2014*	*CBR*	*Life exp. at birth*	*Density of Govt. allo. med. Inst (No./10Km^2)*	*Govt. medical inst./lakh population*	*Bed in Govt. medical inst/lakh pop*	*Pvt. medical inst/lakh pop*	*Bed in pvt. medical inst./lakh pop*
Tvm	6.4	13.4	75.2	0.52	3.5	148	14.8	177
Kol	8.0	10.6	77.1	0.35	3.3	91	14.5	164
Pat	6.4	13.5	76.7	0.25	5.6	163	21.3	308
Ala	8.9	8.1	77.1	0.64	4.2	161	18.2	111
Kot	6.6	11.9	75.6	0.38	4.2	143	22.5	249
Idu	5.5	11.0	72.4	0.15	5.9	99	21.1	440
Ern	3.9	12.9	75.9	0.38	3.5	138	25.5	282
Thr	4.6	14.5	76.4	0.39	3.7	113	10.6	222
Pal	7.5	12.8	76.1	0.25	4.0	93	10.0	84
Mal	7.5	20.8	75.6	0.35	3.0	61	9.2	83
Koz	6.4	17.5	75.4	0.39	3.0	91	12.3	101
Way	11.0	17.5	73.5	0.19	5.1	167	13.6	183
Kan	5.0	16.8	75.6	0.36	4.3	119	12.5	198
Kas	8.2	15.1	75.7	0.31	4.7	83	15.3	120
Kerala	6.5	14.5	75.6	0.33	3.8	114	15.2	179
C V (per cent)	27.2	23.7	1.64	37.13	21.7	28.7	32.3	51.9

Source: Government of Kerala, 2016, 2018; Government of India, 2017b.

Life Expectancy

Kerala recorded the highest life expectancy at birth in India. Demographic data for the period 2011-15 shows that the average longevity of people in Kerala is 75.2 years against the national average of 68.3. The life expectancy of the urban population in Kerala (75.4 years) is slightly more than that of the rural population (75.1 years). The national average stands at 67.1 and 71.9 years for rural and urban populations respectively. Considering rural-urban and male-female components of the population it is found that the female population in rural Kerala has the highest life expectancy of 78.4 years, around 10 years more than the national average of 68.7 years. District-wise life expectancy varies from 72.4 years in the Idukki district to 77.1 years in Alappuzha and Kollam districts. The lowest life expectancy in Kerala (72.4 years) is around 4 years above the all-India average life expectancy. All districts are performing well in this sector with a coefficient of variability of only 1.64 per cent.

Morbidity

Kerala's achievement in the infant mortality rate is commendable. However, Kerala is also noted for a high morbidity rate (perceived ill health), the highest in India, and over the years it is increasing. According to NSSO 60th round data, the rate of morbidity in rural Kerala was

25.5 per cent and in urban Kerala it was 24 per cent against the national average of 8.8 per cent and 9.9 per cent respectively. The NSSO 71st round survey indicated that morbidity in Kerala had increased to 31 per cent in the rural areas and 30.6 per cent in the urban areas (Government of India, 2017b). The national average stood at 8.9 per cent and 11.8 per cent respectively. Comparing these data with that available for the year of 1995-96, it is observed that in the case of Kerala, morbidity increased by 163 per cent in the rural area and 248 per cent in the urban area during the past two decades and at all-India level, morbidity level had gone up by 62 per cent and 119 per cent in rural and urban areas respectively. In the case of Kerala, rural the urban difference in morbidity has reduced from -25 per cent to -1.3 per cent but at the national level, rural urban difference in morbidity level has increased from -1.8 per cent to +32.6 per cent. This vindicates the fact that the difference between rural and urban areas in Kerala is marginal in sharp contrast to all-India situations, and this trend has been noted for other human development parameters also. Similarly, hospitalisation record for one year prior to the survey date showed that, in Kerala, out of 1000 persons 117 persons in the rural areas and 99 persons in the urban areas were admitted in hospitals against all-India averages of 44 and 49 for rural and urban areas respectively.

High morbidity in Kerala drew considerable interest among researchers since the 1980s (Panikar and Soman, 1984; Kannan *et al.*, 1991). The co-existence of low IMR and high morbidity is an important issue widely deliberated. There are several contributory factors, which vary spatially and temporally as disease pattern is changing over the years. The nature of diseases underpinning morbidity varies with socio-economic groups, environmental hygiene, level of literacy, and remoteness of place of residence like that in rural areas or urban areas or urban slums (Gangadharan, 2003). Income is not the only factor for morbidity. Chronic illness is reported more among upper economic groups. However, some chronic illness like asthma is more frequent among poorer section and that of hypertension contributes to morbidity of affluence class (*ibid*). It is observed that the relative share of rich people in total morbidity is more at all-India levels compared to Kerala (Suryanarayan, 2008). While in all-India level morbidity is relatively more among the affluent people, in Kerala this distinction between affluent and non-affluent people in the matter of morbidity is gradually melting down. Deprivation related morbidity is declining in Kerala, however, lifestyle diseases or diseases of affluence like heart disease, diabetes, and hypertension are reported across socio-economic groups in rural and urban areas. There seems to be a process of convergence of the diseases of affluence and deprivation in Kerala (*ibid*). Health consciousness, general awareness, total literacy, easy availability of health care facilities and the urge to avail medical care, all together have also contributed to increasing morbidity of Kerala.

Change in Disease Pattern

Kerala has controlled a number of communicable diseases like malaria and tuberculosis, which were widely prevalent during the 1950s. Recent years witnessed the re-emergence of some of the apparently eradicated diseases like malaria and many new diseases. Dengue, chikungunya, leptospirosis, hepatitis, H_1N_1, and Japanese encephalitis are some of the new diseases that appeared in Kerala, which has led to considerable morbidity and mortality (Government of Kerala, 2018). Dengue fever was unheard of in Kerala till 2003, however, it is now being reported each year from all districts of the state (Government of Kerala, 2014b). Water-borne diseases like varieties of diarrhoeal diseases, typhoid, and hepatitis are reported from different districts. Cholera has also surfaced in many districts. Diphtheria and whooping cough continue to persist in spite of all efforts. In May 2018, Kerala witnessed an outbreak of the Nipah virus, which created considerable tension across the state. The Nipah was deftly controlled, and once again Kerala demonstrated its capacity to consolidate all sections of society to address an emergency situations. Nevertheless, occurrence of such disease from time to time is a big challenge for Kerala. The rise of communicable diseases is partly attributable to deteriorating environmental conditions (Thresia and Mohindra, 2011). Large parts of Kerala, particularly coastal plain and lowlands are now susceptible to communicable disease due to deterioration of water quality of lakes and rivers caused by urban-industrial discharge and agricultural run-off coupled with lack of proper waste management facilities. An increase in cancer cases is also linked to deteriorated environmental conditions.

Non-communicable diseases (NCD) like diabetes, hypertension, cardio-vascular diseases, cancer, and lung diseases, which are life-threatening, are contributing to a high level of morbidity. It caused 52 per cent of total deaths between the productive age group of 30 to 59 in Kerala, and in the case of all-India, NCD related deaths accounted for 42 per cent of total deaths (Government of Kerala, 2018). Lifestyle diseases are increasing due to the change of lifestyle and nature of work, correspondingly, the burden of medical expenses are growing. Kerala also faces problems in mental health. The Census of India, 2011 brought out that 0.20 per cent of Kerala's population suffered from mental illness and another 0.20 per cent suffered from mental retardation compared to all-India average of 0.06 per cent and 0.12 per cent respectively (Government of Kerala, 2018).

SPREAD OF MODERN HEALTH CARE SYSTEM

The introduction of modern medicine or the western health care system coincided with the establishment of colonial power in India. The rulers of Travancore and Cochin, who were forward-looking and introduced several measures for spreading education including girl's education and education for backward communities, took initiative to make the western health care system available to their subjects in the 19th century (Ramankutty, 2000). In 1879, there was a royal proclamation for compulsory vaccination for public servants, prisoners, and students (Panikar and Soman, 1984). Travancore princely state undertook various measures to develop proper health infrastructure, provision for safe drinking water, and even appoint foreign-trained medical professionals in key positions. The appointment of Dr Mary Punnen Lukose, a lady doctor trained in England as the surgeon-general of Travancore in the early years of the 20th century is a very significant step. She was the first woman to be appointed surgeon-general in an Indian state, at a time when women doctors were still a rarity in Europe and America (Ramankutty, 2000). Besides, the establishment of hospitals by Christian missionaries in remote areas and encouraging young Christian girls to take up nursing as career opportunities had a profound impact on Kerala society. Over the years a wide network of health infrastructures—hospitals, dispensaries, etc. was developed to cater to the needs of people in Kerala both in the urban and rural areas. It is noteworthy that all medical facilities were available to all people irrespective of caste and community since 1914. By 1938, separate maternity care centres started, till then both mother and child care were attended by hospitals and dispensaries. There were 140 government medical institutions composed of 32 hospitals and 108 dispensaries, and 20 private medical institutions receiving grant-in-aids of the state at the time of Independence.

Development of Health Sector since Formation of the State

At the time of the formation of Kerala, there was a well-developed health infrastructure. Health facilities grew rapidly under the public sector manifesting the government's commitment to provision health services for people. There are two distinct phases of health care growth. The first three decades since the formation of Kerala was a period of high growth rate. The number of allopathic medical institutions grew from 490 in 1966 to 1014 in 1985 (Fig. 12.4).

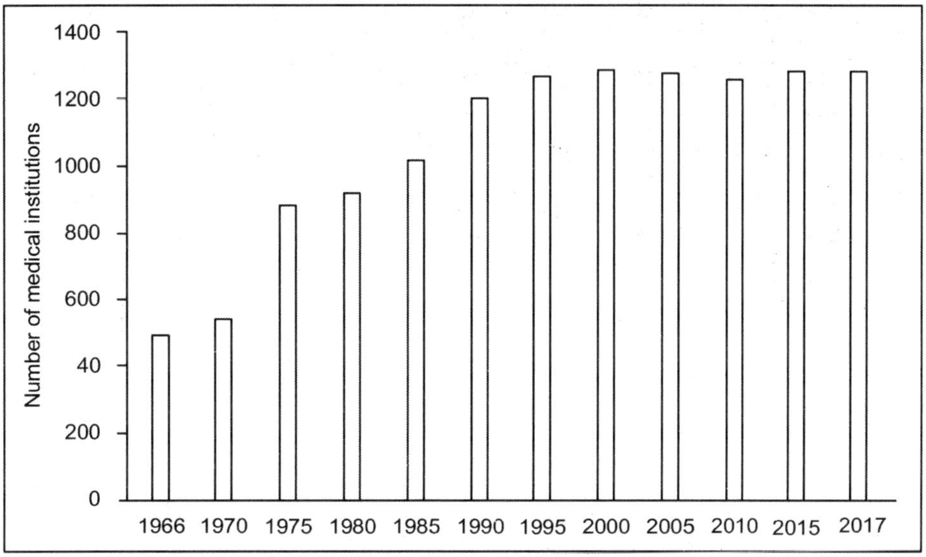

Fig. 12.4: Growth of Allopathic Medical Institutions under Government Sector in Kerala (1966-2017).
Data Source: Government of Kerala: Economic Review—Various Issues.

On average, 28 institutes were added every year during this period of 20 years. Correspondingly, the number of beds also increased from 18,526 to 33,329 during the same period with an average annual addition of 779 beds. The availability of beds per one lakh population increased from 98 in 1966 to 124 in 1985. Thereafter, the growth rate slowed down considerably, and by 2017, there were 1280 Allopathic health institutions with a total number of beds of 38,004. Only 266 institutes and 4675 beds were added since 1985. However, the private sector continues to grow.

PRIVATE HEALTH CARE FACILITIES

The NSSO 71st round survey brought out that around 66 per cent households irrespective of location in rural or urban areas depend on the private sector for medical care in Kerala. There is a steady increase in the number of people seeking health care from private institutions (Dileep, 2009). Available data for private health services covering three-time points, 1986, 1996, and 2004 indicated that by 1986, there were 9663 health care institutions with 49,000 beds, which grew to 12,918 institutions and 64,491 beds by 2004 (Government of Kerala, 2006). The doctor bed ratio was 1: 2.6. in the private sector, whereas in the government sector it was 1: 6.95.

Around 45 lakh inpatients received treatment in private institutions, which was around 2.4 times high than those received treatment in government medical institutions. This, not only, indicates the importance of the private sector, but also, people's growing faith in private sector services. Reasons underlying people's preference for private health care have been investigated from time to time (CPPR, 2017). The fiscal crisis of Government, the huge burden of non-plan expenditure, fall in the quality of services in government hospitals even human behaviour at government institutions are some of the underlying factors governing people's preference for the private sector. Private hospitals have far outpaced government facilities in several cases like provisioning of hi-tech methods of diagnosis and therapy, such as computerized tomography (CT) scans, endoscopy units, magnetic resonance imaging (MRI), neonatal care units, coronary units, etc. (Ramankutty, 2000). A survey in 1987 indicated that only 23 per cent of 10,000 surveyed households used government institutions and this figure varied from 8 per cent among most affluent households to 33 per cent among the poorest stratum (*ibid*). Dependency on the private sector for health services is not limited to the affluent class alone. Even the lower middle class looks for private institutions for avail treatment.

Nevertheless, Kerala stands way ahead of the national average in the matter of public health care facilities. All-India data of 2015 indicated that the average number of people served per government (public) allopathic hospital (health institution) was 27,588 in Kerala against an all-India average of 61,011 persons (Government of India, 2017b). Kerala enjoys strong medical services under ayurvedic, homeopathic, and other traditional medical systems. If all systems are considered the number of available medical facilities will be much higher. However, we would like to limit our discussion to the allopathic system of medicine only.

SPATIAL TREND OF HEALTH CARE FACILITIES

The nature of the settlement system and distribution of population, growth of physical infrastructure, and absence of primate city altogether influenced decentralised distribution of health care facilities in Kerala. There is a conscious decision at the policymakers' level to bridge the rural urban gap in developing health care infrastructure. The NSSO 70th round survey brought out that 94.9 per cent of villages of Kerala has a PHC within 10 km and 99.8 per cent of villages has sub-centres within 3 km whereas the national averages are 71.2 per cent and 71.4 per cent for 10 km and 3 km distance zones respectively. The rural area in Kerala accounts for 89 per cent of institutions and 47 per cent of beds. At all-India level, there are 83 per cent institutions located in rural areas, however, the share of beds in rural institutions is only 27 per cent. Considering these figures in the backdrop of rural-urban population distribution in the ratio of 69: 31 for all-India and 52:48 for Kerala it may be noted that availability of health care services in Kerala is more equitable and rural areas in Kerala are better served compared to rural areas in the rest of the country. The correlation between the rural population and the availability of institutions is high positive and significant (Fig. 12.5).

Distribution of Allopathic Medical Institutions by Districts

The 1280 government allopathic medical institutions in Kerala are composed of 18 general hospitals, 18 district hospitals, 81 taluk hospitals, 680 Primary Health Centres (PHCs), 168 24×7 PHCs, 232 Community Health Centres (CHCs) 18 specialty hospitals, 18 TB Clinics/centres and 47 others (Government of Kerala, 2018). All these institutions are distributed

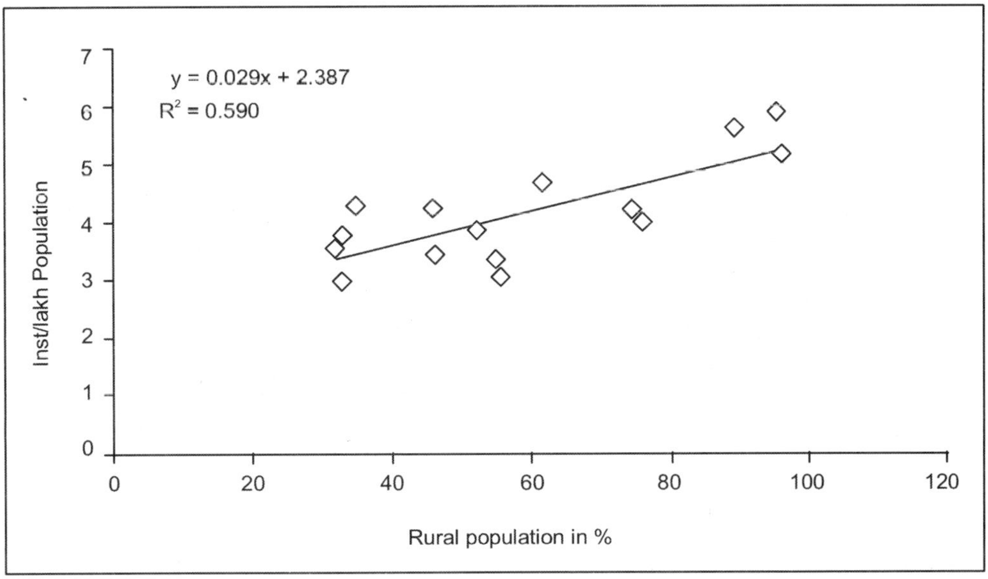

Fig. 12.5: Relation Between Distribution of Rural Population and Number of Government Allopathic Institutions in Kerala, 2017.

throughout the state. Six districts have more than 100 medical institutes each. One medical institution in Kerala serves an average area of 30 km^2. It varies from 16 km^2 in the Alappuzha district to 67 km^2 in the Idukki district (Table 12.3). The density of institutes (number of institutes/ 10 km^2 area) varies from 0.15 in the Idukki district to 0.64 in the Alappuzha district with the state average at 0.33. The scenario changes when the number of institutions per one lakh population is considered. Idukki district tops the list with 5.86 institutions/lakh population and in terms of availability of beds Wayanad district ranks first with 167 beds per one lakh population or bed and population ratio of 1:599. Both these districts are located within the Western Ghats and are dominated by the rural population. The distribution of medical institutions is closely related to the distribution of population density. Private health care facilities have developed parallel to government facilities and boost up overall availability of medical services in the districts. Distribution of private medical institutions is also found related to income variables and literacy. This signifies that income and education determine an individual's preference for private health facilities. Alappuzha, Kottayam, Ernakulam, Kozhikode, and Thiruvananthapuram districts enjoy a high concentration of private medical institutions. All these districts are highly urbanised, record high population density and house medical colleges and other teaching institutes, thereby provide a congenial environment for the proliferation of private medical centres, which prefer to develop in and around market places.

CHALLENGES IN HEALTH SECTOR

Kerala is well known for health care. People from other states and even from abroad come to Kerala for health care services, which is considered affordable and of good quality. Nevertheless, there are new challenges with issues like fiscal constraints, quality of services, aging infrastructure, non-availability of services, underutilisation of available services in Government health institutions, and apathy of service providers. High morbidity, the prevalence of new diseases, and an increase in old age people are another set of issues. However, the most challenging concern is the increasing privatisation of health services. Although government policy in Kerala tried to be inclusive and more equitable than the rest of India, the private sector has emerged as an important player in the health sector and the dependency of rural people on the private sector is also increasing. This has tended to raise household health expenditure making 'health a commodity purchased by ability to pay' (Nabae, 2003). Health-related expenditures are increasing and sometimes it is prohibitive. Demand for private health care is partly an outcome of government policy and several other measures like spreading literacy, awareness building, sensitisation, and felt need for timely intervention in health matters to improve quality of life (Ramankutty, 2000).

There are spatial disparities in health services both under public and private systems. In fact, spatial disparities are more under the private sector as is evident from a relatively higher coefficient of variation (CV). It is observed that CVs in cases of (i) number of medical institutions per one lakh population, and (ii) number of beds per one lakh population are 22 per cent and 29 per cent respectively under government sector whereas under private sector they are 32 per cent and 52 per cent in the same order (Table 12.3). Growing privatisation will increase spatial disparities and gap will escalate. Given the reality of private sector intrusion in the health sector, it is now important to improve quality of services in public health institutions. This will act as a check and balance for the private sector both for services and expenses. Renovation of the public health care system efficiently may be the main strategy to reverse the downward trend (Lekshmi *et al.*, 2014). Many families face a financial crisis due to increased medical expenses. The National Health Policy in 2017 has recommended an increase in allocation under the health sector both at the central and state level. All state governments are advised to incur 8 per cent of budget expenses in the health sector. Kerala now tries to achieve a 5 per cent target. Linking family health expenses with insurance are widely considered among policymakers. This will ease out disease-related expenditure burden of people.

HUMAN DEVELOPMENT INDEX (HDI)

Inspired by Prof. Amartya Sen's idea of capabilities approach UNDP started preparing Human Development report since 1990 and worked out the Human Development Index (HDI) based on a set of indicators mainly taking into account of human well-being as related to health, knowledge and standard of living (UNDP, 2000). Kerala consistently ranks first among all the states in India in the human development index (HDI). Since 1995, Kerala's position has been consolidated with the improvement of HDI from 0.562 in 1995 to 0.773 in 2017 against all-India average of 0.460 in 1995 to 0.639 in 2017. While Kerala falls in the group of high human development and ranks 69 in the UNDP list, India belongs to the medium human development category and ranks 131 (UNDP, 2016). Kerala not only topped in HDI among Indian states, but also recorded the highest progress by 0.211 points against Indian average progress of 0.179 points between 1995 and 2017. This manifests the strong initiatives of the people of Kerala to move forward continuously.

Centre for Development Studies computed district level HDI using four parameters namely (i) life expectancy at birth, (ii) literacy rate, (iii) gross school enrolment, and (iv) real per capita income for preparing Human Development Report of Kerala, 2005 (Government of Kerala, 2006c). HDI varies from 0.753 in Wayanad district to 0.801 in Ernakulam district (Table 12.4). Four districts in the Malabar part and one district in southern Kerala lag behind the state average HDI of 0.773. HDI and distribution of SC/ST population are negatively related highlighting the general trend that the districts with a higher concentration of SC/ST population have relatively low HDI (Fig. 12.6). It can be further inferred that SC and ST population are lagging behind in human development. The fishing community also lags behind in human development. However, their condition is relatively better than people belonging to SC, ST communities. It may be noted that the fishing population in Kerala is mostly concentrated along the coastal tract of nine sea facing districts, which house all major urban centres and are socio-economically well developed compared to midland and highland tracts in Kerala. Fishing people are not so secluded as tribal communities, who live in the heart of the Western Ghats and are away from mainstream activities.

Table 12.4: Distribution of HDI, Per Capita Income and Poverty by Districts in Kerala

Districts/State/CV	*HDI, 2017*	*Per capita income (in Rupees, 2017)*	*Poverty (per cent to total population), 2013*	*SC and ST population ((per cent), 2011*
Thiruvananthapuram	0.773	122,679	7.9	12.1
Kollam	0.787	136,283	7.3	12.9
Pathanamthitta	0.795	96,134	1.2	14.4
Alappuzha	0.794	136,804	3.2	9.8
Kottayam	0.796	126,238	1.3	8.9
Idukki	0.754	132,107	16.6	18.1
Ernakulam	0.801	152,318	2.6	8.7
Thrissur	0.794	129,922	3.2	10.7
Palakkad	0.761	100,128	13.1	16.1
Malappuram	0.749	89,347	6.9	8.1
Kozhikode	0.781	109,632	5.7	7.0
Wayanad	0.753	92,353	32.9	22.5
Kannur	0.783	109,602	7.2	4.9
Kasaragod	0.760	100,198	14.8	7.8
Kerala	0.773	117,811	7.0	10.6
Coefficient of variability (per cent)	2.44	16.8	95.7	41.7

Source: Government of Kerala, 2016, 2018; Government of India, 2017b; Bhandari and Chakraborty, 2014.

POVERTY

Kerala has largely tackled the poverty problem. Only 7 per cent of people in Kerala lies below the poverty line as estimated following the Tendulkar Committee Method (Bhandari and Chakraborty, 2014). People affected by poverty are < 10 per cent in 10 districts out of 14 districts in Kerala. Pathanamthitta and Kottayam districts have remarkable achievements. Little over 1 per cent population in each of these two districts is recorded below the poverty line. However, Wayanad, Idukki, Kasaragod, and Palakkad districts are lagging behind. The difference between the districts of the lowest and the highest incidence of poverty is > 31 per cent point (Table 12.4). The coefficient of variation of incidence of poverty is very large 95.7 per cent, whereas it is relatively low (15.9 %) in the matter of per capita income. The incidence of poverty records wide regional disparity within Kerala. Regional disparity is not so sharp in the matter of district-level per capita income. This is also to emphasize that even though Kerala's overall outcome in social and economic achievements is far better compared to many other states and national average, there exist pockets of deprivation (Chakraborty *et al.*, 2010). Aggregate data help little in poverty eradication. It is important to adopt a spatially disaggregated target oriented approach and capability enhancement initiatives to wipe out deprivation.

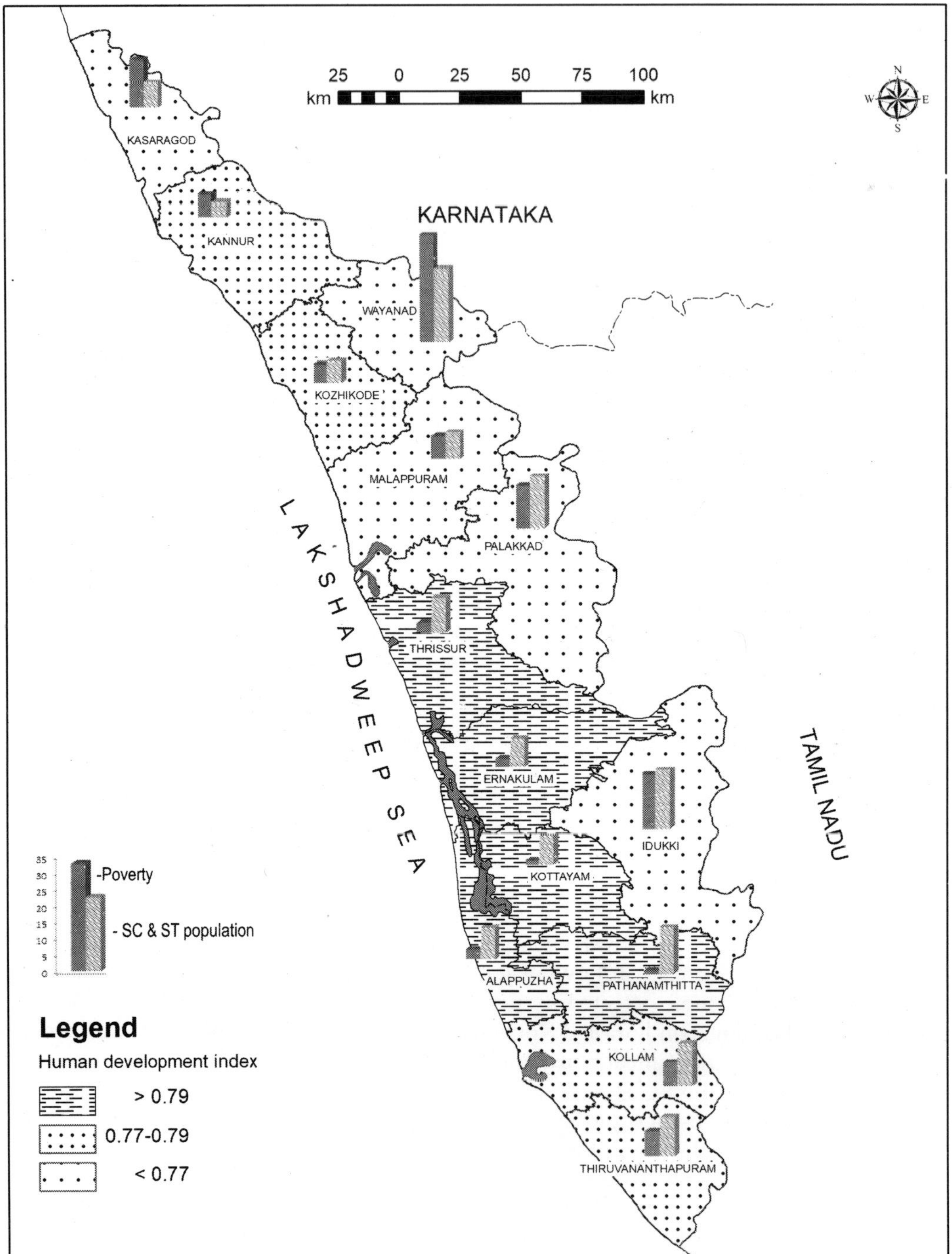

Fig. 12.6: Distribution of Human Development Index, SC, ST Population and Poverty by Districts in Kerala.

Data Source: Government of Kerala, 2006c, 2018; Government of India, 2017b; Bhandari and Chakraborty, 2014.

SPATIAL VARIABILITY OF SELECTED HUMAN DEVELOPMENT INDICATORS

There are spatial variations in health, education, human development index and poverty. The coefficient of variability varies widely across indicators highlighting that spatial disparity and performances of the districts are not the same in all cases (Fig. 12.7). Life expectancy and male literacy are spatially most equitable as evident from a very low coefficient of variability. Six districts, namely Palakkad, Malappuram, Kozhikode, Wayanad, Kannur, and Kasaragod are relatively lagging behind compared to their southern counterparts in the development of some of the social sectors. Among these six districts, Wayanad and Kasaragod are at the bottom. The performances of Kozhikode and Kannur are above average. Seven districts in south Kerala, namely Thrissur, Ernakulam, Kottayam, Alappuzha, Pathanamthitta, Kollam, and Thiruvananthapuram are well placed. Only Idukki lags behind in certain sectors. This district is comparable to Wayanad of the Malabar area. Thrissur, Ernakulam, Kottayam, Alappuzha, and Pathanamthitta districts form a block with high human development.

The spatial variability of human development shows three broad trends (Chattopadhyay, 2019). There is a north Kerala (Malabar) and south Kerala (Travancore) divide. As part of Madras Presidency, the Malabar region was under colonial rule and therefore much neglected in the pre-independence era. Southern Kerala mostly under the princely ruler of Travancore was better cared for education and health. There were differences in human development and

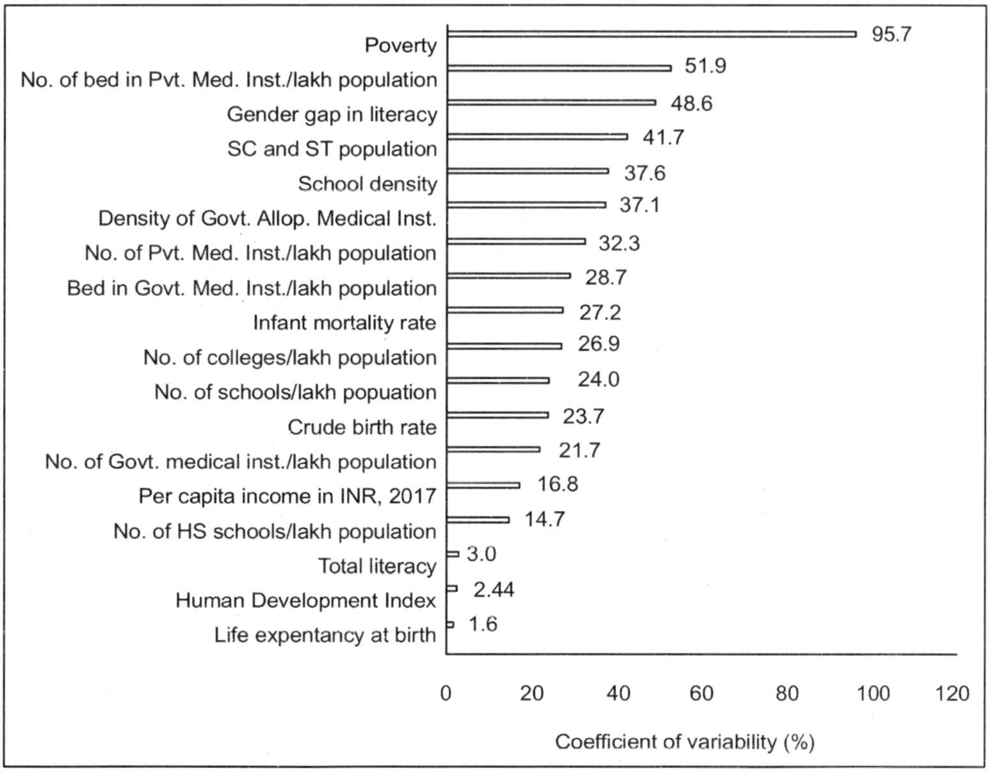

Fig. 12.7: Bar Plot of Coefficient of Variability (%) of Selected Human Development Indicators in Kerala.

also in case of other indicators between these two segments at the time of modern Kerala's journey as a province of sovereign India. In spite of several positive measures by successive Kerala governments, post-1956 reorganisation, the historical legacy of political ecology still persists and spatial variability continues. It is partly a product of political geography.

The second factor is related to terrain. The topographic grain of Kerala is longitudinal. The settlements are ribbon-like and in a continuous chain from one end of the state to the other end along coastal plain and midlands. Performance in human development diminishes from the coast to inland. The location of an area with respect to the coast and altitudinal position play important roles in determining levels of human development. Districts of Wayanad and Idukki, located in the heart of the Western Ghats, are lagging behind. Palakkad district with its interior location also records slow progress. All these districts are characterised by rugged topography and steep slopes.

The third factor underpinning spatial variability is related to the concentration of certain social communities. The fishing community lives mostly along the coast and backwaters. They are deprived compared to other communities living in adjoining areas. Similarly, a high concentration of SC, ST populations is found in the interior districts within the Western Ghats like Wayanad and Palakkad in the Malabar area and Idukki in south Kerala.

Apart from these locational factors, privatisation of education and health sectors has enhanced spatial variability. In one hand, Government services tend to reach the deprived areas and reduce spatial disparity, on the other hand, growing private investment contributes to spatial inequality. This trend has increased specially after globalisation/economic liberalisation. The Government's withdrawal from social service sectors may worsen the situation further.

Kerala has progressed significantly in social and human development. Recently, it is performing well in per capita income and economic growth. Early investment in the human capital building has facilitated this process. Kerala has also turned into a consumer state importing bulk of her consumables from other states of India. The increase of disposable income in the hand of a large section of people in Kerala has contributed to this emerging trend of consumerism and privatisation. Kerala is moving towards a higher trajectory on the development front with a major thrust on knowledge based industries and quality improvement of services. However, there are several challenges including intrastate disparities in social and economic development sectors, the degradation of the environment, financial limitations, and increasing thrust on withdrawal of government from social service sectors. Globalisation has added another level of challenge in the growing privatisation of services and ex-situ decision-making. The geographical advantage that Kerala enjoyed seems to be waning. It is important for Kerala to build on decentralised infrastructure and introduce spatial planning for further consolidation of the Kerala model, minimise spatial variability, and move forward towards a sustainable society.

CHAPTER

13

Decentralised Planning

NEED FOR DECENTRALISATION

Decentralisation is a process of local-level development. Its cherished goals are to overcome regional imbalances of development and ensure distributive justice. The ability of the prevailing democratically elected government and techno-bureaucratic administration to bring distributive justice and equality bridging the development gap following trickle down theory has been doubted (Heller, 2005). It is also argued that decentralisation improves the allocation of resources, delivery and in some cases finance local public goods, cost recovery, and accountability, and reduces corruption in service delivery (Azfar *et al.*, 1999, Rodríguez-Pose, 2008, Jurlina Alibegoviæ *et al.*, 2014).

Over 75 countries in the World have attempted to transfer responsibilities to the lower tiers (Ahmad *et al.*, 2005). The last couple of decades witnessed an emphasis on decentralised planning and development and several decentralised practices have evolved across the continents. It is often argued in the case of India that given the spatial variations of natural resource endowment, development potential, levels of development, and socio-cultural differences, a single development strategy, hitherto followed for the entire country may not bring the desired results (Gadgil, 1967, Misra and Sundaram, 1980, Raj 1971, Ghosh, 1988). The multi-level decentralised planning was suggested to commensurate the spatial variations in resource distribution, and to address the inter-regional disparity in development and heterogeneity in the social structure (Misra and Sundaram, 1980). Kerala is one of the states in India, which has significant achievements in introducing decentralised planning.

THEORETICAL FRAME FOR DECENTRALISATION

Geography's engagement with regional planning has a long tradition. In the post-colonial period, regional planning evolved as part of Government initiatives in many countries around the World and suitable institutions have been built. A new professional field developed and a consensus emerged about the theory and practice of regional planning (Friedmann and Weaver, 1979). Area development, decentralised development, and multi-level planning figured in the deliberations of regional planning and development at the national and sub-national levels (Sundaram, 1983; FAO, 1985; Sundaram, 2017). The geographical political economy remains a highly influential approach in economic geography and local and regional development (Pike, *et al.*, 2006). As a tool for spatial management and planning, the region is conceived as an area or specific territory for which spatial, development, and regulatory plans are made (Klapka *et al.*, 2013). Regionalism and localism are linked with decentralisation (Gopinath, 2009) and this brings the concept of scale, iteration, nesting, and spatial integration as important elements of regional planning and decentralised development.

Pike *et al.* (2006) reviewed the main concepts and theories of local and regional development in the context of all prevailing development theories and observed that 'local and regional development adopts a territorial approach as a means of achieving economic development. Diagnosis of economic, social, and institutional conditions of every territory, and identification of local economic potential are the foundations upon which such development strategies are built' (Pike et al., 2006 : 17). The territoriality is an important dimension of development. It recognises the heterogeneity of geographic space. Absolute or physical space turns into a geographic space through anthropogenic intervention. The nature of anthropogenic intervention depends upon political-economic forces including institutions and infrastructures and level of technological absorption. Apart from the specificity of territorial characteristics, that makes one area different from another area, the other driving forces for decentralisation are the multidimensional nature of employment problem represented through a mismatch of demand and supply of labour, and global economic context allowing greater synergy between the economic and social foundation of the markets (Greffe, 2004).

Decentralisation is also expected to create an opportunity for local people to air their views, to help taking appropriate decisions by the authorities concerned, and to monitor and social auditing. One of the challenges of decentralised planning is local economic development. The ongoing trend of local development, where the government or private agencies working at the local level along with economic, social, and political agencies used to perceive development as interventions is found wanting. The new order that the decentralisation proposes is to set the process of change that affects the local economy's capacity to 'increase economic growth, generate employment, and create new wealth' (Tello, 2010 : 51).

Local economic development strategy is expected to address the issue of (i) mobilisation of local resources and competitive advantage, (ii) locally owned and managed public utility services, (iii) participation and social dialogue, and (iv) multidisciplinary approach to address the problems, (White and Gasser, 2001; Tello 2010). Through decentralisation, it is aimed to optimise utilisation of local natural and human resources, address aspirations of the local people, empower them to participate and take decisions, search for local solutions to address local problems, and also to contextualise local problems in the larger decision-making frame or prevailing political-economic conditions. The process of decentralisation is therefore multidimensional. Broadly, there are six principles of decentralisation. These are (i) territoriality, (ii) mobilisation, (iii) public participation, (iv) local ownership and management, (v) multi-disciplinarity, and (vi) nesting and integration. Strategically, decentralisation process can be initiated through political decentralisation, administrative decentralisation, and fiscal decentralisation (Rondinelli *et al.*, 1984, c.f. Gopinath, 2009; Torrisi *et al.*, 2011). These items are elaborated here in brief:

- Political decentralisation refers to holding of regular party-based elections to constitute local self-governments below the state level, draw marginalised groups in the policy-making and delegate legislative power to share governance functions.
- Administrative decentralisation means delegation of administrative/bureaucratic powers to the local bodies, empower them to take administrative decisions and transfer responsibilities to deliver social services like education, health, social welfare, etc.

- Fiscal decentralisation envisages transfer of budgetary and other financial powers with authority to act in implementing revenue and expenditure policies and raise revenues.

There is also ideological contestation about the meaning of decentralisation. While the policy literature and multilateral donors usually focus on the efficiency dimensions of decentralised local government, civil society groups and social movements have tended to emphasise participation to create wider space for 'expanding the role of public powers to underwrite social citizenship' (Heller, 2005; 2007). This philosophical positioning is significant in understanding Kerala's decentralisation, which stresses empowerment, entitlement, and welfare.

DECENTRALISATION PROCESS AT NATIONAL LEVEL

Decentralisation in India, particularly rural decentralisation through Panchayati Raj Institutions has generated widespread interest as it proposes to address some of the vital issues concerning basic necessities of the human society—poverty alleviation, provisioning of public services, change in social, economic, and cultural lives, improvement of quality of life, empowerment, and ensuring human dignity and meeting aspirations. Academicians, scholars, even international funding agencies showed interest in decentralisation experiences of India (Johnson, 2003; Gopinath, 2009; World Bank, 2000; UNDP, undated).

Gandhiji's vision of 'Gram Swaraj' through universal education, economic self-sufficiency, and village democracy forms the bedrock of India's attempt of decentralised development in rural areas. Since Independence, there had been various committees/initiatives for decentralisation (Table 13.1).

The first major initiative for decentralised development was the launching of the Community Development Programme (CDP) in 1952 to triclke down growth impulses. It started as a pilot project to boost agriculture, improve rural health, education, and communication, and finally to transform the social, economic, and cultural lives of rural India (Dube, 1958). Administered through Block Development Officer (BDO), this programme brought development administration close to the people. The pilot project brought perceptible changes and therefore by the end of the 1st Five Year Plan in 1956 there were 248 Blocks in the country and by 1964 Block Development Offices were established throughout the country (Karunaratne, 1976). However, doubts were raised from the very beginning of this programme and there were contrasting views about its usefulness (Nayar, 1960; Karunaratne, 1976). While examining the work of CDP initiated in 1952 and National Extension Services introduced in 1953 to suggest improvement Balwantrai Mehta Committee recommended the scheme of democratic decentralisation, which set the ground to launch three-tier Panchayati Raj Institutions at the village, Block, and District levels in 1957. The aim was to decentralise the process of decision making and encourage people's participation (Hedge, 2000). Rajasthan was the first state to adopt this scheme in 1959. The following years witnessed the formation of various committees to look into different aspects of Panchayat Raj Institutions (PRI).

The Ashok Mehta Committee formed in 1978 to review the PRI identified bureaucratic apathy, lack of political will to devolve power to the lower tiers, capturing of power in the Panchayats by local elites, and ambiguity about the functions of PRIs to undermine various attempts of decentralisation and recommended several measures including that of the districts to serve as administrative units for PRIs (Johnson, 2003). The states of Andhra Pradesh,

Table 13.1: Decentralisation—Chronology of Attempts and Committee Reports in India

Year	*Item*	*Ideas and concepts*
First Plan 1951-56	Community Development Blocks (CDBs)	To break up the planning exercise into national, state, district and local community levels
Second Plan 1956-61	District Development Councils	Drawing up of village plans and popular participation in planning through the process of democratic decentralisation
1957	Balwantrai Mehta Committee established	Village, Block, District panchayat institutions established
1967	Administrative Reforms Commission	Resources to be given/local variations accommodated, purposeful plan for area
1969	Planning Commission	Formulated guidelines; detailed the concept of the district plan and methodology of drawing up such a plan in the framework of annual plans, medium-term plans and perspective plans.
1978	Ashok Mehta Committee M.L. Dantwala report	Review of activities of PRI. Block-level planning to form link between village and district-level planning. Block-level planning to form link between village and district-level planning
1983-84	Centrally Sponsored Schemes/Reserve Bank of India	Strengthen district plan/district credit plan
1984	Hanumantha Rao Committee	Decentralisation of functions, powers and finances; Setting up of district plan bodies and district plan cells
1985	G.K.V. Rao Committee	Administrative arrangements for rural development; District Panchayat to manage all development programmes.
1992-93	73rd and 74th Constitutional Amendment	Constitutional status of Local Self Governments

Modified after Planning Commission of India (2008b).

Karnataka, and West Bengal followed some of the recommendations and passed new legislations to strengthen PRI. In subsequent years, while the G.V. Rao Committee (1985) recommended Block Development Offices to assume more powers, the L.N. Singhvi Committee (1986) recommended that local self-government should be constitutionally enshrined and Gram Sabha (Village Assembly) must form the foundation of decentralised democracy. All these commissions acknowledged the importance and necessity of PRIs and at the same time highlighted the weaknesses and recommended measures to overcome these shortcomings. Observations of all these committees prepared the ground for the 73rd and 74th constitutional amendments to provide Panchayats the constitutional status (Jha, 1999).

CONSTITUTIONAL AMENDMENT

According to the Article 40 of the Constitution of India (Directive Principles of State Policy) distributing powers between the Union of India and states, the subject of local bodies and

Panchayati Raj was vested with the states, although it did not elaborate on the relations between the state and further lower tiers of government. The 73rd and 74th constitutional amendments provided a legal basis and defined details of powers and functions of various organs of Panchayats and Municipalities, respectively. The Article 243G envisaged development functions of Panchayat as '(i) preparation of a plan for economic development and social justice and (ii) the implementation of schemes for economic development and social justice as may be entrusted to them including those in relation to the matters listed in the Eleventh Schedule.' Mandatory provisions included that: (a) every state should constitute local governments at the village, intermediate, and district levels through periodic and direct elections; (b) some of the positions to be filled in local governments through direct election are to be reserved for individuals belonging to identified categories (such as the Scheduled Castes and Tribes, women, etc.); (c) every state should constitute a finance commission from time to time to review the financial position of local governments; (d) every state should ensure that these mandatory provisions are incorporated into state legislation (Government of India, 1993).

Apart from provisioning of a 5-year term democratically elected body at the Panchayats to maintain continuity of activities one of the most important salient features of the constitutional amendment is the gram sabha (Village council), which is mandated to meet twice a year and examine (i) annual statements of account and audit report, (ii) report of administration, (iii) tax-related matters, and (iv) selection of schemes, beneficiaries, and location. Thus Gram Sabhas are given a key role in the process of the PRI system. Both the central and state governments are expected to play the role of facilitator by establishing and nourishing the village panchayats as effective self-governing institutions.

Since the 73rd and 74th constitutional amendments in 1992-93 there had been high expectations about democratic decentralisation and empowerment of local people in furthering their cause. In reality, questions are raised about states' initiatives to implement the PRI system in its full spirit, the operational imperatives and how far democratic decentralisation has taken root in India (UNDP, undated; World Bank, 2000; Oomen, 2004). The states, which have taken serious efforts in PRI, mostly follow the top-down approach as noted in the case of Karnataka (Satish Chandran 2000). In spite of constitutional mandate, elected Panchayat system and commitments of the Central and state governments to channelize various development programmes through Panchayats, the basic premises of decentralisation—territoriality, local people's participation in decision making and dialogue, provisioning of local public good and services, creation of local economic opportunities, and a multidisciplinary approach are found missing in most cases. Re-examination of structural aspects of decentralisation and further institutional reform to bring efficiency and nurturing of human capital are necessary to effect decentralisation process (Kalirajan and Otsuka, 2010).

Lessons from Kerala are important in this context. Perhaps Kerala is the only state that has taken a significant step towards decentralisation following a bottom-up approach and tried to address the basic premises for decentralisation. Even West Bengal, which installed elected Panchayats way back in 1978, started with land reforms and pioneered in empowering Panchayats well before Constitutional Amendments hardly took any lasting initiative to set the process of bottom-up planning and provisioning for human development at the local level. In fact, Panchayats in West Bengal continue to follow the top-down approach in planning, and in recent years it has been rebureaucratised with Block Development Officers playing decisive roles.

DECENTRALISATION IN KERALA

Kerala, although technically a late starter, has out-shone all other states in the country in the matter of implementing the Panchayati Raj system and ushered a new dimension in democratic decentralised by launching the People's Plan Campaign (PPC) in 1996. The decentralisation as espoused through PPC is a 'means of carrying class-struggle forward, of buttressing the class-strength of the rural poor by developing institutions where they can, in principle, assert themselves directly and hence more effectively' and this is in contrast to the market-driven decentralisation that attempts to blunt class-struggle and encourage 'a scenario of obedient-and-supplicant-villagers-being patronized-by-NGOs, and of substituting the concept of the Rights of the people by the concept of Self-Help' (Patnaik, 2004). How did Kerala succeed to chalk out its own course of action and achieve many of the goals of decentralisation? A question often debated at the national and international deliberations. Kerala's geographical setup, environmental resource distribution and settlement pattern facilitate decentralisation (Chattopadhyay and Franke, 2006). Apart from this natural foundation, there had been several initiatives at the state level contributing to decentralisation some of which can be traced during the pre-Independence period. We shall limit our discussion mainly to the post-Independence period. Broadly there are two phases in decentralised planning in Kerala. Phase-I may be considered as a preparatory phase during which a series of initiatives directly or indirectly contributed to lay the foundation for decentralised planning, beginning with administrative reforms. Phase-II begins with the launching of the people's plan campaign (PPC) and includes post-PPC activities.

ADMINISTRATIVE REFORMS

The general election of 1957 witnessed the installation of an elected communist-led government for the first time anywhere in the World. It created the ground for political decentralisation based on the Report of the Administrative Reforms Committee (ARC) (1958), which envisaged the formation of two-tier local government structures in Kerala, one at Panchayat and Municipal level and another at the district level. There were proposals to transfer certain power and responsibilities to the elected district councils from the state, and in stages, these district councils 'will assume full authority over development work in the district except the very important schemes which may be specifically retained in Government's direct control and function as full-fledged local self-governing units' (Government of Kerala, 1958). As the government was dislodged in 1959, these administrative reforms could not be implemented; however, the recommendations of ARC formed the basis of Panchayat Acts in the 1990s. Since the formation of the state in 1956 till 1996, Panchayati Raj in Kerala passed through many uncertainties (Charvak, 2000). It was during the 1987-1991 period that the Panchayats were given untied money to take up a programme of their own. The amount of money was meagre, however, it had sown the seed among the panchayat people that they had to devise their own plan.

LAND REFORMS

Land is a capital resource. Providing access to land is a basic step to empowerment. The land reform bill introduced after the formation of the elected government in 1950s and 60s could finally be passed in 1969 as Kerala Land Reforms (Amendment) Act and implemented in the 1970s put an end to the feudal system and ensured the rights of the tenants. Thus, a large section of people got access to land resources facilitated by the government both through legislation and cost-sharing (Franke and Chasin, 1989). Land reforms removed landlordism, provided ownership to a majority of people, helped reduce of the difference between economic classes, and even minimised caste inequalities (Franke, 1992, Jaydev and Ha, 2015). Given the nature of land in Kerala and its productivity, access to land resources had far-reaching impacts on the socio-economic conditions of people. Land reforms by giving land rights ensured dwelling places for a large section of people, distribution of wealth, social security, and redistribution of the power structure.

SETTING UP OF ENABLING INSTITUTIONS

The 1970s is an important decade for Kerala. Apart from the implementation of land reforms this decade also witnessed the setting up district planning offices in all districts linked with the technical division of Kerala State Planning Board. Another significant development was the adoption of the Science Policy Resolution by the state government and the establishment of a number of research and development institutes dealing with health, industry, economics, earth science, water resource, forest resources, and fisheries. These institutes were set up to help Kerala in its planning process and bring frontier knowledge for the development of the state. Subsequently, new institutes were established under the Government of India, and the Government of Kerala. Besides, there are new Universities, Professional Colleges, Management and Technology institutes. All these have created an ambiaence of professionalism and the Government of Kerala is able to draw expertise from these institutes as and when required. Decentralised planning, which considers territorial heterogeneity as opportunity requires location-specific precision science and technology. Generalised observations are not sufficient to provide inputs for decentralised decision making.

CIVIL SOCIETY

Civil society is a loaded term and is subject to interpretation. There are different perceptions, meanings, and manifestations associated with the term 'civil society' in the contemporary context (Tandon and Mohanty, 2000). The common denominator of all the formulations is that civil society provides the space for citizens to collectively act for common public goods. Kerala has one of the strongest civil societies (Tornquist, 1999). With high level of literacy, activism, and awareness, people are organised voluntarily in a multitude of organisations ranging from socio-religious associations to environmental groups and women organisations. Many of these organisations/associations were involved in the education and health sectors since the early part of the 20th century. Kerala's achievements in health and education can be partly attributed to these initiatives.

Roots of many of the organisations, particularly socio-religious groups and educational bodies can be traced to pre-Independence era. Many organisations espousing the causes of the environment, the welfare of the underprivileged groups, women, and other socially marginalised groups emerged after Independence. A strong people science movement (PSM) gathered momentum in Kerala since 1960 pivoting around Kerala Sasthra Sahitya Parishat (KSSP) with a slogan "Science for Social Revolution" (Ekbal and Thomas Isaac, 1988). It succeeded to provide a platform for people across the society of different political spectrums and also the intellectual capability for voicing their concerns about different aspects of social development. As part of people's science movement, it propagates science awareness, rationality, and works in tandem with other organisations and government machinery. Apart from these, there are various associations and trade unions formed under political patronage. What is significant in Kerala is that a socio-political-technical space evolved for continuous dialogue and discussions about the welfare of people and emerging national and global issues. People are socially conscious and voice their concerns even in remote areas.

Civil societies in Kerala succeeded to create common platforms to express an individual's concern, develop consensus among different sections of people, spread awareness, promote discussions, and helps mediate and negotiate with the authorities for common public goods. The state can hardly ignore the vibrancy and the energy that the civil societies bring. As these forces are distributed across the length and breadth of Kerala, the local issues find their way in mainstream deliberations of policymakers, and setting up of development agenda. The idea that the relationship between civil society and government in a democracy should be mutually reinforcing (Tandon and Mohanty, 2000) is perhaps maturing in Kerala. The International Congress on Kerala studies organised every five years since 1995 provides an opportunity for the cross-section of society to articulate their views, which through a series of deliberations can find their way in setting the development agenda of the state.

LIBRARY MOVEMENT

The library movement in Kerala is a unique feature. It manifests the contribution of voluntary effort. Since the establishment of the first public library in Trivandrum in 1829, there had been a steady growth and now Kerala can boast of 8182 libraries affiliated to Kerala State Library Council. There are 248 libraries for every one million people in Kerala. The library management is decentralized through the District library council and Taluk library council under the overall guidance of the State Library Council. Activities of the libraries, particularly rural libraries include helping students in studies, career guidance, book distribution for women, and also nurturing of social and cultural lives of women. More than 1000 libraries opened children's wing and started book reading competition among children. Libraries as statutory bodies are turning into focal centres for the cultural exchange of local people and help grooming concerned and responsible citizens. It is necessary to equip libraries technologically for attracting the young generation and serving as a local knowledge centre.

ENVIRONMENT MOVEMENT

Environment movement in modern Kerala can be traced to the protest organised by local people of Vazhakkad, a village in the Bank of river Chaliyar, and local fish workers against the Grasim industries at Mavoor in 1964. Industrial effluents from the plant were causing pollution problems resulting in resource depletion and health hazards. Although initial attempts were not successful, there were protests in subsequent years and finally, the Kerala State Pollution Control Board ordered the closure of the plant in 2001. The Silent valley movement was one of the most significant environmental movements not only in Kerala but in the whole of India. A hydro-electric project was proposed by Kerala State Electricity Board (KSEB) over the river Kuntipuzah, a tributary of Bharathapuzha. The proposed reservoir would have inundated virgin tropical rainforest. People from all walks of life gathered together and lodged a strong protest pivoting the project. The movement, started locally, soon gained national and international momentum. The project was finally abundant at the instance of the Government of India and the silent valley area was declared as a national park in 1985.

Some of the other notable movements are: (i) Stop the use of Endosulfan pesticides, which was prevalent in the cashew plantations in Kasargod district causing great harm to human health. The use of endosulfan has been banned in Kerala by the order of the Department of Agriculture, Government of India; (ii) the Plachimada movement in Palakkad district in 1998. It is one of the finest examples where local people dominated by *Adivasis* dared Multinational Corporation and agitated against contamination of groundwater caused by the Plant. The plant was closed down in 2006. The Plachimada movement drew national and international attention; (iii) The Muthanga Adivasi movement in 2001 claiming property rights on the land. A subject, yet to be addressed properly, is a potential source of agitation and also conflict between settlers and *Adivasis*. The land distribution has not been fully materialized. Nevertheless, *Adivasis* started organizing themselves and asserting as a collective social force with increasing consciousness about their rights to live in the society (Seethi, 2017); (iv) Save river campaign. There were several initiatives in Kerala for peoples' movement to save rivers. *Nadi Samrakshana Samities* (Save River Society) have been constituted for various rivers to espouse the cause of river protection. Some of the notable *Samities* pertain to the rivers of Pamba, Periyar, Chalakudy, and Bharathapuzzha. Besides, there are other initiatives undertaken by the government to revive rivers through people's participation particularly in south Kerala; and (v) Save Western Ghats campaign. There are various groups raising their voices to protect the Western Ghats, particularly after the 2018 devastating flood.

All these movements testify how the struggles of local people bloomed as mass movements. This was possible due to the awareness of people and their eagerness to participate in deliberations espousing a larger cause of social development. Public participation is an important component of decentralized planning and development.

TOTAL LITERACY CAMPAIGN

Kerala has remarkable achievements in spreading literacy. It was the first state to be declared total literate. The new education policy (NEP) of the Government of India adopted in 1986 stressed decentralization, social mobilization, and integration for removing illiteracy (Government of India, 1995). It also constituted the National Literacy Mission to spearhead the programme for universalizing literacy. Districts were considered as units in place of states for implementing the programme and social mobilization was found necessary over the departmental approach to come out of the curse of illiteracy. It was felt that social mobilization will catalyse the involvement of larger society. Accordingly, the NEP kept provision for actively involving voluntary agencies, groups, and interested individuals to spread elementary education. The new policy also considered it necessary to integrate adult literacy and non-formal education programmes with socially relevant themes like family planning, health care, environment, and nutrition.

The situation in Kerala was ripe to take advantage of this new policy. Apart from various initiatives so far taken up by the Government of Kerala for spreading literacy, there were other initiatives also. Already KSSP launched its own literacy mission in 1977 and prepared ground for mass mobilization, while organised science procession covering 900 villages and meeting 5000 people through length and breadth of the state and a series of district and local level initiatives (KSSP, 1991). KSSP along with district administration launched the total literacy programme under the aegis of Total Literacy Mission for Ernakulam district in 1989. The Kerala Association for Non-Formal Education and Development (KANFED) was formed to implement the programme. There were 42 voluntary organizations and groups participated in the total literacy campaign in Ernakulam district (Sivadas, 1991). Trainers and instructors drawn from KSSP, political parties, teachers' union, and local voluntary orgnisations were trained by resource persons. The programme was taken up after a detailed field survey and dividing the entire districts into local units. It took the form of a mass movement with people voluntarily participating not in the programme alone but also in meeting local logistics. The atmosphere was socially charged both for illiterates to learn and for literates to avail the opportunity to teach (Tharakan, 1990).

By 1990, Ernakulam district attained total literacy and was declared as the first total literate district in the country. Soon, the Ernakulam experiment was considered as a model for the state and also for the nation to eradicate illiteracy. It is important to note that the total literacy programme in Kerala was not only limited to enable people to read and write, but also to spread awareness about other social and environmental issues and to make people conscious and cultivate self-dignity. It was thus a silent social revolution that brought attitudinal change among the illiterates (Rajan, 1991). The regional and localized approach followed in Ernakulam district and subsequently in the entire state of Kerala was a key element for the success of the total literacy programme as distribution of illiterate people varied spatially and also across the social strata. Total literacy campaign manifested the importance of synergy of Government, Civil Society and common people, power of mass mobilization, and necessity of decentralized approach in accomplishing the task of executing socially useful programmes.

PANCHAYAT RESOURCE MAPPING (PRM) PROGRAMME

While the total literacy campaign was underway, one of the Research and Development Centres in Kerala, Centre for Earth Science Studies (CESS) now rechristened as Nationa Centre for Earth Science Studies (NIESS) was experimenting with micro-level resource mapping in Ulloor Panchayat in the suburb of Thiruvananthapuram city to generate biophysical resource data for local level planning (CESS, 1991). This experiment was finally perfected as Participatory Panchayat Resource Mapping (PRM) programme in collaboration with social science professionals and people's science movement organisation like KSSP. The detailed methodology has been developed spelling out step by step activities (CESS, 1991; 2000; Chattopadhyay *et al.*, 1999; Chattopadhyay *et al.*, 2004). This programme was executed through a series of activities (Table 13.2).

The first complete survey was accomplished for Kalliasseri Panchayat in Kannur district of north Kerala by CESS and KSSP under a project sponsored by the Department of Science and Technology, Government of India. Subsequently, additional data were collected and a total development report was prepared for this Panchayat supported by the Kerala Research Programme for Local Level Development (KRPLLD) project of CDS. Kalliasseri Development Society, formed to spearhead the development of the Panchayat, mapped all their resources, reorganized schools, installed high-efficiency cooking stoves and solar street lights, set up a shrimp farm in a swampy area, and set up dry season vegetable co-operatives. It may be reasonably argued that the Ulloor exercise followed by the Kalliasseri experiment inspired much of Kerala's decentralised planning effort in subsequent years. The PRM has been completed for all Panchayats in Kerala under the guidance of the Kerala State Land Use Board (KSLUB). Cadastral scale mapping has emerged as an important tool for involving local people, conscientisation, and documentation of plot-level basic resource data and drawing up local plans. The PRM exercise has been repeated in several instances according to local requirements involving local volunteers (Fig. 13.1). There were some modifications in methodology and nature of maps being prepared during the course of work to suit project requirements,

Table 13.2: Activities for Executing PRM Programme in Kerala

Phase/Type of activity	*Details*
Environment building	Interaction with the panchayat and NGO. Campaigning in the panchayat, preparing ground for initiating the programme. Formation of a PRM committee. Conducting PRA as entry programme.
Training materials and guide book	Preparation of materials for training volunteers, like hand outs, sample maps and other required materials. Guide book for master volunteers and other professionals.
Selection of volunteers	Selection of at least five volunteers from the local people in each Ward/Village/hamlet/unit area of 1-2 km^2.
Training of Volunteers	Training of master volunteers, who will lead the team and coordinate the work at the panchayat level, preferably in a R&D Institute. Training of other volunteers in the respective panchayat.
Mapping by volunteers	Landuse and asset mapping in cadastral scale follows training of other volunteers. Consolidation and re-plotting of mapped data continue.
Mapping by Professionals	Mapping of land form, surface material/soil, water potential and other relevant details in desired scale (1:12,500 or in cadastral scale for smaller area).
Finalisation of maps	Finalisation of all maps, preparation of short write up and preparation of environmental appraisal map for discussion at the panchayat.
Action planning	Presentation of maps in the panchayat and developing an action plan map through interaction with the panchayat. Appraising PRM committee/Panchayat about the resources, potentials and problems of the panchayat.

Source: Compiled from CESS, 1991, 2000.

nevertheless, the basic framework of plot-level data generation, people's participation, and location-based intervention remained the same. The PRM methodology has been widely applied for watershed management not only in Kerala but also in other states. Riverbank mapping programme now being executed by ILDM, Department of Revenue and Disaster Management, Government of Kerala has been developed following PRM methodology. It is learnt during field-work and subsequent interaction with local people that Panchayats require proper technical support to use resource maps for planning purposes and drawing intervention measures.

PANCHAYATI RAJ ACT (1994)

The 73rd and 74th constitutional amendment provided mandatory provision to set up a democratically elected three tier Panchayati Raj System in all states in India. Accordingly, the Legislative Assembly of Kerala passed the Panchayati Raj Act, 1994. It conformed to the mandatory provision of the Constitution. Gram Sabha is the basic unit. The three-tier system consists of Gram Panchayat, Block Panchayat, and District Panchayat. Besides, there are urban areas. At present, there are 1200 local self-government institutions comprising 941 Grama Panchayats, 152 Block Panchayats, 14 district Panchayats, 87 municipalities, and 6 corporations. This is in the variance of all previous attempts to decentralize through two-tier system of the district council and pachayats and conceiving district governments with jurisdiction over rural and urban areas (Thomas Isaac and Franke, 2000). The Government of Kerala also appointed Finance Commission for the Panchayati Raj Institutions as mandated by the 73rd Constitutional amendment.

Fig. 13.1: Women Volunteers Mapping Panchayat Resources in Thanneermukkom Panchayat of Alappuzha District in Kerala. Photo by Srikumar Chattopadhyay.

PEOPLE'S PLAN CAMPAIGN

The year 1996 was a watershed year for decentralised planning not only in Kerala but also in India and many other developing countries experimenting on decentralisation. People's campaign for the preparation of the 9th Plan (1996-2001) was launched. Kerala's many movements and experiments in democracy, equality, and participation culminated in this remarkable experiment in local democratic planning and decentralised development (Chattopadhyay and Franke, 2006). There had been several scholarly deliberations on this programme at the national and international levels (Thomas Isaac and Franke, 2000; Heller, 2005; Tornquist, 1999; 2001; Franke, 2008; Gopinath, 2009; Jafar, 2014 among others).

Apart from the mandatory provisions of elected Panchayats and State finance commission, the 73rd and 74th Constitutional amendments provided optional provisions for the state aimed at strengthening local institutions through innovation in the matter of delegating power, preparing development plans, allocating financial resources, and to use services of professional/experts. The newly constituted State Planning Board (SPB) in Kerala in 1996 made full use of all constitutional provisions and spelt out an innovative course of action. The brochure issued by Kerala State Planning Board in 1996 stated that:

> '35–40 per cent of the Ninth Plan programme will consist of schemes formulated and implemented by the local bodies within the respective areas of their responsibilities. In order to empower the Panchayats and municipalities to undertake this task in a scientific and participatory manner the board resolved to organize a people's campaign for Ninth Plan'.

To implement this programme, that Bandyopadhyay (1997) termed as innovative, audacious, and original, the details were meticulously worked out and acted upon. It was initiated through various phases having clear objectives, involving various actors, and bringing about a concrete output (Table 13.3). Details pertaining to all these steps have been elaborately discussed in Thomas Isaac and Franke (2000). This programme substantially decentralised the functions of the government bureaucracy and introduced decentralised planning while mobilizing the energy of hundreds of thousands of activists and volunteers to go beyond what government-funded projects could accomplish. It was not simply to draw a plan from the below, rather it strived to bring about the attitudinal transformation of the participants. This was not possible through an executive government order. It required continuous engagement, creativity, and the social logic of the movement.

Table 13.3: Major Phases of People's Plan Campaign for Democratic Decentralization in Kerala, 1996

Phase	*Dates*	*Objective*	*Activities*
1. Grama Sabha/ward convention	August-September	Identify felt needs of the people; generate participation and interest in the Campaign	Meet at local school or other location; plenary sessions and small group discussions, listing of local problems
2. Development Seminars	October-November	Assessment of resources, development problems and tentative solutions	Collect data at local offices; carry-out transect walks; write local development reports; conduct development seminar
3. Task Forces	December	Turn proposals from Development Seminars into formal projects	Meet, discuss, and draft project proposals
4. Village and Municipal Councils	January	Finalize local plan	Prioritize projects; draft and approve local plan
5. Block and District Councils	February	Finalize block and district plans	Analyse local plans; integrate them and develop appropriate complementary projects
6. Volunteer Technical Corps-Expert Committees	February-March	Appraisal and technical corrections to projects	Read, evaluate, suggest improvements to projects. Assist Panchayats to make viable projects
7. Release of Funds, Project Implementation, Monitoring, and Evaluation	March-June	Implement and monitor the projects in the local plan	Open accounts, draw checks from treasury, set up beneficiary committees, hire contractors, activate monitoring committees

Note: Dates during the first year of the Campaign, 1996-97, were significantly delayed.
Source: Thomas Isaac and Franke 2002.

The campaign generated procedures to insure delivery of project funds to the former untouchable Scheduled Caste and Scheduled Tribe populations who, previously often, did not receive them. It brought about special projects for women and simultaneously encouraged women elected officials and activists to participate more fully and effectively in public life. It created conditions for eliminating or drastically reducing corruption in several areas of government spending. It rationalized procedures for all kinds of projects and programmes. It promoted and supported the spontaneous release of local initiatives in areas such as neighborhood groups that hold promise for even deeper decentralization and for strengthening of community and of civil society. The campaign gave training to thousands of local officials and activists in what must be one of the most extensive adult education and empowerment programmes conducted anywhere in the world. The campaign raised consciousness about the local resource base, problems encountered by local people, and environmental issues in development planning in Kerala. It built upon the PRM we discussed before and inspired several local communities to create innovative projects in environmental protection and sustainable resource management. Panchayats started preparing their own development reports—'*Vikasana Rekha*' and identify projects. These Development reports comprised eight chapters—introduction, development strategy, resource mobilization, sectoral programmmes, integrated development, the welfare of Scheduled Castes and Scheduled Tribes, women's development, and monitoring. All Panchayats in Kerala have their own plan and development reports.

Plan Integration and Technical Support

Plan integration at the hierarchic level is an important aspect of decentralized development. Plans prepared by the Panchayats were supposed to be integrated at the block and district levels. This integration was supposed to be organized in such a way that the plans prepared for the higher tier units complemented those of the lower tiers. Integration was also planned for the ongoing centrally sponsored schemes implemented through the community development blocks. At this point, it was realized that the District Planning Committees did not have sufficient technical staff to undertake proper scrutiny of the vast number of projects that emerged from the Grama Panchayats. People in the Grama Panchayats had difficulties in understanding the technical issues in project formulation, and especially in budgeting, comprehending engineering drafts and formulas, and presenting statistical tables required by the state Government.

To deal with this problem, campaign organizers came up with the idea of appealing to Kerala's many retired but active experts. Four thousand volunteers were mobilized as Volunteer Technical Corps (VTC) who gave at least one day per week of free time to help evaluate and improve the project proposals prepared by the Grama Panchayats. Around 50 per cent of VTC were Graduates, Post Graduates, M.Phil., and Ph.D. degree holders and 41 per cent were professionals. The VTC later merged with other expert groups to become the Expert Committees that helped bypass contractors and reduce corruption. The Voluntary Technical Corps were expected to provide the expertise required at the Block or Corporation or District levels. The VTC had no right to change the priorities set by the local councils. Their functions were confined to technical and financial appraisals of the projects and providing suggestions for modifications so that the projects could be made viable. Local Governments have absolute freedom in formulating and implementing projects based on their priorities, subject to plan guidelines issued by the state Government. There was a massive transfer of functions, functionaries, and funds within ten months from the beginning of the Campaign during August-September, 1996. Capacity building through training followed giving responsibilities; and accountability followed the transfer of power and financial resources. It was one of the most significant institutional reforms in governance.

FINANCIAL DEVOLUTION

Financial decentralization is one of the three pillars of decentralization. Kerala has successfully implemented the devolution of Plan funds based on a non-discretionary and equitable formula. The local self-governments now receive funds to implement the schemes devised by them. Since 1997-98 an average of 25 per cent of the state's investible resources have been devolved to local governments as development funds (Government of Kerala, 2018). Another 9 per cent of funds are shared with the LSGs from the states' tax revenue. LSG receives funds under the general category, Scheduled Caste Sub-Plan, and Tribal Sub-Plan. The outlays are fixed based on recommendations of the State Finance Commission as envisaged in constitutional amendments. However, adequate fiscal autonomy is yet to be granted to the panchayats who remain heavily dependent on transfers from Central and State Governments.

TWO DECADES OF PEOPLE'S PLAN CAMPAIGN (PPC)

Two decades have passed since the introduction of PPC. There had been perceptible changes in plan formulations at the ground level and several milestones have set. There were also changes in modus operandi with the change in Governments and their priority. Nevertheless, the process set by PPC continued with varying momentum. One of the issues related to project implementation by the Grama Panchayats is the procedural delay in plan preparation, approval, and implementation as the financial year ends by March 31st. This issue is being addressed by the Government and new procedures are set so that formulation of annual plans could be completed by the first quarter of the year, so that, there is sufficient time for plan implementation. There is also a problem with the integration of local plans and annual plans, which is being addressed. Participation at the Grama Sabha level is found to be waning and the enthusiasm is missing in many cases. To address these issues, beneficiary committees are being introduced. It is expected that the beneficiary committee can halt degeneration and also weed out the corruption that has crept in over the years for plan implementation at the lower level. One of the significant events in Kerala is the unprecedented growth of urbanization. Around 48 per cent of the total population in Kerala now lives in urban areas. The growth of urban infrastructure is not at the same pace. Formulation of plans for Urban Local Governments is an important issue in the decentralized development of Kerala. The state-level mission mode activities are also being integrated with the activities of Local Self Governments.

DISTRICT PLAN

The preparation of district plans is mandatory as envisaged in 73rd and 74th constitutional amendments. There are several attempts in various states across the country to workout district plans as a part of the exercise of decentralized planning. However, the approach was mostly centralised and top-down. Constitutional amendments clearly spelt out that the district plan will be a consolidation of the plans prepared at the lower tiers of Block and Gram Panchayats. It was also constitutionally mandated to establish the District Planning Committees (DPCs) to accomplish the task of consolidation. There were various initiatives undertaken to work out a biophysical data base for all the districts in Kerala as a part of PPC. Incidentally, it may be noted here that terrain and drainage maps were prepared for all the districts in Kerala as part of a project on Terrain Analysis of Kerala (Chattopadhyay and Mahamaya, 1995). Block-level maps are also prepared to consolidate Block level data. In 2000, the development perspective report was prepared for the Trivandrum district as a first step towards formulating a district plan (Mridul *et al.*, 2000). However, it had not progressed further.

While the importance of preparing the district plan is well-realized, there are debates on approach and methodology. The Planning Commission of India (2008b) has brought out a manual for 'Integrated District Planning'. It provides a guideline for working out the district plan by the states. It also indicates the reforms that are required at the state level in order to prepare a participative district plan. Town and Country Planning Department, Government of Kerala prepared Kollam district plan following guidelines prepared by the Planning Commission, Government of India. Recently, The Government of Kerala took a renewed interest in the preparation of district plans to promote multi-level planning and to ensure greater integration between the development needs and priorities of various tiers of local governments and different socio-economic classes in the state. The task has been entrusted to the District Planning Committee.

The necessity of a bottom-up approach, i.e. preparation of plans at the Gram Panchayat level and aggregation of those plans at the subsequent higher levels is well accepted, however, there is a lack of adequate ground-level data to accomplish this task. This is another problem that warrants due attention. In absence of the required ground level data base that this exercise requires, it is imperative to work out a methodology that will help designing districts plans till grassroot level data for each village are fully generated.

KUDUMBASHREE PROGRAMME

This is one of the programmes manifesting the strength of the local need-based approach, innovation, and people's participation. All three are essential components of democratic decentralization and local level development. From a modest beginning as Community Based Nutrition and Poverty Alleviation Programme (CBNP) in Malappurm district in the early 1990s, up-scaled in 1997 as Kudumbashree and turned as the main vehicle for poverty eradication in 1998 this programme has come a long way and succeeded to 'alter lives of poor women in the state, change their perception, rebuild their confidence, boost their morale, rediscover their dignity and honour, empower them economically, socially and politically' (Jose undated). The Kudumbashree programme, acclaimed nationally and internationally, is considered a model programme worth emulating by other Indian states. Its functions can be broadly grouped under three sectors: social, economic, and women empowerment. Its activities are wide ranging including that of micro-finance, micro-enterprise, skill development, mobilization of people, health, education, and environment management.

Kudumbashree mission is a comprehensive self-help group-based activity. It can boast of 4.4 million women participants or 25 per cent of the total female population in Kerala in 2017. There are 277,175 Neighbourhood Groups (NHGs), 19,854 Area Development Societies (ADSs), and 1073 Community Development Societies (CDSs) distributed in all the districts (Table 13.4). Even Neignbourhood Groups (NHGs) are formed for elderly people, disabled persons, and transgender people. Besides, there are 66,743 children's NHGs known as *Balasabha* covering 1.06 million children. Kudumbashree mission has penetrated all parts of Kerala. Kudumbashree executes programmes in close cooperation with local self-government organizations. This positive linkage has great potential in facilitating women to enter local self-government institutions (John, 2009) and thereby further strengthening and deepening democratic decentralization.

Table 13.4: Distribution of Different Groups under Kudumbashree Mission by Districts in Kerala, 2018

Districts	*NHGs*	*CDSs*	*ADSs*	*Balasabha*	*No. of Children*
Thiruvananthapuram	29900	83	1546	4686	72492
Kollam	25112	74	1420	4022	56584
Pathanamthitta	9976	58	920	2654	43469
Alappuzha	20418	79	1384	3523	53054
Kottayam	15333	78	1344	3105	45236
Idduki	12381	54	861	1738	18636
Ernakulam	23197	101	1833	7110	129375
Thrissur	24245	100	1794	4880	77264
Palakkad	22468	96	1730	14100	248750
Malappuram	26582	110	2257	7525	136682
Kozhikode	27772	82	1566	5666	71496
Wayanad	9584	26	512	1426	16687
Kannur	19610	81	1545	4150	67750
Kasaragod	10597	42	777	2158	21808
Kerala	277175	1964	18489	66743	1059283

Data Source: www.kudumbashree.org/pages/518, June, 2018.

FUTURE CHALLENGES OF DECENTRALISATION

Kerala's development history demonstrates 'that critically examining one's context and making, by mass protest if necessary, the appropriate adjustments leads to a radically more healthy, egalitarian, and meaningfully informed society' (Brown, 2013), which is the goal of decentralisation and sustainable development. There are several lessons to note from Kerala's experience. These are useful to understand the process of decentralised planning and development. Other states in India striving to implement democratic decentralisation may also find them applicable. Since the formation of Kerala, there were attempts to introduce political and administrative decentralisation from time to time. However, most of these attempts were of limited success till the launching of PPC in 1996. With the constitutional mandate, the state could initiate political, administrative, and fiscal decentralisation and has succeeded to a considerable extent. A significant achievement of this programme is that the agenda of decentralisation has been forced into the public discourse on development (Kannan, undated).

If the six principles of decentralisation—territoriality, mobilisation, participation, local ownership and management, multidisciplinarity, and nesting and integration are considered, the most spectacular achievements are in the matter of mobilisation, participation, and multi-disciplinary approach. The architects of the programme could successfully infuse a belief among academicians, researchers, politicians, and people spearheading science movements, that decentralisation is necessary for the sustainable development of Kerala. Existence of such a group in Kerala is a product as much as of social and political movements and reforms in pre-Independence era as that of post-independence, more particularly, post-state reorganisation radical political initiatives, land reforms, labour reforms, and the emergence of concerned, informed and educated citizenry, who would like to assert their citizenship through participation and by questioning the government. Existence of 'a differentiated civil society—both an effect and cause of movements—more than anything else that explains the cumulative and in particular democratising impact of social movements in Kerala'(Heller, 2005 : 100). When PPC was launched, Kerala society was perhaps ready for such a big bang in participatory development.

The programme was able to galvanise disparate groups' interests create scope and space for participation, and build institutions in lower tiers. There were synergies among different stakeholders and power groups. Bureaucrats, technocrats, academicians, research and development professionals, unions, associations, civil societies, party workers, and people science movements extended sustained support in various stages. As the programme was chalked out meticulously, there were proper opportunity structure, and institutional design for the participation of all stakeholders. Poor people and marginal social groups were involved in the process of planning through Grama Sabhas. Kudumbashree mission with emphasis on NHGs, CDSs, and ADSs have further strengthened the process of participation of marginal and poor groups. The PPC was introduced from the top, however, with progress, it turned into a continuous process of learning-by-doing with adequate space for dialogue and participation even before institutional set up was firmed up. Panchayats in Kerala succeeded to produce annual plans, design projects, and improve project execution. Panchayat plans emerged as a negotiated priority followed by data base analysis. However, over the years many of these positive developments had slid down. Spontaneity, participation, and innovativeness waned. This necessitated a re-examination of methods followed and the introduction of remedial measures.

In the matter of territoriality, ownership, and nesting, and integration there are several steps yet to be taken to strengthen the decentralisation process. Emerging tasks include characterisation of Panchayats with respect to territoriality, synthesis of panchayat plans along with the hierarchic order, and empowering Panchayats for ownership. Spatial integration of panchayat plans at the Block and District levels are important steps towards nesting and integration. This requires regionalisation. An integrated district plan as is being proposed now may help in this process. Decentralisation as a development strategy has advantages and disadvantages. While there are many advantages, specially, in the matter of efficient delivery of services and public goods and better matching of policy and needs of local people, questions are often raised about the capacity of the local bodies to address larger economic issues and overcome territorial disparity. It is necessary to introduce multilevel planning and decentralisation simultaneously to address these concerns. Decentralisation is a continuous process and requires proper nurturing, technical input, capacity building, and institutional structure. Kerala is entering into the 3rd generation of development initiatives with a major thrust on knowledge economy and innovations. Priorities and local aspirations are changing. The decentralised planning needs necessary reorientation to cater to these changing circumstances. The campaign mode had served its purpose by creating a socio-political-technical space for discourse on development, participation, and local action. The next phase is to build over that base and take necessary steps to strengthen decentralisation for transiting to sustainable development.

CHAPTER

14

Natural Hazards and Disaster Management

NATURAL HAZARDS AND DISASTER MANAGEMENT: AN EMERGING CONCERN

Natural hazards and disaster management have drawn considerable attention from scientific and academic communities and decision-makers during the last couple of decades as almost all countries face one type of disaster or the other and loss due to disaster is increasing over the years. Flood, cyclone, drought, earthquake, tsunami, volcanic eruption, forest fire, sea surge, landslide, soil erosion, etc. all are natural processes and form an integral part of our geo-system. But for these events, present-day distribution of land, water, and air would not have been possible. A natural event turns into a disaster when it causes loss to the economy and human lives. Accelerated changes in all human activities including population number and indiscriminate extraction of natural resources have disturbed the ecosystem balance, pushed a section of people to the marginal areas, and thereby increased the risk of human suffering, death, and destruction. Human occupancy of hazard-prone areas is increasing and thereby rendering more and more people vulnerable to mounting economic losses. The situation is further complicated due to man-made disasters mostly associated with various human activities and inappropriate development decisions.

India with its sub-continental nature and having wide variations in geographic, geologic, and climatic conditions is prone to various types of hazards. Hazard affected areas overlap and most of the states in India experience more than one type of hazard or are multi-hazard prone. It has been estimated that 68 per cent of the land in India are drought-prone, 60 per cent are prone to earthquake, 12 per cent are flood-prone and 8 per cent of area experience cyclones (Sharma and Asutosh, 2012). The average annual economic loss in India is INR 1000 crore and 70 per cent of this loss are on account of the flood. Disaster affects around 50 million people in our country every year. As a whole, around 85 per cent of the land area in our country is vulnerable to natural hazards and 22 states have been marked as hazard prone states.

NATURAL HAZARDS AND DISASTER: SOME CONCEPTUAL ISSUES

Natural hazards, as the term implies, occur naturally. It is an extreme event that causes harm to society. Extreme events are unusual phenomena and not necessarily always harmful. Disaster occurs when an extreme natural event coincides with a vulnerable situation surpassing a society's ability to control or survive the consequences (Kreimer and Munasinge 1990). The Government of India (2011) defined disaster as a "catastrophic situation in which the normal pattern of life or ecosystem has been disrupted and extraordinary emergency interventions are required to save and preserve lives and or the environment'. Another two terms, vulnerability, and risk are also used in disaster management literature. Vulnerability is defined as the state of exposure to the possibility of being harmed or the degree to which people, property, resources, ecosystems and cultural, economic, environmental and social activity is susceptible to harm due to an extreme event. Risk is the possibility or chance of loss due to external and internal vulnerability. It is, usually, worked out based on the likelihood of natural hazards and vulnerability to natural hazards. The intensity of risk is directly proportional to the incidence of hazard and vulnerability. The terms like disaster, vulnerability, and risk are largely anthropocentric and are used mostly from human perspectives. Kapur (2008, 2010) deliberated on disasters and vulnerability of India in detail and provided certain important insights having policy implications.

TYPES OF HAZARDS

Hazards are usually categorised into four types:

(i) Geological hazards—earthquake and volcanic eruptions;
(ii) Meteorological hazards—cyclone, hurricanes, heat waves, cold waves, freezing rain, lightning;
(iii) Hydrological hazards—flood, drought, landslides, soil piping and tsunami; and
(iv) Biological hazards—various types of diseases.

Besides, there are man-made hazards like accidents, suicides, drowning, terrorism, and war, etc. causing the death of people and loss of property. The High Powered Committee, constituted by the Government of India identified 32 types of disasters, categorized under five groups based on generic considerations and departmental responsibilities (Government of India, 2011). These are: (i) Water and climate-related disaster, (ii) Geologically related disaster, (iii) Chemical, industrial, and nuclear-related disaster, (iv) Accident related disaster, and (v) Biologically related disaster. We shall follow this classification while discussing natural hazards in Kerala.

NATURAL HAZARDS IN KERALA

Kerala is not exposed to highly devastating natural hazards like many other parts of the country. In fact, Kerala enjoys a sheltered condition due to the presence of the Western Ghats along the eastern border and passive continental margin along the western boundary. Nevertheless, there are several hazards that affect different parts of Kerala from time to time, and the incidence of hazard is increasing in recent years. Due to geographical location, weather conditions, and high population density, Kerala faces both natural and man-made disasters and is multi hazard-prone. Around 14 per cent of the total area of Kerala is flood-prone; 10 per cent area are landslide prone; nine districts bordering coastline are exposed to various coastal hazards; lightning kills several people throughout the state; low-intensity earthquake occurs from time to time; and there is the recurrence of drought severely impacting agriculture and drinking water conditions. Soil piping is emerging as a menace causing land subsidence in various parts of Kerala. Besides, there are manmade calamities like suicide, road accident, and drowning. Each of these hazards affecting Kerala is briefly discussed here.

WATER AND CLIMATE RELATED HAZARDS

FLOODS

Flood is a natural hazard that affects most countries around the world. There are four basic types of floods: flash flood, standing flood or waterlogging, riverine flood, and sea surges. Our discussion here will be limited to the first three types of floods as they are interrelated, rainfall dependent, and mainly concerned with a drainage problem. Sea surge has been discussed under coastal hazards. The tropical countries with seasonally concentrated high rainfall are subjected to recurrent flooding and suffer from heavy economic and human losses. India is one of the most flood-affected countries in the world that witnessed some of the deadliest floods. Floods invade some parts of the country almost every year during the monsoon. There were 12 major floods affecting large parts of the country between 1980 and 2009. The last five years, 2014, 2015, 2016, 2017 and 2018 witnessed devastating floods causing several deaths in each case. The states like Assam, Arunachal Pradesh, Bihar, Gujarat, Orissa, and West Bengal, suffered from a flood in 2017 with a total death toll crossing 700 persons and around 11 million people were affected. The 2018 flood was one of the most devastating floods in the recent history of Kerala. Analysis of national-level data for the period of 1953-2009 reveals that on average, annually, 7 million ha of croplands are affected, 1464 lives are lost, 86288 heads of cattle die and damage to crops, houses and public utilities in the country amounts to around Rs. 16,500 million (Sharma and Asutosh, 2012). The average annual loss is increasing over the years as investments in the flood-affected areas are multiplying with growth and development.

Geological Survey of India estimated that 12.5 per cent of the country's total area falls under the category of major flood-prone area distributed in the Bramhaputra river valley, the Ganga river valley, North West riverine tract, and Central India, and Deccan region. Flood prone areas increased from 25 million ha in the 1960s to 34 million ha by the end of the 1970s and to 40 million hectares by 2015. Kerala does not figure in the category of devastating floods like those affecting other states in the country as rivers in Kerala are small. However, high rainfall and undulating topography cause severe flooding in many parts of Kerala. The Central Water Commission (CWC) estimated that an area of about 8,700 km^2 or 22.37 per cent of the total geographical area of Kerala is flood-prone.

Incidence of Floods in Kerala

Historical records indicate that an extremely high flood in 1341 AD in the Periyar river led to the silting up of the Cranganore port that had been a major trading center for centuries. It opened a new outlet to the sea at Cochin. Since meteorological recordings began in 1872, the most devastating flood affecting Kerala was in 1924. This incident is popularly known as 99 Flood due to its occurrence during the Malayalam Era of 1099. The next incidence of the severe flood was in the year 1961causing considerable loss of property. In recent decades, the floods of 1989 and 1992 were severe, although rainfall in both the years was lower compared to that in 1961 and 1924 (Fig. 14.1). High rainfall in consecutive days and geo-environmental status of river basins, both upstream and downstream are significant factors contributing to flood incidence. This point has been elaborated further in a separate section. Kerala lost more than INR 7,750 million due to the 1989 flood. More than 2 million people were affected; about 135 thousand houses were damaged and crops of more than 65 thousand hectares were destroyed (Chattopadhyay and Franke, 2006). During the 1992 devastating flood, 988 villages were affected and 60 lives were lost.

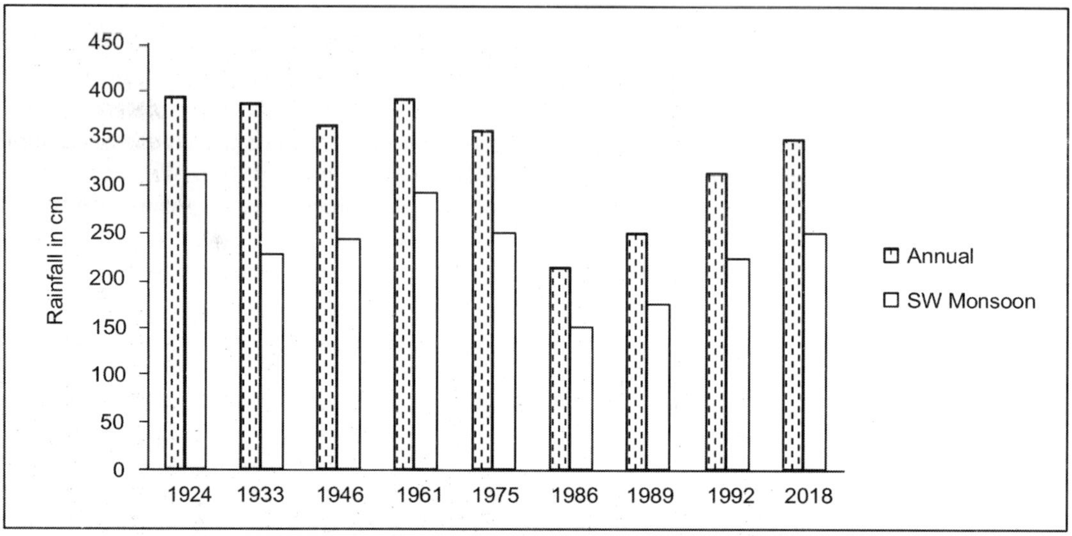

Fig. 14.1: Annual and South-West Monsoon Rainfall during High Flood Years and Selected Non-Flood but High Rainfall Years since 1924 in Kerala. (Flood years: 1924, 1961, 1975, 1989, 1992, 2018)

Data Source: IMD.

In the current millennium, the year 2009 witnessed widespread flood affecting 9 districts and causing 13 deaths. Kannur district alone incurred a loss of INR 52 crore. Flash flood and landslides affected Kannur and Kozhikode districts in 2012 with the death of 9 persons. In the following year of 2013, flash floods and landslides left 20 people dead and 20,000 people displaced. Kuttanad was the worst hit in July 2013 with the repeated incursion of the flood. The second paddy crop was completely destroyed and embankments protecting paddy fields were on the verge of being collapsed. Heavy rains soaked Kerala in mid-September 2017, considered to be the biggest rainfall event since 2013. The districts of Kozhikode, Wayanad, Palakkad, Kottayam, and Alappuzha witnessed widespread flood. Waterlogging in urban areas like Thiruvananthapuram and Kochi was reported. The flood of 2018 was an unprecedented event in Kerala since 1924. We provide a separate discussion about the flood event of 2018.

Flood of August 2018

The recent flood in Kerala has drawn national and international attention for more than one reason. Kerala apparently insulated from large scale natural disaster was suddenly exposed to unprecedented flood havoc claiming 488 lives and affecting 56 lakh people across the state in August 2018. Heavy rainfall and floods are intertwined. The average annual rainfall in Kerala is around 300 cm. About 70 per cent of this rainfall precipitates during the south-west monsoon spreading from June to September. Like all previous years this year also, monsoon broke on 1st June 2018. Rainfall crossed normal rainfall by 2nd week of June itself. Heavy rainfall continued for more than two and a half months except the 1st week of July. Rainfall received in Kerala from 1st June to 18th August was 42 per cent more than the normal rainfall usually received during this period. In a span of seven days, from 9th to 15th August, Kerala received 35 cm rainfall, whereas normal rainfall for the entire month of August used to be 32 cm. Idukki district, source of several major rivers in Kerala, received 568 per cent more rainfall than the normal between 13th and 19th August. Volume of river discharge was 12,300 million m^3 in just three days from 15th to 17th August. This was around 16 per cent of the total volume of water received by the state annually. Kerala was completely rain-soaked by the first week of August. Soils were saturated; wetlands were flooded and reservoirs were full to the brink. Shutters of 24 reservoirs were opened by 10th August and during the next 2/3 days, shutters of several other reservoirs were also opened. Flood and landslides intensified since 8th August turned into deluge during subsequent days.

Ten out of 14 districts experienced flood fury. Fourteen lakh people have to be moved to temporary relief camps and around 60 million people were affected. There were heavy losses of infrastructure like roads, bridges, schools, colleges etc. apart from loss of agriculture. UNDP pegged the total loss at INR 31 thousand crore (Table 14.1). This amount had not taken into account the ecosystem damages like loss of topsoil of large tracts of land, unsorted sediment deposit over productive agricultural land, sediment washout due to strong surface run-off, and biodiversity loss to indicate a few. The flood of 2018 was often compared to the flood incidence of 1924 when the entire state of Kerala was inundated. However, in 1924, the flood occurred during 16th–18th July with total water yield of 12.6 billion m^3, whereas, in 2018, the incidence of the flood took place from 15th to 17th August with a total water yield of 12.3 billion m^3. Occupancy of flood plain had increased manifold during this intervening period of 94 years. Accordingly, losses have also been multiplied.

Table 14.1: Loss due to August 2018 Flood in Kerala

Human lives lost	:	488 persons
Villages affected	:	774 nos.
PWD roads damaged	:	16,000 km
Regional roads damaged	:	82,000 km
Bridges damaged	:	182 nos.
Cattle lost	:	46,000 heads
Poultry lost	:	2,00,000 heads
Total estimated loss	:	INR 31,000 crore

Data Source: Computed from Newspapers, August-September 2018.

Flood Prone Area

The term flood plain is used to denote the land adjacent to a stream or river stretching from its bank to the base of the valley wall and experiences flooding during periods of high drainage discharge. All flood plains are flood-prone. Kerala landforms are characterized by hills, undulating terrain, plains, and valleys. Around 14.5 per cent of the total area of Kerala is prone to flood (Mathai, 2009). District level distribution indicates that the Alappuzha district has the maximum area of 763 km^2 under the category of flood prone area closely followed by Ernakulam (719 km^2) and Thrissur (688 km^2) (Table 14.2). These three districts along with adjoining

Kottayam district account for 47 per cent of the total flood prone area in Kerala. Geographically, the area is more or less contiguous around the Vembanad lake and spreads over greater Kuttand in Alappuzha and Kottayam districts and the "*Kole*" (depression) lands of Trichur and Ernakulam districts. Greater Kuttanad is the most susceptible to floods due to its peculiar landform. This area is a deltaic formation bordering the southern and eastern margin of the Vembanad lake, characterized by 'saucer shaped islands' lying mostly 1 to 2 meter below mean sea level. Every year, during the monsoon, the entire landmass gets waterlogged. Kuttanad taluk within Alappuzha district having 282 km^2 or 92 per cent of the total area of taluk as flood-prone is the most flood-prone taluk in the state. It is usually flooded during every monsoon. Floods of a 10 year return period are generally severe in Kuttanad. In fact, people living here consider flood as part of their lives and they have traditionally adjusted to live with flood and evolved agricultural practices suitable for waterlogging. In recent years, these practices are being discontinued in favour of modern cultivation methods.

Besides the Vembanad lake basin, the flood plains of Bharatappuzha, Chaliar, Vallapatnam, Kallada, and Vamanapuram are flood-prone. Broad valleys within the plateau areas in the Western Ghats experience flash flood. Wayanad district has around 11 per cent of its area under the flood-prone category. Spatial distribution of floods indicates that flooding does not depend on absolute elevation rather on relative height with respect to the river bed and the sudden rushing of surface run-off (flash flooding) that is generated in the surrounding highlands and slopes. Flash floods in the hilly districts are invariably related to landslide/landslips causing loss of human lives and damage of property.

Table 14.2: Distribution of Area Prone to Different Types of Hazards by Districts in Kerala

District	*Flood prone (km^2)*	*Landslide prone (km^2)*		*Lightning incidence (No.)*	*Storm surges (km^2)*	*Coastal erosion (km)*
		High	*Low*			
Thiruvananthapuram	268.0	46.0	115.0	59	10.5	59.0
Kollam	284.0	76.0	191.0	257	31.5	40.0
Pathanamthitta	213.0	170.0	426.0	125	-	-
Alappuzha	763.0	-	-	51	44.0	34.0
Kottayam	461.0	62.0	191.0	218	-	-
Idukki	39.0	388.0	874.0	92	-	-
Ernakulam	719.0	61.0	229.0	104	58.0	33.0
Thrissur	688.0	108.0	217.0	36	15.0	21.0
Palakkad	567.0	325.0	367.0	83	-	-
Malappuram	602.0	198.0	268.0	249	21.0	22.0
Kozhikode	289.0	109.0	207.0	242	19.0	44.0
Wayanad	215.0	103.0	197.0	19	-	-
Kannur	339.0	169.0	273.0	353	31.0	27.0
Kasaragod	199.0	34.0	206.0	108	34.0	34.0
State	5646.0	1849.0	3761.0	1996	264.0	314.0

Data Source: Mathai, 2009.

Alteration of Basin Ecology and its Impact on Flooding

Flooding is associated with excess rainfall. However, a study analysing rainfall and flood incidence over a period of 37 years (1960-1998) indicated that flood has occurred even during the years of below-average rainfall (Chattopadhyay and Franke, 2006). The occurrence of flood incidence has increased and with slight unusual rainfall, which is now common in Kerala, there are floods as flood plains and spill areas have significantly reduced and thus lost their capacity to accommodate excess water. Human intervention alters the basin ecology, contributing to increased flooding and changing flood patterns. The main human activities influencing basin ecology are reclamation of wetlands, shrinking of water bodies, floodplain occupancy, embankment construction, and changes in the land use and land cover patterns in the upper catchment. In Kerala, like in many other places around the world, these activities continue unabated. Even in developed countries like the United States of America, loss from flood continues to increase primarily due to expanded use, and occupation, and development of downstream floodplains.

Conversion of wetlands to paddy fields and to drylands for settlement and tree crop culture are widespread phenomena throughout Kerala. River catchments have lost most of their natural vegetation cover due to the expansion of plantation agriculture and the spread of cultural landscape within the erstwhile forest tracts. This has reduced the water holding capacity of the catchments and increased surface run-off. By the end of the 1990s, there was only 58 per cent of 5,736 sq.km of wetlands and paddy fields, which had retained their wetland condition and functioned as flood cushioning area (KSI, 1994). This has further diminished as is evidenced by the reduction of area under paddy cultivation. Gross area under paddy covered only 1960 km^2 in 2015-2016. Paddy is cultivated in three seasons—autumn, winter, and summer. Considering the maximum coverage under paddy in anyone of these three seasons for each district, it emerged that the net area under paddy would be around 1144 km^2. Kuttanad, the most flood-affected area in Kerala experienced considerable change in land use patterns. Paddy fields, flood spill area, are limited to less than one-third of the total area of Kuttanad of 1400 km^2 (Fig. 14.2). This indirectly indicates the nature of reduction in open space or flood spill area. As a result, the height of the flood-water in

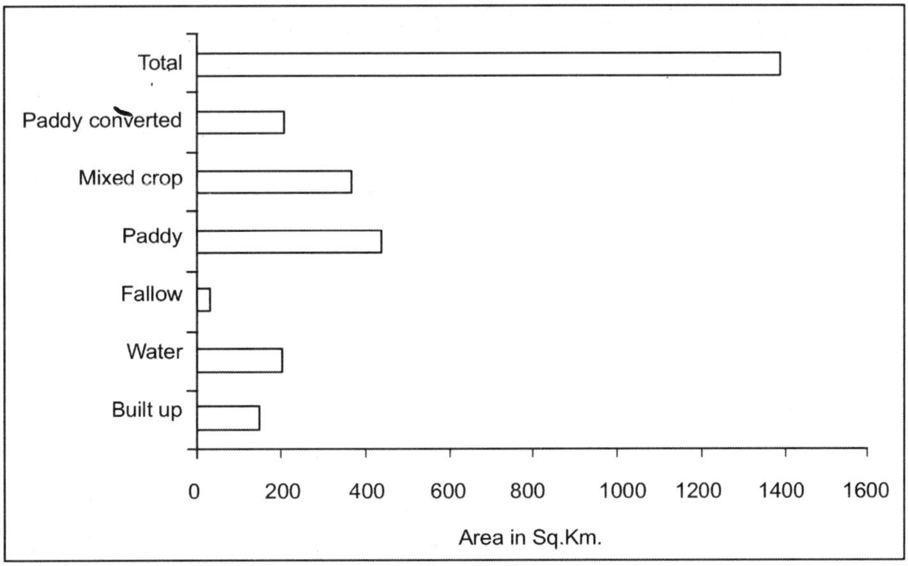

Fig. 14.2: Land use in Kuttanad of Alappuzha District in Kerala, 2018.

Data Source: Kerala State Land Use Board, Government of Kerala, 2018.

the rest of the open areas increases, and the flood moves upstream causing an increase in flood prone areas. It also causes a breach of embankments in many places.

Drastic reduction and clogging of canal and drainage systems in and around urban areas like Kochi and Thiruvananthapuram have compounded local flood problems further. The duration of "standing floods" has increased significantly. Even during normal monsoon, these urban centers experience prolonged and deep waterlogging conditions (Fig. 14.3). An increase in population concentration and other infrastructures erected within the flood plains and alteration of basin ecology resulting in the reduction of flood cushioning area and shorter duration of the residence time of water in the upper catchments have contributed to enhanced flood proneness and damage. Therefore, flood management is no more a hydrological analysis of rainfall-run-off behaviour. It is important to consider a larger perspective and address several other parameters, which are mostly anthropogenic, controlling water movement within the basin.

Fig. 14.3: Waterlogging in Parts of Thiruvananthapuram City in Kerala, 2015. Photo by Shaji Johnson.

Strengthening of Non-Structural Measures

There is a growing realization globally that structural measures of flood mitigation like the construction of reservoirs, raising of embankments' height and provisioning of drainage facilities may not be the appropriate solutions. Therefore, it is necessary to look for non-structural alternatives and adapting to flood incidence by managing our activities particularly in the flood plains and flood cushioning areas. Non-structural flood management activities include

(i) channel maintenance — thalweg, river bank, and river gradient; (ii) flow maintenance — moderation of flood flow, and maintenance or improvement of dry season flow or base flow; (iii) catchment conservation; (iv) flood plain conservation; (v) restriction on flood plain occupancy; (vi) restoration and rejuvenation of palaeo channels and water storing facilities; (vii) zoning of flood plains with different risk factor; and (viii) limiting the conversion of wetlands for other purposes. Such measures can be taken up with a holistic view considering the entire river basin as a unit of planning. It helps to identify problems, setting a proper boundary for activities like bank protection, catchment conservation, erecting of check dams, riparian vegetation, flood plain conservation, room for rivers, and devising activity zone based on the flood risk factor. A micro-level watershed-based approach is necessary to understand the process of flood incidence, to prioritize, and to undertake proper intervention measures. The thrust should be on the area planning approach for flood plain management and adaptation to the flood incidence and not merely flood water dispersal and structural mitigation measures. Proper institutional arrangements delineating the role of different tiers of administration, Panchayats, and technical support are necessary and finally, people's participation in flood management should be ensured at the ground level.

OTHER METEOROLOGICAL HAZARDS (CLIMATE RELATED)

Meteorological hazards cause the maximum loss of lives and destruction of property. It is of two types: (i) those on-setting gradually like droughts, and (ii) those occurring suddenly like tropical cyclones and the associated storm surge, lightning, severe thunderstorms, and associated tornadoes and large hail, torrential heavy rains that cause floods and landslides, and heat and cold waves (Joseph, 2009).

Meteorological Drought

Droughts are broadly classified as (i) Meteorological drought, (ii) Hydrological drought, (iii) Agricultural drought, and (iv) Socio-economic drought. The last three drought conditions are largely the result of meteorological drought, which is being computed based on rainfall. We shall discuss here only meteorological drought. The India Meteorology Department (IMD) declares drought in any area when rainfall deficiency in that area is ≤ 26 per cent of its long term normal rainfall. Drought condition is further classified into moderate and severe depending on rainfall deficiencies between 26 per cent to 50 per cent and > 50 per cent respectively (Shewale and Shravankumar, 2005). However, administratively, in India, a situation is considered drought when rainfall deficiency exceeds 10 per cent, and the affected area crosses 20 per cent of the total area under plains in the country (*ibid*). Kerala is a wet state, nevertheless, it experiences drought conditions from time to time. Recently, IMD has classified Kerala under the mild to moderately drought category. The country-level analysis of long-term (1875-2005) data indicated that Kerala has an 8 per cent probability of moderate to severe drought incidence and consecutive drought years were 1951-52, 1965-66, and 2002-2003 (Shewale and Shravankumar, 2005).

The State-level study covering 109 years from 1901 to 2009 indicated 53 drought years of which 13 were of severe nature and only four droughts were disastrous (Gopokumar, 2011). KSDMA (2010) listed 66 drought years from 1871 to 2000. On average, Kerala experiences one year of deficit rainfall every two years. The decadal distribution shows that the incidence of droughts increased since 1961 and each of the last five decades witnessed 6/7 drought years of various intensity. Since 1980, the officially declared drought years were 1983, 1986, 1987, 1992, 1997, 1998, 2002, 2003, 2004, 2010, 2012, and 2013 (KSDMC, 2016). The period spanning from September 1982 to June 1983 experienced the maximum long-duration dry spells griping most parts of Kerala. It was a late drought experienced during March, April, and May of 1983. About 3.23 lakh hectares of paddy cultivation were lost during the 1983 drought. Coconut, rubber, coffee, and tea crops were also affected. The 1987 drought was also severely impacting the entire state of Kerala and affecting 9.82 lakh hectares of crop land and 6 lakh cattle population (Nathan, 2000). The estimated loss was INR 1000 crore in that year. A similar drought situation occurred during 2003-2004, which was early drought with water deficiency in the months of October, November, and December of 2003 and January and February months of 2004 (Rao, 2010). Recently, 2012-2013 was a major drought year and perhaps the worst in 115 years when the Government of Kerala declared all 14 districts as drought-affected (KSDMC, 2016).

In general, the climate in Kerala is turning towards dryness, with a 3.32 per cent increase in aridity index (Gopokumar, 2011). The epochal behavior of the Indian monsoon with alternate dry and wet periods of 30 years duration seems to be interrupted as expected wet epoch of the 1991-2020 period already witnessed several drought years (Joseph, 2009). Generally a drought year is followed by a year of normal or excess rainfall. The phenomenon called Tropical Biennial Oscillation is part of a larger global meteorological dynamics in which the summer monsoon of India and Australia and the sea surface temperature of tropical Indian and Pacific Oceans play an active role (*ibid*). The Nepha (or cloud) analysis based on satellite data product indicated the shifting of clouds far over southern India during drought situations like in 1966 and there was no high-level moving system of waves in the upper tropospheric easterlies (Nathan, 2000). Orography plays little role in drought incidence in Kerala. Combating drought may not be possible as it is related to global climatic circulation however, adequate preparedness for drought situations can be an important disaster management initiative.

Cyclonic Storms

The Probability of the occurrence of the cyclonic storm in Kerala is low. The state is grouped in the category of IIA, low vulnerability zone, in all-India perspective. During a period of 116 years, from 1891 to 2007, Kerala experienced 23 cyclonic storms or an average of one storm every five years. Among these 23 storms, 18 were severe cyclonic storms and the rest five were cyclonic storms. Two systems formed in the Arabian Sea and struck the Kerala coast as a severe cyclonic storm. During the last 35 years, a total of eight systems with an intensity of Severe Cyclonic Storm originated over the Bay of Bengal, crossed over the east coast of India, and reemerged in the Arabian Sea as depression (IMD, 2017). Ernakulam and Malapuram districts are mostly affected by storms. All 14 districts in Kerala have been listed as moderately vulnerable to cyclone wind, and fall in the lowest category of cyclone proneness in all-India perspective (Mandal and Mohapatra, 2010).

Cyclone Ockhi

The tropical cyclonic storm of a very severe category, named Ockhi by IMD, affected parts of the Tamil Nadu coast, Kerala coast, and Lakshadweep during the last few days of November and early December of 2017. Cyclonic storm in the North Indian Ocean is a regular event occurring during the period from April to December with peak incidents reported in the month of May and November. The year 2017 witnessed 10 depressions of which 6 were deep depressions, which intensified in the category of cyclonic storms (3 nos.), severe cyclonic storms (2), and very severe cyclonic storm (1) (Table 14.3). The cyclone Ockhi was the most intense storm after cyclone Megha which occurred in 2015. The origin of this cyclone can be traced to the Gulf of Thailand where a low pressure formed. The depression moved to the south-east coast of Sri Lanka, gained strength, and intensified from depression to a very severe category and affected the Indian coast. The maximum wind speed rose to 155 km/hr.

Table 14.3: Cyclonic Storms in North Indian Ocean in 2017

Type of depression	*Wind speed (Km hr-1)*	*Number of occurrence*
Depressions	31-50	10
Deep depressions	51-62	6
Cyclonic storms	63-88	3
Severe cyclonic storms	89-117	2
Very severe cyclonic storm	118-165	1
Extremely severe cyclonic storm	166-220	-
Super cyclonic storm	≥ 221	-

Data Source: IMD, 2017.

Hundreds of fishermen were reported missing and several of them lost their lives. By the end of the year, 2017, Kerala alone reported 76 death and 144 fishermen missing. The fact-finding team of the Government of India proposed Rs. 404 crore relief to Kerala. Loss of lives was mainly due to infrastructural weakness and communication gaps. The fishermen had already moved to the sea by the time information about cyclones could be disseminated. The rehabilitation package announced by the Government and religious organizations included compensation, building up of infrastructure, promotion of community resilience to disaster, and also setting up of Ockhi township to rehabilitate displaced people. Loss due to Ockhi was a manifestation of a natural event like a tropical cyclone turning into a disaster.

Storm Surge

The coastal zone spreading over nine districts experiences unexpected flooding due to storm surge, which is an abnormal rise of seawater generated due to cyclone. According to IMD, the mean height of the storm surge in Kerala varies from 2.3 m to 3.5 m. If the storm surge coincides with high tide, the maximum surge height in the state will cross 4 m and the minimum will be 3 m. The low-lying areas adjoining tidally active estuaries and lagoons are inundated due to storm surge resulting in damages of properties and other installations in these areas and also corrupting a thin freshwater lens. Several coastal stretches are vulnerable to storm surge. The preliminary estimate indicates that a 263 km^2 area is likely to be affected by this mode of inundation (Mathai, 2009). The central Kerala coast-extending over Alappuzha and Ernakulam districts have the maximum area exposed to storm surge (Table 14.2). Thiruvananthapuram coast experiences frequent storm surge in certain stretches regularly (Fig. 14.4). The incidence of the storm surge depends upon various factors like bathymetry, coastal configuration, geomorphology, elevation, and composition of the coastline. Part of the Kerala coast, characterized by promontories and hard rock cliffs is less affected. The coastal tract in Kerala

Fig. 14.4: Storm Surge along Thiruvananthapuram Coast in Kerala, 2019.

Note that the coastal laboratory of the National Centre for Earth Science Studies collapsed. Photo by K.V. Thomas.

is densely populated, and settlements have sprung up right along the coastline, therefore storm surge creates several socio-economic problems, and warrants proper attention.

Thunderstorms and Lightning

Thunder and lightning, some of the dramatic weather phenomena are caused by the formation of cumulonimbus clouds. They are part of the global system called 'Global Electric Circuit' associated with the layer of 'Electrosphere' at about 60 km above the ground. During lightning, the charging current flows from the top of the cloud to the Ionosphere. The Electrosphere and the earth form a capacitator which is charged to a voltage of 3,00,000 continuously by the lightning discharges. It is estimated that lightning strikes somewhere on earth surface approximately 44 times every second, a total of 1.4 billion lightning hits every year. Thunderstorms are created by intense heating of the earth surface therefore they are more frequent in hot and humid tropical areas than higher latitudes, and landmass experiences more storms compared to the ocean. According to the Met Office, the United Kingdom, Java island of Indonesia is the most thundery place in the world with 220 days of thunderstorm every year (www.metoffice.goc.uk/learning/library).

Kerala experiences thunderstorms and lightning causing loss of lives and property. The majority of lightning incidences in Kerala are generated by convective cumulonimbus (Cb) cloud, the formation of which commences in the forenoon with the onset of convective cycle initiated due to heating by the solar radiation (Murali Das *et al.*, 2009). Usually, they mature by the afternoon. Most of the lightning strikes from this type of cloud occur in the afternoon hours. In some years like 2002, there were frontal thunderclouds causing lightning at almost all times of the day (*ibid*). Charging mechanisms of the cloud, the number of charged cells formed and the regional meteorology decide the thundercloud formation in a region. Temporal distribution of lightning hit in Kerala indicates that 72 per cent of the lightning strike in the months of April-May (pre-monsoon) and October-November (North-East monsoon), and 83 per cent of event happen in the afternoon between 1500 hr to 1900 hr (KSDMA, 2010). Based on felt reports available with village offices and old newspapers for 17 years between 1986 and 2002, it has been estimated that on average about 70 people in the state die due to lightning every year (Murali Das *et al.*, 2009). The foothill zone and midlands in Kerala reported the maximum number of incidence. The lightning strike rate was moderate in the coastal plain, whereas it was low in the highland zone.

District level distribution indicated that the highest number of events (353) was reported from the district of Kannur followed by Kollam (257) and Malappuram (249). The districts of Kozhikode and Kottayam also recorded more than 200 incidences each. Wayanad district reported the least number of events (19) and Thrissur with 36 incidences occupied the second-lowest position. At Taluk level distribution, Kadakkal and Ochira in Kollam district, Kuttiayadi in Kozhikode district, and Taliparamba in Kannur district experienced very high incidences (Mathai, 2009). A conspicuous gap in lightning hit may be observed in the Palakkad Gap and the adjoining area towards the west covering the Thrissur district (Fig. 14.5). Similarly, Wayanad, situated on the plateau top and sloping towards the east, received few lightning-hits. This distribution pattern clearly manifests orographic influence, which plays important role in the formation of a cumulonimbus cloud causing thunder and lightning.

Hailstorm

Hilly areas, particularly Idukki and Wayanad districts, experience hailstorms regularly. In 2010, a severe hailstorm damaged cardamom plantations and several houses in the Idukki district. Sultan Battery area of Wayanad district was hit by huge hailstorm during 1st May night, 2017. Hails covered roads and rooftops for several hours. Mild hailstorm happens often but severe hailstorm like this event is a rare phenomenon.

SEA EROSION

Coastal erosion is one of the natural hazards affecting the coastline of Kerala severely. Nine districts in Kerala adjoining coastline experience erosion of various intensities (Table 14.2). There are stretches of high erosion and low erosion. Wave action, winds, tides, nearshore currents, and storm surges cause coastal erosion. Erosion and accretion are part of the coastal processes. However, human activities such as the construction of harbours, and jetties, groynes, river training works, beach sand mining, and dredging of the outlets may lead to erosion. Kerala coastline is composed of Recent alluvium, Tertiary formations, and Precambrian crystalline rocks. Rocky coasts and enclosed pocket beaches are less eroded compared to the coastline composed of laterites and alluvium. Cliff sections composed of laterites like that in Varkala experience undercutting of the soft clay layers due to wave action at the base and ground-water action along the cliff face. These two actions together accelerate the slumping of

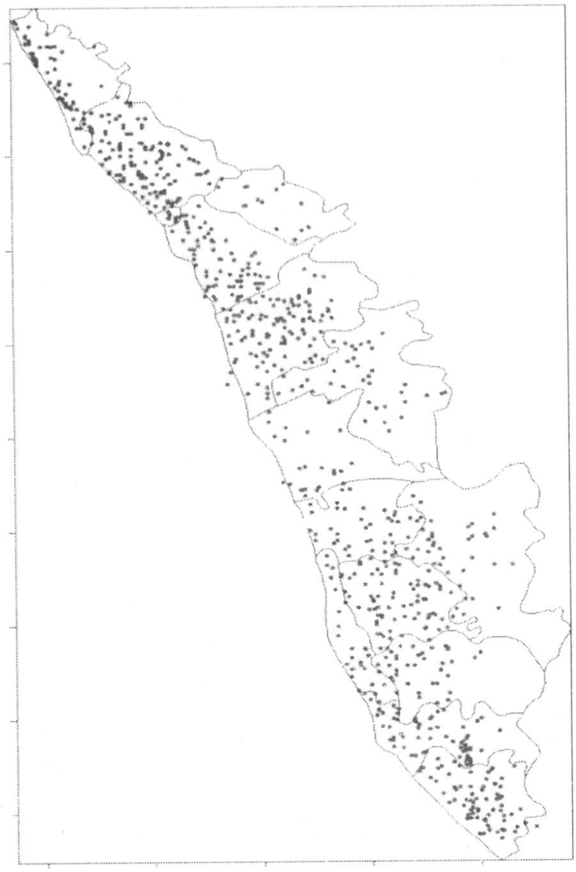

Fig. 14.5: Distribution of Lightning Events in Kerala (1986-2002).

Source: Muruli *Das et al.*, 2009.

top layers resulting in net landward migration of the shoreline. Some segments in the sandy areas show long term erosion and thereby contribute to change in coastal configuration. Alappuzha coast experiences both erosion and accretion. The presence of mud banks impacts the coastal process in this area. Several stretches that underwent erosion earlier are now either accreting or in a state of stability. The presence of sea walls inland and a long accreted area towards the western (sea) side of the sea wall manifests this change from erosional to accretional coast. The Ambalapuzha coast presents one such example.

Out of 560 km long coastline, 313 km are erosion affected at present. Only a 17 km stretch or little over 3 per cent of coastline is highly eroding and devoid of the sea wall. Thiruvananthapuram district has the maximum length of 12 km under this category. Another 199 km stretch is also under the high erosion category but protected with sea walls. The rest 97 km of eroding coastline falls under the low erosion category. Construction of a sea wall is considered as final protection measure in common parlance, although its effectiveness is being questioned. Groynes (a structure perpendicular to the coastline) are being constructed in several stretches with encouraging results. However, there is also a problem with end erosion in the case of groynes. Management of coastal erosion is a high priority in Kerala as the coastal stretch is densely populated and highly urbanised with several other installations. The existing coastal protection measures have not yielded the desired results. The use of coir geotextiles to protect eroding coast is being considered in recent years. Implementation of integrated coastal zone management plans may provide long term solutions to the problems of coastal erosion.

TSUNAMI

The earthquake with a magnitude of 9.3 in the Richter scale and releasing the energy of around 5 megatons struck the north Indian Ocean on 26th December 2004 and caused the unprecedented tsunami alarming the world. More than 2 lakh people lost their lives, millions were affected for livelihood, economic growth was derailed and even the geography of some parts of the world has changed. In India, the Andaman Nicobar islands, and coastal stretch of Tamil Nadu were seriously affected followed by Andhra Pradesh and Kerala. Recorded history indicates that the Indian subcontinent experienced the tsunami in 1881 (Nicobar island earthquake), 1883 (August 8th Krakatoa volcanic explosion in Indonesia), 1941 (June 26th earthquake in Andaman Nicobar islands), and 1945 (November 27th Karachi earthquake), prior to the 2009 tsunami (Manimaran and Chacko, 2006).

Incidence of 2004 Tsunami in Kerala

The first tsunami waves hit the east coast between 7.30 am and 9.00 am and the second hit was around 11.00 am. The Kerala coast experienced the first tsunami wave between 11.00 am to 11.30 am. The time lag is due to Kerala's location in the shadow zone with respect to the direction of propagation of the tsunami. Kerala coast received diffracted wave bounced back from the Lakshadweep oceanic ridges. The tsunami waves attained heights of 3 to 5 m and inundated various parts of coastal areas at different times. Coincidence with high tide compounded the inundation level (Kurian *et al.*, 2006). Among the affected coastal stretches, the damage was severe in the low coastal land around Alappad-Arattupuzha Panchayats on both sides of the Kayamkulam outlet at a distance of about 100 km north of Thiruvananthapuram. Alappad part is a barrier bar of 200 m to 500 m wide bordered by 100 m-150 m wide backwater and parts of Arattupuzha is also a barrier bar of 500 m to 1000 m wide. According to local people, there was a vast sheet of water spreading over the land and the backwater. For a couple of minutes, there was no distinction between the backwater and the sea.

The intensity of tsunami inundation (spread and depth) varied spatially and temporally. There are two hydrodynamic aspects of the tsunami: (i) modulation of wave power by the initial thrust of energy release and seafloor topography, and (ii) rise of sea water level along the coast (Mahadevan, *et al.*, 2005). The spread of tsunami wave, level of inundation and impact depended upon bathymetry, coastal configuration, geology, coastal geomorphology, presence of backwater, and elevation of the land. Once the flood water reached the coast it was the landscape configuration that had controlled the flow of water. While coastal vegetation and coastal protection measures like a sea wall, groyne, etc. moderated tsunami inundation, the nature of destruction was linked to settlement pattern and construction material.

Loss due to Tsunami

Seawater initially receded and part of the nearshore sea bed got exposed during the tsunami.

This unprecedented phenomenon attracted many people, some of whom walked down the exposed sea bed but caught by the onrushing tsunami wave of high velocity and could not survive. The presence of the backwater in Kerala reduced flood fury to some extent. The backwash caused much more damage in many parts of Tamil Nadu. It was reported that about 176 people were killed and 1600 persons were injured in the coastal belt of Kerala. Further, the tsunami pounded 187 villages affecting nearly 25 lakh persons in the state. As many as 6,280 dwelling units were completely destroyed, 11,175 were partially damaged and nearly 84,773 persons were evacuated from the coastal belt adjoining the sea and accommodated in 142 relief camps after the tsunami. Many concrete houses, fishing boats, canoes, and automobiles of this area were uprooted and thrown at a distance of 100 m to 200 m. Even blocks of rocks of the seawall were thrown ashore to a considerable distance by the rushing waves. Tsunami has affected coastal geomorphology, the bottom configuration of the outlets due to inflow and outflow of tsunami waves, groundwater, land use, infrastructure, settlement system, and economy in various measures. People were not only dislocated but also lost their livelihood opportunities.

GEOLOGICAL HAZARDS

Earthquake

Earthquakes are movements of the surface of the earth resulting from the sudden release of energy, stored in the lithosphere, which creates seismic waves. It is detected through the Seismometer measuring vibrations and is measured using the Richter scale, a logarithmic scale where values are expressed in exponential terms. Earthquake is one of the most dreaded natural disasters causing immense harm to human lives and properties. National Institute of Disaster Management (2017) estimated that 59 per cent of the area of India is prone to earthquakes of various magnitudes. According to the Bureau of Indian Standards, the Seismic zoning map of India divides India into four seismic zones from zone II to zone V depending on seismic activity.

Seismicity of Kerala

In all-India perspectives, Kerala falls in Seismic Zone-III, whereas most parts of the Himalayas come under Seismic Zone IV and V. The Zone III is a zone of moderate damage risk, where the maximum expected intensity is VIII in modified Mercalli scale which roughly corresponds to the magnitude of 5.6 in Richter scale. Being part of stable Peninsular shield Kerala is relatively safe with little chance of occurrence of high magnitude earthquakes. So far, there had not been any major earthquake reported from the state causing large-scale damage to human lives and properties. A seismic hazard map with a 10 per cent probability of exceedance in 50 years has placed Kerala in low level hazard region (Mathai, 2009). However, Kerala experiences high ground acceleration in pockets of Central Kerala, and several minor tremors occur from time to time throughout the state. The Latur (Killari) earthquake in Maharashtra in the year 1993 in an area otherwise considered to be least vulnerable raised several questions. It is now being increasingly felt that no region is safe. The sudden release of accumulated strain energy along planes of weakness in the earth's crust may generate earthquakes of high magnitude. The landscape of Kerala is structurally controlled as manifested in the disposition of landforms and orientation of drainage channels. There are deep-seated major faults like Periyar fault, Idamalayar fault, Muvattupuzha fault, Bhavali fault, and Kuthuparamba fault and several minor faults that act as weak planes and are potent to generate tremors as a result of crustal readjustment.

Incidence of Earthquakes

Kerala houses a number of stations to monitor seismicity. The IMD Observatory in Thiruvananthapuram, KSEB stations in various hydroelectricity project sites, and the station at Peechi, Thrissur district operated by NCESS provide a fairly good network for the seismic records. Available data indicates that only four earthquakes with a magnitude of >4 occurred in recent times (Rajendran, *et al.*, 2009). These were earthquakes at Nedumkandam (in the year 1988), Wadakkancheri (1994), and Erattupettah (2000, 2001). Another earthquake of similar magnitude was reported 150 km south-west of Thiruvananthapuram in 2001. Among all these earthquakes, the 2000 Erattupettah recorded the highest magnitude of 5 (Bhattacharya and Dattatrayam, 2002). Considering earthquakes of all magnitude, Kerala can be divided into three zones: <3 M, 3 M to 4 M, and >4 M (Mathai, 2009). Central Kerala spreading over Kottayam, Idukki, Thrissur, and Palakkad districts experiences a higher level of seismicity. The midland part in this area has generated a number of significant earthquakes, the most prominent being the multiple events of 1856 and 1953, the magnitude of which was comparable to 2000/2001 events in central Kerala (Rajendran *et al.*, 2009). Districts of Thiruvananthapuram in south Kerala and Kozhikode and Kannur in north Kerala recorded a moderate level (3 to 4 M) of seismicity.

Are Minor Tremors in Kerala Hydroseismic?

The hypothesis of hydroseismicity, proposed recently, suggests that a natural increase in the hydraulic head along permeable fractures caused by a transient increase in the elevation of the water table in recharge areas may trigger earthquakes (Costain *et al.*, 1987). It attributes most of the intraplate and near-intraplate earthquakes to the dynamics of the hydrological cycle, and consider rainfall-induced pore fluid pressure at depths as the triggering mechanism in the areas already under stress and near failure (BSSA, 2010).

The temporal distribution of micro-tremors in Kerala indicates a pattern. More than 60 per cent of micro tremors occur after peak rainfall and these events can possibly be explained based on the hydroseismicity hypothesis (Rajendran *et al.*, 2009). The continuing low-level seismicity in the vicinity of the Idukki reservoir in recent years after decades of impounding has partly been considered as a response to the hydrological cycle (Rajendran *et al.*, 2012). Kerala receives more than 300 cm of rainfall and the foothill zone in some places receives more than 400 cm of rainfall. More than 60 per cent of this rainfall precipitates during the monsoon season spanning over four months. Widespread land use change, landscaping and rock quarrying have altered hydrological pathways, opened up concealed fractures, and also facilitated high percolation of water in the localities dominated by well-terraced plantations. Thus anthropogenic activities emerge as a major contributor to increased soil pore pressure and fluid movement along pre-existing faults thereby catalyzing the occurrence of minor tremors.

Landslides

Landslide is one of the natural hazards affecting large parts of India spreading over the Himalayas, the North East hill ranges, and the Western Ghats. Covering scarp slope of stable continental block, the south Indian Shield, the Western Ghats comes second in rate of landslide incidence after the Himalayas. By definition, a landslide is the mass movement of rocks and unconsolidated materials down the slope under the influence of gravity. Based on the type of material and movement characteristics, landslides are classified as rock slump, sediment slump, rockfall, soil fall, soil creep, debris flow, avalanche, rotational slide, planar slide, etc. (Thampi, *et al.*, 1998). Debris flow is the most common type of landslides in Kerala. It causes landscape change, destruction of property, and even threatens human lives.

The Western Ghats occupies more than 40 per cent of the landmass of Kerala. The topography is youthful and highly erosion-prone. Rivers draining the slopes of the Western Ghats are fast flowing. They erode plateau margins continuously through headward erosion. Removal of a large quantity of material along the foot slope leaves the overburden unstable and in the process, the entire plateau margin particularly the areas having >25° slope are highly susceptible to a mass movement. Rocks, particularly gneisses in parts of the Western Ghats slope, are unstable due to prolonged weathering and consequent friable nature. Due to heavy rainfall and consequent infiltration, the soil column in the laterite dominated areas is oversaturated resulting in movement of the overburden along the contact zone or litho-merge. There is also mudflow in some cases.

Both natural and man-made factors cause landslides. Factors like rainfall, rock strength, geological structure, slope steepness, relative relief, altitude, the thickness of overburden, the water-bearing limit of soil, disposition of streams, and micro-level landform, singularly and in combination render landscape unstable and susceptible to slide. Slope and rainfall are the most important causative factors in Kerala. Rainfall of >20 cm a day has been the triggering factor for landslide in most cases. Man-made factors causing slope instability or disturbance to the natural equilibrium mostly include land use change, slope modification for construction of roads, and raising plantations and land shaping to accommodate various other anthropogenic activities.

Landslide zonation attempted for Kerala based on parameters like slope, lithology, structure, relative relief, land use/land cover and hydrological conditions brings out that 5610 km^2 area or 14.5 per cent of the total area of Kerala are prone to landslides, of which 1849 km^2 area (4.7 %) falls in the class of high landslide hazard category and the rest 3761 km^2 area (9.8 %) are in the category of low landslide hazard (Mathai, 2009). All 14 districts in Kerala except the coastal district of Alleppey extend eastward up to the hilly Ghats sections and are affected by landslides. Idukki district has the maximum area under the high hazard category followed by Palakkad (Table 14.2). The taluks of Devikulam, Vythiri, Nilambur, Mannarkad, and Rani are the most landslide-prone in the state. Landslides are reported during every rainy season. Sometimes, it turns out to be fatal.

In the recent past, the landslide incidence in Amboori panchayat of Neyatinkara Taluk in Trivandrum district on 9th November, 2001 was perhaps the most tragic. It claimed 39 lives besides, devastating several houses, displacing 100 people, and destroying crops. A part of the Kurisimala hill crushed down due to heavy rainfall. The entire soil column was lost causing loss of agricultural land and resulting in loss of means of livelihood of the people living in this area.

Kavalappara Landslide, August, 2019

A fatal landslide was reported during August 2019 from Nilambur valley and Puthumala area of Mepadi in Wayanad plateau. Kavalappara village of Pothukal Panchayat near Nilambur town witnessed devastating landslides. This area was settled relatively recently. There was originally a tribal colony. Settlers moved a couple of decades back and brought agricultural practices including rubber plantations, for which the slopes have been reworked. The landslide occurred here wiped out a number of houses located on the hill slope, and the entire soil column was lost (Fig. 14.6). The incident took place on 8th August 2019 and killed 59 persons. People were buried under the soil. Several bodies could not be recovered from the huge debris piled up in the foothills. It is told that none of the tribal people was affected. They could sense the impending disaster and flew away before the devastation broke out.

Fig. 14.6: Landslide in Kavalappara Village, Nilambur Valley, Malappuram district in Kerala. Note the piled up debris in the foreground. Exposed rocks can be seen in the crown part of slide. The stream, blocked due to human activities, has occupied its position after the slide. Photo by Srikumar Chattopadhyay.

The landslide took place on the slope of a 260 m high hillock. The crown of the slide is located at an altitude of 220 m amsl. The country rock is horneblende biotite gneiss. It appears that unconsolidated materials were deposited in the hollow of a talus slope, both by stream action and gravity flow. The entire hillslope has been washed away and dumped over the stream flowing down the foothills. The debris has a slope of 18/19° whereas the natural slope as observed in the undisturbed area is 28/30°. In all probability, the movement started at the bottom of the break-in slope and the upper part moved down exposing the bedrock, which has a gradient of 40° to 45° below the crown. A first-order stream established itself at an altitude of 100 m down the slope. Debris flow was in two segments leaving part of the interfluves undisturbed, which was a spar. A patch of settlements and crops remained intact in this spur backed land. There were reports of three incidents taking place in the vicinity, partly in the forests and also in the habited area. The rainfall was very high. The nearby Nilgiri station recorded 216 cm rain precipitated during four days—6th to 9th of August 2019. While excessive rainfall in consecutive days is the main triggering mechanism of this landslide, the contribution of anthoropogenic factors is equally important. People occupied not only the vulnerable slope but also settled in the runout zone of this slope.

Various human activities including settlement construction block natural channels (mostly first-order streams) along hill slopes in most cases. These channels, mostly ephemeral, are natural hollows filled with debris material, which is loose. During heavy downpour surface and subsurface water tend to accumulate along these concave slopes which are often blocked. As water cannot flow freely, it saturates the entire unconsolidated material and with excess rainfall, the whole slope moves down. Kavalappara and Puthumala are two such cases. These areas are also structurally weaker zones experiencing mild earth movements during debris movement, causing further slides.

Documentation of deaths due to landslides since 1961 brings out that the number of fatal landslides is increasing over the years (Kuriakose, *et al.*, 2009). This trend perhaps indicates the growing human occupancy of vulnerable areas. There is a general trend of migration from the lowland and midland to the high lands in Kerala. Wayanad and Idukki are two main centres of attraction. Zonation of vulnerable area, planning for appropriate land use in conformity to slope, and arrangement of proper drainage along hill slopes are important precautionary measures to manage landslide hazards in Kerala.

Soil Piping

The soil piping is a process of internal erosion manifested by subsurface migration of soil particles by the suffusion of piping, which is induced by the regressive erosion of particles from downstream to upstream. It involves hydraulic removal of subsurface soil causing the formation of underground passage (Ingles, 1968). Many instances of land subsidence reported from various parts of Kerala during the last decade are most probably linked to soil piping (Sankar and Tomson, 2017). While investigating land subsidence in Thirumeni village of Kannur district in 2005, a number of subsurface tunnels were discovered and it was ascertained that these features had developed due to soil piping (Sankar, 2005). Since then, several such cases are reported. The general conditions initiating soil piping in Kerala can be listed as (i) enough water saturating soil column, (ii) sufficient hydraulic head for the subterranean flow of water, (iii) presence of permeable/erodible soil above the base level, and (iv) an outlet for flow (Sankar, and Tomson, 2017). Soil piping is a natural process. However, in many cases, human activities like alteration of natural vegetation, land shaping, and such other interventions accelerating water infiltration in soils initiate subsurface pipe formation. Decomposed tree trunks and roots also form conduits for water movement in the soil. So far, land disturbances due to soil piping have been reported from all districts in Kerala except Alappuzha, Kollam, and Thiruvananthapuram. Soil piping is severe in Kasaragod and Kannur districts, where formations of the lateritic mesa are widely noted. This phenomenon is also widespread in the mountainous districts of Wayanad and Idukki. The list of natural hazards at the state level of Kerala includes soil piping.

ACCIDENT RELATED HAZARDS (MAN-MADE HAZARDS)

There are hazards like terrorism, war, etc., causing loss of human lives and property. These incidences are directly linked to human activities. Kerala is free from such hazards however there are losses of human lives due to suicides, road accidents, and drowning. The suicide rate is relatively high in Kerala. During 2014, Kerala recorded a suicide rate of 23.9 (number of persons per one lakh population). Although the suicide rate in Kerala came down from 28.8 noted in 2001-02, still it was well above the national average of 10.6. The High suicide rate in Kerala is attributed to larger socio-economic issues including social tension, financial problems, and family discontent.

Road accidents are emerging as a big challenge. Available data from Kerala Police (www.keralapolice.org) showed that in 2016, there were 39, 420 road accidents in which 44,108 persons were injured, and 4287 persons succumbed to injuries. The 16 year period from 2001 to 2016 witnessed 58,903 deaths due to road accident. On average 3681 persons lose their lives every year. Accident death rate is 11 persons per one lakh population, and the trend is increasing (Fig. 14.7). One of the underlying reasons for accidents is overcrowding of roads.

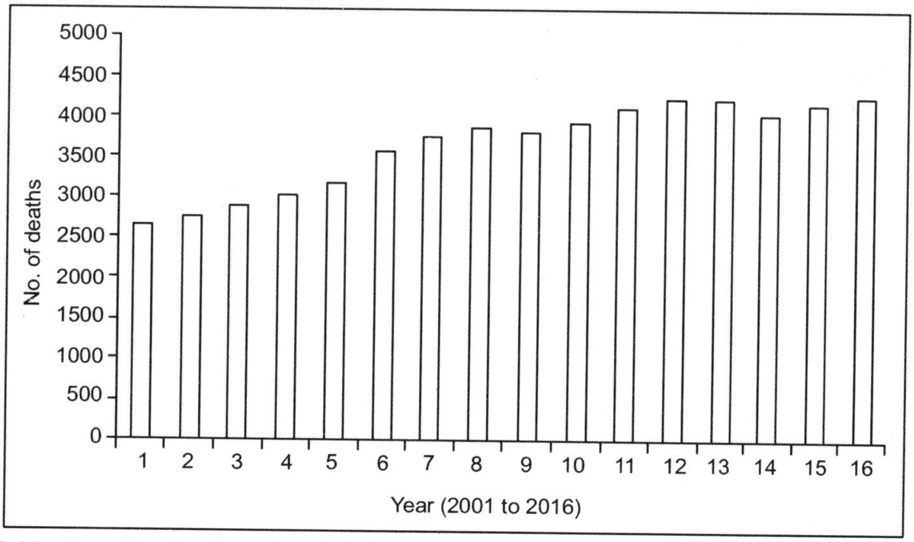

Fig 14.7: Number of Deaths due to Road Accident in Kerala (2001-2016).
Data Source: www.keralapolice.org

Number of registered vehicles has increased tremendously, whereas road development could not keep pace. In 2004, Kerala had 2.79 million registered motor vehicles, which rose to 12.04 million by 2018. Registered motor vehicles outnumbered the total number of households (7.85 million) in the State. Kerala has around 66,000 km of road excluding panchayat roads. Average number of vehicles for each kilometer of non-panchayat road is around 182. Roads are generally narrow and quality is not up to the mark.

Drowning death is another tragedy affecting Kerala society regularly. In fact, drowning comes second after the road accident in claiming human lives. There were 1380 water-related deaths in 1334 incidents by 2015 (Antony, 2017). Although an average of more than 4 persons dies every day due to drowning, the unfortunate incident mostly goes unreported as there is no single agency to follow such occurrences in Kerala (Thummarukudy, 2017). Apart from rivers, lakes, and sea, even the abandoned rock quarries are found to turn into watery graves. The places where frequent incidents are reported are Payyambalam beach in Kannur district, Kappad beach in Kozhikode district, Banasurasagar dam in Wayanad district, Adyanpara waterfalls in Malappuram district, Walayar dam in Palakkad district, Chalakudy river in Thrissur district, Periyar river in Ernakulam district, Meenachil river in Kottayam district, Sasthamkotta lake in Kollam district and Kallar river in Thiruvananthapuram district. World Health Organisation (WHO) has recommended a set of action plans to reduce drowning deaths. Kerala may consider executing some of the action plans suitable for Kerala and find ways to ensure water safety.

DISASTER MANAGEMENT

Famous American geographer Gilbert F. White, well known as Father of Flood Plain Management in the USA said, “Hazard always arises from the interplay of social and biological and physical systems; disasters are generated as much or more by human actions as by physical events” (White, 1978). This observation perhaps forms the foundation of modern-day disaster management. It is well realized that the advancement of science and technology has enabled us to understand many of the underlying causes of the disaster however; we are not yet adequately equipped to predict or prevent disaster. Damage caused by extreme events escalates over the years due to increasing occupancy of the vulnerable areas and extensive development in the high-risk sites. Apart from massive human loss, natural and man-made disasters lead to environmental degradation and jeopardise the social and economic fabric of the society particularly vulnerable societies throughout the world. Millions of people lose their livelihood security. The nexus between disaster and poverty is increasingly evident. Disaster management is, therefore, an exercise to control human activities, identify safe operating space, and devise precautionary steps, so that losses are minimized.

Disaster management is a multidisciplinary activity with a number of stakeholders. While understanding the natural disasters and its prediction lie in the purview of natural sciences, the management issues, institutional arrangement, habitat and economic rehabilitation, warning system, awareness, and information dissemination are the subject matters to be addressed by various other disciplines including geography, economics, engineering, informatics, medical science, social science, and psychology. A comprehensive approach is necessary. Disaster management practices can be conceived in two parts: (i) scientific analysis to understand the causes and consequences of disaster, and (ii) managerial part to prepare the society facing eventualities with minimum loss to reduce risks and to build capability. The disaster management activities can be classified under three broad heads: pre-disaster, during the disaster, and post-disaster. Each of these phases consists of several activities to be performed by the individual and institutions involving communities and Government. Besides, there are strategic issues, development initiatives, vulnerability/risk management, and coordination at various hierarchic levels. All these issues are important for understanding disaster management and making adequate preparation.

Generic or Agent Specific Approach

Researchers debate on the approach to disaster management—generic or agent-specific. The agent–specific approach assumes that each type or class of disaster has its own distinct characteristics/peculiarities, conversely, the generic approach assumes that there are more individual and organizational features that are similar for occurrences of all disaster (Baum, Fleming and Davidson, 1983; Quarantelli, 1987). The thrust towards a generic approach is primarily due to the realization that an event, natural or man-made, does not automatically result in a disaster and a disaster can be identified only in terms of a social happening, by the characteristics of an individual and groups responding to a situation. This socially oriented conception of disaster shifts the focus to the common or similar properties of the social consequences and away from the physical features of natural and technological agents and their effects (Quarantelli, 1991). As socio-behavioural features of disasters are not agent-specific and many of the human and organizational problems are common to almost all disasters, social science studies advocate a generic approach. The generic approach is also preferred from a management point of view as it requires considerable institutional planning and efficient and effective handling of machinery without duplicating the efforts. However, an agent specific approach is necessary to understand the nature of disasters.

Disaster Management Initiatives in Kerala

The Government of Kerala has initiated several measures for disaster mitigation following the guideline suggested by the Government of India. The Revenue Department has been renamed as the Department of Revenue and Disaster Management. The Institute of land management under the Revenue Department is now known as the Institute of Land and Disaster Management (ILDM). It imparts regular training for disaster management. There is also a crisis management group at the highest level under the Chief Minister. Besides, the State Disaster Management Authority (SDMA) has been constituted as an apex body to coordinate disaster management activities. Every district has a disaster management cell. SDMA initiated several activities including collation of data gathered from all sources and creation of awareness. Preparation of State Disaster Management Plan Profile is one such initiative. It has provided the state of art situation and spelt out various activities including preparedness, capacity building measures, and response plan. Dispersed settlement and high density of population right along the coastline are major challenges in Kerala. The coastal people are most vulnerable and there is hardly any buffer zone left between the coastline and the settlement. The occurrence of most of the natural hazards is site-specific. While landslides take place in the hilly terrain, sea surges affect coastal plain. Flood is normally associated with flood plains and the spread is limited to certain elevations above the river level. All 14 districts in Kerala report flood-prone areas and suffer from lightning. Landslides affect 13 districts extending up to the Western Ghats, whereas coastal erosion is confined to the 9 coastal districts. Among these nine districts, eight districts face problems of landslides as well as coastal hazards. Assessment of the vulnerability of people living in hazard-prone areas and apprising people about the nature of risk associated with these areas are necessary. At the same time, it is important to develop a robust early warning system and required infrastructure to face emergencies as learnt during the floods of 2018. Floods and landslide incidences repeated in 2019 also. Subsequently, a committee appointed by the Government of Kerala to study the Extreme Rainfall Event in the state has analysed flood and landslide incidences in detail and prescribed a set of mitigation measures (KSCSTE, 2019). There is a need for a deep science-based approach, and to involve local people in disaster management activities. Kerala has demonstrated its resilience during the August 2018 flood and outbreak of the Nipha Virus the previous year. All sections of the society contributed to relief and rescue operations. The same spirit may be harnessed to strengthen Kerala's disaster management system.

POSTSCRIPT TO CHAPTER 14: COVID 19 PANDEMIC

Novel Coronavirus 19 has shaken the world. All countries, irrespective of physical, cultural, or economic status, are exposed to an unprecedented health emergency. Novel Coronavirus 19 was first reported from China by December 2019, when pneumonia of unknown causes was detected in Wuhan, China, and it was informed to the country office of the World Health Organisation (WHO). The disease is infectious and fatal with hardly any known treatment. It swiftly spread worldwide through direct and indirect contacts. The entire world is caught unaware. WHO announced the name of COVID-19 for this new coronavirus disease and declared the outbreak as Pandemic, public health emergency of international concern. According to WHO, 215 countries, areas, and territories around the world are in the grip of this fatal disease by 9th May 2020. Historical data indicates that a similar global health emergency occurred 100 years back (1918 to 1920) when the world experienced an incidence of Spanish Flu-lethal influenza-affecting large parts of the world.

Available data till 9th May 2020, indicates that the number of confirmed cases in the world crossed 3.82 million, and over 2.63 lakh lives were lost (covid19.who.int). The USA alone has more than 1.24 million confirmed cases with a death toll of over 78 thousand. Spain, Italy, UK, Germany, France, the Russian Federation, and Turkey, each, reported more than one lakh confirmed cases and several thousand deaths (*ibid*). Medical science is facing one of its greatest challenges in recent years. Various experiments are going on in different countries and scientists are deeply engaged to develop vaccine and design courses of treatments for COVID 19. Fighting COVID 19 is a global challenge.

India is one among the countries that are suffering from COVID 19 disease. By 9th May 2020, the death toll in India reached 1981 with confirmed cases of 59,662 and recovered cases of 17,487 (corona.mygov.in). Thirty-two states and union territories in the country have reported the incidence of COVID 19. The number of cases and deaths are increasing with every passing day. The Government of India declared COVID 19 as a national disaster, invoked the Disaster Management Act 2005, and took several steps including information dissemination to contain the spread of the disease. Lockdown has been considered as the best course of action to contain this epidemic. Accordingly, there is a complete suspension of all activities, barring essential services, in the country since 25 March 2020. The lock down period has been extended in three phases till 18th May 2020. People are advised to confine themselves to their homes, refrain from attending gatherings, take necessary precautions not to expose themselves to this contagious disease, and maintain social distance to break the chain of virus spread. Along with medical service protocols, guidelines have also been framed spelling out behavioural norms at personal and public levels to cure this disease and arrest its spread. Some of the activities are expected to begin in phases under strict surveillance, however, it is not yet clear when full normalcy can be restored.

Suspension of all activities due to lockdown has severe negative economic and livelihood consequences. Industry, business, and commerce are almost stopped. People in unorganised sectors, daily wage earners, and migrant labourers are facing extreme hardship. There are certain government measures to assist people financially, and also by providing food items through the public distribution system. Nevertheless, there are institutional weaknesses to cover all people in the country. The marginal people are suffering. It is an unprecedented situation.

Kerala was the first state in India to report an incidence of the COVID 19 by January, 31, 2020, when one student from Kerala, who was studying in Wuhan, China, returned from there and tested positive for the novel coronavirus. By 9th May, 2020, Kerala reported a total number of 503 cases of which 16 are active cases; 484 persons are discharged and 3 persons have lost their lives (*The Hindu*, 2020b). Not only the number of cases in Kerala is the lowest among the major states in India, but also, there is a progressive decline in active cases. Kerala has succeeded to contain the spread of disease and to keep the death toll at a low level. Indian Council of Medical Research (ICMR) has lauded Kerala for her unparalled achievement and referred to it as the 'Kerala Model' (*The Hindu*, 2020c). In fact, Kerala's success to flatten the COVID 19 curve has been widely acknowledged across the world. Kerala's legacies of egalitarianism, social rights, and public trust have paid dividends in this hour of crisis (Heller, 2020).

During the last three years, Kerala faced four emergency situations beginning with the outbreak of the Nipah virus followed by flood disasters of 2018, and 2019 and now the incidence of COVID 19. Kerala has handled all these emergencies commendably and is able to provide important lessons for disaster management. Preparation to tackle novel coronavirus in Kerala started almost immediately after receiving the news of the outbreak of the disease in China, as there were a number of Malayali students studying there. When the first case tested positive in Kerala by January end, 2020, protocols were already planned. This early planning has helped the state to have a head start in tackling this health emergency. By March 10th, 2020, the Government of Kerala acted decisively and ordered the closing down of all schools, banned social gatherings, and requested people not to congregate in religious places. Kerala used its experience gained during fight against the Nipha virus in 2018. Along with medical care, the Government of Kerala rolled out financial packages even before the Government of India declared lockdown, started supplying food grains through PDS outlets, activated entire government machinery, maintained complete transparency in informing larger society about the daily status of the disease and called for citizen's participation in arresting this pandemic. The legacy of the state that has been created by strengthening inherited social infrastructure through various people-centric welfare measures, introduced in Kerala during the last six decades, irrespective of political formations, a vibrant civil society, and people's consciousness has provided the foundation for such government actions. In 1996, the Government of Kerala launched the People's Plan Campaign, emphasised decentralisation, activated Panchayati Raj Institutions down to the Gram Sabha level, and thus facilitated the involvement of people in planning process and enlargement of operational space for civil society. Subsequently, there were various movements/initiatives concerning the environment and social development including the Kudumbashree programme. All these have helped building up people's capacity, involving them in planning and development process, and deepen social democracy in Kerala. Chapters 12 and 13 of this book have deliberated on all these issues.

Good and well-penetrated health care infrastructure, timely and affirmative action of the government, the synergy of different departments, decentralised approach, the involvement of all stake-holders, people's participation and above all people's faith in state initiatives and their willingness to strictly adhere to the prescribed protocols have helped Kerala in time of emergencies. Investments in health and education and the empowerment of people increase the resilience of society and equip it to withstand critical situations. This is amply demonstrated through Kerala's efficient management of disasters.

CHAPTER

15

Changing Geography of Kerala

Kerala's geography is undergoing change. Like all other parts of the World, clearly, the landscape that we see in Kerala today was not the same that people in Kerala knew even fifty years back. Today's landscape of Kerala is a mosaic of dense settlements, both in rural areas and in urban agglomerations, fishing people along the coast, subsistence agriculturalists distributed along with tree crops, planters, tribal settlements (in the Western Ghats), and modern information technology centres in and around towns and cities. Apparently, the emerged functional landscape is similar to that of the rest of the country and Kerala's uniqueness is slowly being eroded. Chattopadhyay and Franke (2006) discussed a couple of these issues as part of environmental change in their book 'Striving for Sustainability: Environmental Stress and Democratic Initiatives in Kerala'. In this chapter, we intend to highlight some of the changes which have altered or have the potential to alter the face of Kerala. Some of the issues raised here have been discussed in the concerned chapters, nevertheless, they are repeated here for presenting a comprehensive picture of the changing geography of Kerala.

POPULATION

Demographic change is one of the most spectacular geographical facts. Data since 1951 indicated that, in the past six decades (1951-2011), Kerala's population grew by 2.46 times against all-India growth of 3.35 times. However, if the entire 20th Century (1901-2001) is considered, it is found that Kerala's population grew by five times—from 6 million in 1901 to 32 million in 2001, whereas India's population grew little more than three times from 238 million in 1901 to 1027 million in 2001. Decadal growth rates in Kerala were higher compared to all-India average since the beginning of the 20th Century till 1971. It started falling down drastically in subsequent decades, and in the 2001-2011 Census period, Kerala recorded population growth of 4.88 per cent against an all-India average of 17.8 per cent (Fig. 15.1). Now, the decadal growth rate in Kerala is 3.6 times lower than that of the all-India average. This phenomenon of low population growth rate in Kerala during the last four decades has raised considerable interest among social scientists, particularly demographers across the world.

The high growth rates in Kerala compared to the all-India average in the early decades of the 20th century indirectly suggests that people in Kerala enjoyed better health condition compared to the rest of India. Female literacy, the consciousness of general people about child health care, a substantial decline in infant mortality rate, and couples' preference for a small family in order to provide their children the best of livelihood opportunities have contributed

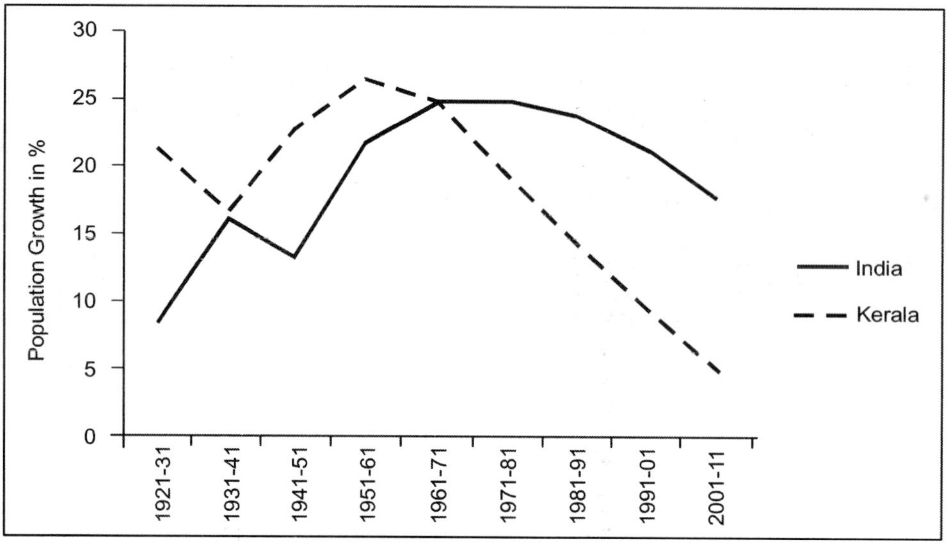

Fig. 15.1: Decadal Growth of Population in India and Kerala (1921-2011).
Data Source: Census of India, Kerala—Various Issues.

to declining population growth. With the present trend, Kerala is likely to achieve zero population growth in the coming two to three decades (Planning Commission of India, 2008a). Kerala is turning into an aging society. Age-wise growth rate indicates that people below 35 years of age are decreasing whereas the proportion of elderly people, senior citizens, is increasing with the dominance of the female population, whose longevity is higher than that of the male population. The state is in the 3rd phase of demographic transition.

URBANISATION

According to the 2011 Census, the urban content of the total population in Kerala rose to 47.72 per cent against an all-India average of 31.16 per cent. Kerala is now in the trajectory of high urban growth. There are some districts each of which has recorded more than 60 per cent urban content. Urban growth is considered a positive sign in development trajectory as urban areas provide efficient and more opportunities for livelihood, help raising income and improving quality of life, and generate better out-turn per unit of capital investment. Urbanisation also provides scope for the growth of informal sectors. The other side of this high growth of urbanization is increasing pressure on existing amenities like housing, health, education, communication, transport, drinking water, sewage system, etc. There is also a demand for new amenities in the newly annexed urban centres. Often, these demands are not fully met and by the time new facilities are created, demand exacerbates. So there is always a gap between demand and supply.

Urban landscape in Kerala experiences perceptible change. The distinctiveness of urban centres of Kerala is fast melting down in recent years. Traditionally, Kerala's urbanization was spreading horizontally maintaining a broad rural-urban continuum. Most of the urban services used to be available in rural areas also. This trend is slowly changing. Differences between rural and urban areas are increasing. Growth of real estate and spatial and vertical expansion of cities like Thiruvananthapuram, Kochi, and Kozhikode and other medium and small cities during the last two decades following globalization and liberalization are gradually but steadily altering Kerala's unique urban landscape, both along the skyline and over the ground. The style and architecture of buildings are also changing.

Inter-settlement space is reducing in Kerala like the rest of urban centres in India, thereby, adding to more and more built-up lands. The main cities like Thiruvananthapuram, Kochi, and Kozhikode are evolving like any other major modern urban centres in the country with a high concentration of population, high rise buildings, high vehicular traffic, congested roads, dominating built-up lands, waste dumps and deteriorated environment. Population migration from rural to urban areas was limited in Kerala. However, this trend is changing. Migration from rural to urban areas is going up in recent years. Attitudinal change among the young generation in favour of urban life is well evident. This is in contrast to earlier trends registering people's preference in Kerala to settle in rural hinterlands for living a quite life.

ECONOMIC GROWTH

Kerala is experiencing an upward trend in economic growth. During the period from 1956 to 1986-87, the economy witnessed slow growth. Since then, particularly following globalization, Kerala's economy grew by 6 per cent and from 2000-01 onwards the growth rate was 8 per cent, whereas India recorded a growth rate of 7 per cent. By 2014, Kerala had recorded a growth rate of 6.49 per cent against the national average of 4.04 per cent (Government of Kerala, 2014a). In south India, Kerala occupied the 2nd position next to Tamil Nadu in the matter of growth. The service sector is the dominant contributor to economic growth. Transport, communication, trade, hotels, banking and insurance, and real estate have performed well. Per capita Gross State Domestic Product (GSDP) of Kerala in 2017-18 was US$ 3100 against the national average of US$ 1800. Kerala's per capita income was well below the national average during the 1980s.

Kerala had a strong linkage with the international economy, historically, with spices trade. In recent decades, there is large-scale emigration of people strengthening the linkage further. Kerala's economy is heavily dependent on foreign remittance. Kerala tops among the states in India by receiving 19 per cent of total foreign inward remittance. With opening up of the Indian economy, it is expected that Kerala having its traditional cultural and trade linkages with the outside world will be able to profit from the new opportunities (George, 2011). Although, there are slow changes in food habits in urban areas and not much technological and managerial skill has been infused into Kerala economy, there is a definite impact due to large-scale inflow of funds through foreign remittances. Investment in the real estate sector is very significant. It has effected a large-scale change in Kerala's biophysical environment. There is also appreciable increase in material consumption. Kerala has turned into a consumer state.

AGRICULTURE AND FOOD SUPPLY

India witnessed a food shortage during the 1960s and started importing food. There were all-out efforts by the Government to produce food. This had culminated in the green revolution and by the 1970s, India become self-sufficient in food grain production. Kerala was not in the belt of the green revolution, however, it recorded the introduction of high yielding varieties of paddy during the era of the green revolution. There was a steady increase in the area under rice till 1981. An increase in rice cultivation was also supported by the expansion of surface irrigation. Subsequently, rice acreage started declining. It covered only 6.63 per cent of the total cropped area in 2016-17. Another important food crop, tapioca, also recorded a reduction after reaching the maximum coverage in the year 1976-77. Percentage shares of food crops like rice and tapioca have decreased by 69 per cent in 2016-17 compared to that in 1960-61. The area under non-food crops like rubber, coffee, areca nut, and pepper recorded a steady increase. Rubber and coffee gained the maximum by 328 per cent and 405 per cent respectively. Rice was 1st ranking crop till 1980-81. By 1990-91, it was second-ranking crop after coconut and by 2000-01 rice ranked third below coconut and rubber. Rubber has also replaced degraded forests, tree crops, and tapioca.

Modern Kerala was food-deficit from its birth in 1956. The demand-supply gap was 40 per cent in 1960-61 and by 2015-16, the gap widened by 85 per cent (Government of Kerala, 2017c). The thrust on food grain production in the 1960s and 1970s was linked to the food shortage in India. Subsequently, easy availability of food grain from other states as a consequence of India's self-sufficiency in food grain production, well-entrenched public distribution system, increasing cost of cultivation and lower return, and slow change in food habits in urban areas have shifted farmer's interest from food grain production to cash crop cultivation in Kerala. Rice lands are further under pressure due to several demographic and economic factors. These lands are even kept fallow as many farmers consider rice cultivation is a non-profitable venture. It has been so alarming that, recently, there are several government initiatives to revive rice cultivation. Kerala, due to its physiographic settings, has limited land suitable for growing rice. The present demand cannot be met with the current level of productivity and population pressure, even, in case all rice fields are brought under rice cultivation. Nevertheless, it is important to protect and enhance rice cultivation for subsistence farmers and rural economy, and also as a measure of ecosystem conservation.

CLIMATE CHANGE

Climate is one of the overarching natural endowments which contributed to Kerala's speciality among the states in India. Two monsoon seasons, spatially well-distributed rainfall, and limited seasonal and diurnal variations of temperature, long sunshine, and orographic influence have enriched the potentiality of primary production of Kerala. However, there is a perceptible change in climate, particularly in rainfall characteristics and temperature during the last couple of decades. Moisture and thermal regimes are changing, as a result, climate shifts from B4-B3 to B2-B1 class, moving Kerala from wetness to dryness within the humid type of climate (Rao *et al.*, 2009).

Weakening Trend of Monsoon

Kerala is experiencing a weakening trend in monsoon rainfall, which characteristically occurs in strong and weak spells with varying intensity. The change is taking place in these spells and in rainfall intensity. Increasing dry spell causes a change in climate character. The duration of weak monsoon spells had increased by 30 per cent in the period from 1950 to 2002 (Joseph and Simon, 2005). During the south-west monsoon season (1 June to 30 September), the number of days with daily average rainfall more than 8 mm day^{-1} increased by 45.4 per cent, and the rainy days receiving rainfall more than 12 mm day^{-1} decreased by 78.1 per cent during these 53 years' period. The decreasing trend is prevalent all-over Peninsular India. During the period from 1871 to 2005, Kerala witnessed more number of deficit years (24) compared to the number of excess years (Krishnakumar *et al.*, 2009). Rainfall in south-west monsoon declined by 233 mm in 135 years (from 1871 to 2005) indicating an average reduction by 1.7 mm year^{-1}. However, this trend was not uniform all through these years. While in the early years rainfall was more or less steady or even increasing, particularly during non-monsoon periods, the declining trend in all seasons is evident during the last couple of decades. Annual rainfall fell by around nine per cent from 1990 to 2005. Monthly rainfall data indicated that the contribution of rainfall during the month of July remained stable with lighter variations. In contrast, the contribution of rainfall during August, September, October and November is decreasing (Krishnakumar *et al.*, 2009). However, rainfall in recent years recorded an increasing trend. Multi-decadal variations are evident. Temporally, the month of peak rainfall shows a shift from July to August.

Spatial Trend in Rainfall Change

Change in rainfall pattern in Kerala varies spatially both from north to south and also from coast to the hill in the west to east direction. The coastal area reported a marginal increase in annual rainfall. Precipitation decreases from the coast to foothills. The extreme rainfall for 1-day duration is found to have decreased by 10 per cent to 15 per cent over large parts of the state (Soman *et al.*, 1988). The foothill area spreading over Ernakulam, Kottayam, and Idukki districts of south Kerala showed a 15 per cent to 20 per cent decrease for the 5-day and 10-day extreme rainfall; the decrease is around 10 per cent over the adjoining hilly regions of north Kerala. The north-east monsoon shows a significant decreasing trend over Thrissur and Palakkad districts and Ghat sections of south Kerala. Both the monsoons appear to be weak in south Kerala (Singh *et al.*, 1989). North Kerala shows a decrease in rainfall in recent decades. Wayanad district experiences deficit rainfall for consecutive years. However, the last two years (2018, 2019) witnessed high rainfall. Apart from global climatic factors, local level disturbances affecting aerosol distribution might also have contributed to the rainfall distribution pattern.

Temperature

Surface air temperature is increasing over the years in Kerala with the mean maximum temperature rising by 0.8 °C, mean minimum by 0.2 °C and the mean average by 0.5 °C as evident from IMD data during the period from 1961 to 2003 (Gopokumar, 2011). The rate of temperature increase is the maximum across the high ranges (1.46 °C) as noted in Wayanad and Idukki districts followed by the coastal tract (1.09 °C) and it is marginal (0.25 °C) in the midland segment (*ibid*). The annual mean maximum temperature in Trivandrum city records an increasing trend from 2000 to 2010. The Bharatapuzha basin in Central Kerala experiences an overall upward trend in annual and daily temperatures (Raj and Azeez, 2011).

The projection of the Indian Network for Climate Change Assessment (INCCA) indicates that the Western Ghats region bordering Kerala may experience an average temperature rise by 1°C to 3°C by 2050 and the maximum air temperature rise may touch 4.5 °C (Government of Kerala, 2014b). Large temperature variations in the high ranges can be attributed to several factors ranging from global climate change to deforestation and land use change at the local level. Increased temperature causes high evapo-transpiration loss and loss of soil moisture resulting in loss of productivity particularly of shallow root crops which depend on soil moisture in the upper soil horizon. The rise in temperature and decline in rainfall feed each other and together affect agricultural scenario and livelihood of people depending on the primary sector.

Sea Level Rise (SLR)

Globally, the sea level is on the rise. According to the 5th IPCC report global mean sea level rise is estimated to be close to 1.7 mm yr^{-1} for the period of 1901-2010 and 1.8 mm yr^{-1} for the period of 1961-2003. The estimated average sea level rise for the India coast stands at 1.3 mm yr^{-1}, comparable to the global trend (Unnikrishnan and Shankar, 2007). Long term data analysis for selected tide gauging station on Indian coast indicates that the net sea-level rise is the highest in Diamond Harbour, Kolkata (4.96 mm yr^{-1}), and the lowest is noted in the case of Visakhapatnam (0.93 mm yr^{-1}), while for Kochi it is estimated at 1.81 mm yr^{-1} (Unnikrishnan *et al.,* 2015). However, short term data (1992-2007) reveals an elevated trend of 3.7 mm yr^{-1} for Kochi (*ibid*). One metre rise in sea level may inundate 169 km^2 area around Kochi (Govt. of Kerala, 2014b). The coastal low lands particularly Kuttanad wetlands, which lie 1 to 2 metre below mean sea level will be critically affected by sea water ingression. Not only the human settlements have to be rehabilitated, but also the existing ecological conditions will be altered due to the upstream migration of the saltwater/freshwater wedge along all watercourses. Besides affecting agriculture and related activities, this change will impinge upon freshwater availability and drinking water supply. The socio-economic and financial cost to manage this emerging situation will be substantial.

COASTAL CONFIGURATION

Various human interventions in the coastal tract like a sea wall, groyne, settlements, construction of barrage and spillways, removal of coastal dunes, loss of mangroves, and reclamation of coastal wetlands have altered the natural coastal landscape of Kerala. We shall discuss here some of these changes.

Shoreline Change

The shoreline itself has undergone considerable modification during the last four to five decades. There is an overall balance between erosion and accretion for Kerala if the loss and gain of materials are accounted for. The orientation of the coastline remains largely the same. However, there are micro-level modifications of coastal processes resulting in loss of land in some areas and built up of land in some other places. Parts of the coastline are affected by intense erosion regularly during monsoon months for a couple of years and subsequently, beaches are found to have built up. Intense human activities in most parts of the Kerala coast interrupt the normal process of the erosion-accretion cycle. Available studies indicate that around 53 per cent of the coastline in Kerala is artificial coast protected by the sea wall, a structure parallel to the coast, riprap revetments, and or groynes, a structure perpendicular to the coast. Some of the districts like Kollam and Ernakulam have nearly 80 per cent of their coastline made artificial (Table 15.1). Around 30 km of coastline experiencing a high rate of accretion are distributed throughout the state, in patches. Construction of sea walls along the coast, either eroded or erosion-prone, is a major intervention in the beach ecosystem (Fig. 15.2).

There are 106 groynes distributed all through the coastline of Kerala. Twenty-five-break waters impact coastal processes and affect sediment distribution patterns. Besides, there are one major port (Kochi), 16 medium and minor ports, 11 harbours, and 90 fish landing centres dotting the coastline. All these structures and construction of roads and settlements adjacent to the coastline have altered the character of the Kerala coast. Beaches bordered by coconut

Table 15.1: Status of Kerala Coastline by Districts, 2010

Districts	*Status of coast (Figures in per cent)*			
	Erosion	*Accretion*	*Artificial*	*Stable and rocky coast*
Thiruvananthapuuram	23.33	16.08	29.06	31.53
Kollam	6.89	5.47	79.92	7.72
Alappuzha	9.30	23.19	61.98	5.53
Ernakulam	3.83	13.39	79.74	3.04
Thrissur	1.46	40.84	50.67	7.03
Malappuram	9.78	24.51	61.31	4.40
Kozhikode	10.96	23.68	50.41	14.95
Kannur	1.95	24.66	55.71	17.68
Kasaragod	17.49	33.84	32.34	16.33
Kerala	10.37	23.91	52.69	13.03

Source: Ministry of Environment and Forest, Government of India: Shoreline Change Assessment for Kerala Coast.

Fig. 15.2: View of Sea Wall Along Kerala Coast. Note that settlements have been constructed right on the coastline. Photo by Srikumar Chattopadhyay.

groves, a trademark feature of the Kerala coast have now yielded place to man-made structures. This restricts the normal functioning of coastal processes, natural beauty of sea coast, beach development and also hinders the free movement of people to the sea.

KUTTANAD WETLANDS AND VEMBANAD ESTUARINE SYSTEM

The coastal plain of Kerala is characterized by estuaries and lagoons commonly known as backwaters or *Kayals* in vernacular language. Most of these brackish water bodies are being reclaimed for various purposes. There are three major interventions in the Kuttanad area: Embankments and Block reclamation, Thottapally spillway, and Thanneermukkom barrage across the Vembanad estuary (Fig. 15.3). We shall highlight some of the salient features.

Embankments in Kuttanad

All landscapes, today, are anthropogenically impacted. However in the case of Kuttanad, the human impacts are much more intensive and the landscape here is truly a hybrid formation, where human actions contributed in good measures to land building processes by way of constructing embankments, removal of silts from the river beds, and reshaping the lands. Criss-crossed by drainage channels, this subsiding landmass is a conjugation of several riverine islands locally known as *Padasehkarams.* There are around 1270 *Padasekharams* (polders), which are altitudinally at lower levels than the water levels in the surrounding rivers and therefore encircled by embankments or man-made levees. These lands remain waterlogged during monsoon months and all agricultural practices take place after dewatering.

Block Reclamation

Block reclamation in the southern part of the Vembanad lake was completed in the 1950s with the aim to increase rice cultivation as part of the 'grow more food campaign'. There are six major Blocks, namely Rani, Chittira, Marthandam, R Block, C Block, and D Block (Fig. 15.3). The reclaimed area was brought under cultivation. In some blocks, human settlements have also colonised. However, several problems cropped up over the years. The R Block which witnessed good agricultural practices, particularly coconut cultivation is now mostly abandoned. Similarly, other reclaimed areas are also left uncultivated and now there is a proposal to develop tourism projects in some areas as part of backwater tourism. Block reclamation has changed the shape of the lake and modified hydrodynamic conditions, water flow, and sediment movement. Outer embankments around the Blocks are now being replaced by concrete structures under the Kuttanad package, proposed by the M.S. Swaminathan Committee for the overall development of the Kuttanad area.

Thottapally Spillway

The Thottapally spillway cum regulator was constructed to check saline water inflow from the sea through an inlet at Thottapally and to divert part of the floodwaters in Kuttanad along with an easier and more direct outlet to the sea, otherwise, these floodwaters used to exit through the Kochi mouth after traveling about 75 km long distance towards the north. This spillway, 1290 m long and 60 m wide, fitted with 40 gates was completed in 1955 (Fig. 15.4). It is an important landmark along the National Highway no. 47 before reaching Alappuzha town from Haripad along the coast. The shutters are left open from June to December to facilitate flood water discharge and kept closed for the rest of the year to arrest saltwater ingression.

Thanneermukkom Barrage

The Thanneermukkom barrage across the Vembanad estuary commissioned in 1975 was intended

Fig. 15.3: Satellite View of the Pamba Delta, Block Reclamation, Parts of the Vembanad Lake and adjoining Areas in Kerala.

to prevent saltwater incursion during summer months to ensure paddy cultivation in the Kuttanad area spreading over Alappuzha and Kottayam districts (Fig. 15.5). The barrage is operated through 62 shutters, although originally planned for 91 shutters. At present rest of the shutters are being constructed under the Kuttanad package. With the commissioning of the Thanneermukkom barrage the southern part of the Vembanad estuary lying to the south of the barrage and covering a water spread area of 69 km^2 was ecologically separated from the main estuarine system and turned into a freshwater body. It has divided the estuarine ecosystem and altered its hydrographic as well as ecological characteristics. The shutters remain closed for a considerable period ranging from a minimum of 81 days in 2001-02 to the maximum of 181 days in 1986-87.

There is a conflict of interest between paddy cultivators and the fishing community in the matter of duration of the closure. Paddy cultivators press for keeping the shutters closed for a longer duration, whereas the fishing community and aquaculturists demand closure for the minimum period to allow saline water inflow during summer months. Significant adverse impacts surfaced due to the operation of the barrage on fisheries and the aquatic environment. There is a reduction in the upstream migration of marine fishes and prawns and an increase in weed growth upstreams. The structure has severely restricted the natural flushing of pollutants. With the development of cities and townships on its banks and within catchments, industrial, domestic and municipal sewage flow into the backwater without proper treatment, besides

Fig. 15.4: View of Thottapally Spillway in Alappuzha District of Kerala.
Source: Photo by Srikumar Chattopadhyay.

Fig. 15.5: View of Parts of Thanneermukkom Barrage in Alappuzha District of Kerala.
Source: Photo by Srikumar Chattopadhyay.

agricultural run-off of fertilisers and pesticides. In view of the environmental deterioration, and the fact that the area under paddy cultivation came down from 609 km^2 in the pre-Barrage period to 376 km^2 by 2003 (MSSRF, 2007) the demand for closure of the shutters for a longer period become feeble. At present, the closure period is proposed for 90 days from December 15th to March 15th. Human interventions have altered lake ecology in several dimensions ranging from physical configuration to water quality and biodiversity. We shall discuss the water quality issue in a separate section in this chapter.

Configuration of Vembanad Estuary

The Vembanad lake system, the largest estuary along the west coast of India with typical salient characteristics of tropical estuary runs parallel to the west coast from Alappuzha in the south to Munambam in the north with a total length of 113 km. The main body of the lake from Kochi to Alappuzha is 83 km long and its width ranges from a few hundred metres to 15 km in the widest stretch along the southern boundary. However, this width has reduced substantially due to reclamation all around the lake and block reclamation of the lake bed in the southern part. Spreading over Ernakulam, Alappuzha, and Kottayam districts, the Vembanad lake has been included in the list of Ramsar sites in 2002.

The erstwhile princely state of Travancore encouraged the farmers to reclaim parts of the Vembanad lake in the Kuttanad region for paddy cultivation. The area of the Vembanad lake in 1834 was 365 km^2 which has reduced substantially over the years mainly due to reclamation and encroachments (Gopalan, 2002; Nair *et al.*, 1987). Around 162 km^2 area has been reclaimed from the backwaters for the purpose of paddy cultivation, paddy-cum-shrimp culture, coconut husk retting, and housing during the period between 1834 and 1984 (Table 15.2). As a result, the Vembanad estuary shrunk horizontally by about 44 per cent in this 150 years' period. In additon to construction, port development, and urban development project including the creation of Welingdon island claimed a little over 7 km^2 area.

Table 15.2: Reclamation of the Vembanad Estuary in Kerala (1883 to 1984)

Period	*Area Reclaimed*		*Purpose*
	Area in hectare	*Per cent to total area of the lake*	
1883-1903	2226.72	6.1	Agricultural
1912-1931	5253.15	14.4	Agricultural
1941-1950	1325.00	3.6	Agricultural
1970-1984	5900.00	16.2	Paddy-cum-shrimp culture
1900-1984	1500.00	4.1	Housing, agricultural and traditional industries
Total	16204.87	44.4	

Source: Gopalan, 2002.

Loss of Water Holding Capacity of Vembanad Lake

The water holding capacity of the Vembanad lake is coming down due to a reduction in water spread area and loss of depth. The average depth of the Vembanad lake south of the Thanneermukkom bund is 3 m while the maximum depth is 8.5 m. Consequent to the reduction in area and depth, the total volume of the backwater lying between Alleppey and Azhikkode has reduced. It has been estimated that the Vembanad lake which had a capacity of holding 2.4 billion m^3 of water prior to large-scale reclamation started at beginning of the 20th century, can now accommodate only 0.6 billion m^3 of water (CWC, 2018). It has been estimated that the excess flood water (1.634 billion m^3) generated by the rivers of Meenachil, Pamba, Manimala, and Achankovil during the 15th-17th August 2018 flood could have been very well consumed by the Vembanad lake had it had its original water holding capacity (*ibid*). Accelerated siltation of the lake bed took place due to excessive erosion in the provenance zones of the river catchments on account of landscape alterations such as deforestation, land use change, and intervention in river regime which has altered the flow and circulation pattern of the water. Dredging in the Cochin harbour to clear the shipping channel also impacts the siltation pattern.

ALTERATION OF TOPOGRAPHY

Alteration of topography is a widespread phenomenon in Kerala. The state, well known for undulated topography right from the coast, is experiencing drastic changes that have long term impacts on biophysical setup, particularly on soil and water. Topography in Kerala rises step by step from the coastal plain. Valleys along rivers and streams are indented within a hilly terrain. They are mostly narrow due to the dominance of hard rock. The laterite dominated areas, which cover 60 per cent of land area in Kerala, is characterised by valleys, mostly formed due to parallel slope retreat. The traditional arrangement of settlement along the ridges and upper slopes and agricultural activities along the valleys are yielding to more compact settlements. Low lying paddy fields are filled up. Hill slopes are terraced resulting in increasing length of overland flow and high rate of infiltration. Construction of buildings in all types of lands across Kerala results in modification of landscape configuration. Growth of population, socio-economic condition, and real estate business along with ex-situ factors of foreign remittance are the principal drivers. The construction industry is growing at a faster rate. Its share in state income increased from 10 per cent in 1987-88 to 16.8 per cent in 2008-09, and, during this period, India experienced growth from 5.7 per cent to 9.2 per cent (George, 2011). By 2011-12, the construction sector contributed around 12 per cent of GSDP (Gross State Domestic Product) (Government of Kerala, 2013). In a span of seven years from 2004-05 to 2011-12 contribution of construction sector grew by 69 per cent. The bulk of the construction materials are extracted from Kerala itself.

Mining and Quarrying

Removal of sand dunes in the coastal plain, flattening of lateritic hillocks, cutting of laterite bricks from hard crust laterite surfaces specially in north Kerala, excavation of clay from the flood plains, sand mining from flood plain and also in-stream and quarrying of hard rocks for extracting building materials are widely observed. Hills are continuously being excavated. Besides, there are laterite quarries, and clay quarries. Data available from the Mining and Geology Department, Government of Kerala indicated that by 2011, the Malappuram district reported 1754 quarries consisting of building stone (granite) quarries, laterite quarries and ordinary clay quarries. This indicates that on average there is one quarry in every 2 km^2 in this district. Ernakulam district reported 776 quarries. Out of these, 691 were hard rock quarries and 55 were laterite quarries. There are 818 quarries in the Trivandrum district (Table 15.3). In recent years, several licenses are issued for starting new quarries. Besides, several unauthorized quarries operate in Kerala as observed by the Gadgil report on the Western Ghats. A study based on satellite image has brought out that there are 5924 granite quarries in Kerala distributed in all districts (Sajeev and Alex, 2017). Palakkad and Ernakulam districts have more than 800 granite quarries each.

Excavation of Hillocks

Kerala's topography, particularly, along the lower midlands is changing fast with the flattening of hillocks. This is well evident in the areas adjoining highways all through the state. Although proper surveyed data are not available, a traverse across the state will bring out this reality.

Table 15.3: Distribution of Quarries by Talukas in Thiruvananthapuram District of Kerala, 2011

Sl. No.	Taluk	Total no. of Quarries	No. of quarry/10 km^2	Type of quarries			
				Granite/hard rock	Laterite	Brick clay/clay	Sand
1.	Chirayinkeezh	221	6	173	29	17	2
2.	Nedumangad	209	2	199	-	9	1
3.	Neyyantinkara	275	5	189	-	53	33
4.	Trivandrum	113	4	96		16	1
	Trivandrum district	818	4	657	29	95	37

Source: dmg.kerala.gov.in/docs/pdf/MIS/district/quarries-Thiruvananthapuram.pdf

Undulated topography is yielding to flatlands. The hills/undulated topography, which has attained slope stability due to the prolonged weathering process, are now grossly altered (Fig. 15.6).

The hard rocks are excavated for building materials and the hard crust (duri crust) laterites are cut as laterite bricks, which is an age-old practice. The soft lateritic hills are excavated to fill-up low lands and paddy fields. This process has been accelerated during the last two decades with a change in the settlement system in favour of high-rise buildings particularly around major centres. While building stones and laterites are excavated from topographically elevated areas, clay, and sand are being mined from flood plains, and also in-stream. In coastal areas, sand dunes are removed for various purposes. Removal of material alters topographic configuration. When rocks are mined and removed, a positive topography (elevated area) gives place to a negative topography (hollow area). Local base levels are altered, water stagnates, sediment balance changed and even the groundwater regime is affected. All these trigger a host of immediate and long term changes. Abundant quarries are seen throughout Kerala.

Fig. 15.6: Excavation of a lateritic hill in Thiruvananthapuram district of Kerala. Photo by Srikumar Chattopadhyay.

Rain-water accumulates in these structures and, if unattended, these structures may turn into mosquito breeding grounds leading to health hazards. There are also reports of drowning deaths in these abandoned mines. Proper planning is necessary for the wise use of these quarries.

SOIL

Soils in Kerala suffer from erosion and witness change in composition and have been impacted by high-intensity human activities like any other densely populated area. Agriculture, cultivation in the up-slopes, draining of wetlands, and overfertilization of soils in the lowlands and in the slopes modify the natural soil qualities. It has been estimated that 9.5 lakh ha of cultivated land are affected by accelerated erosion (Government of Kerala, 2014c). The average annual rate of soil loss from the Western Ghats is around 22 tons $ha^{-1}yr^{-1}$. Sample studies indicated the rate of erosion in some pockets is even >200 tons $ha^{1}yr^{-1}$. High fertilization has resulted in available phosphorus of even >100 kg/ha in some areas (Dinesh *et al.*, 2014). Filling up of lowlands by laterite soils, brought down from the hillocks, change the soil character both in the place of excavation and also in the areas of dumping. Low-lying areas filled up by laterite soils will now show a profile that has coarse-grained laterite soil in the top layer and the bottom layers are composed of fine-grained clay dominated alluvial materials. This alteration of soil is most intense around urban centres and coastal plains. This is reflected in the soil survey, which reports more and more area under the 'unclassified' category.

RIVER ECOLOGY

Rivers in Kerala are undergoing physical and biogeochemical changes due to large scale human interventions, which can be categorised into two: (i) direct, and (ii) indirect interventions (Chattopadhyay and Mahamaya, 2014). Construction of dams and the impounding of reservoirs in the upstream sections for irrigation and hydroelectric projects are the most common human interventions on river systems. Water is diverted through main and branch canals impacting larger areas in the case of irrigation projects. Encroachment on the river bank, flood plain occupancy, removal of floodplain materials, river bed mining, and destruction of riparian vegetation, check dams, and similar activities altering the configuration of land along the river corridor and surrounding areas are some of the direct interventions. Discharge of domestic solid and liquid wastes in the river, industrial effluents, excessive use of fertilizer and pesticides in agriculture, and other such activities causing water quality problems also fall under the category of direct intervention. It has been brought out that land use change in the catchment area impacts water quality (Chattopadhyay, *et al.,* 2005) and increases nutrient load (David *et al.*, 2015) Indirect interventions include population growth, urbanization, socio-political and socio-religious activities. The impact of globalization is also considered as an indirect intervention.

Dams and Reservoirs

The natural flow of the river is interrupted due to the construction of dams and impounding of reservoirs. While a new base level is created for the drainage system upstream of the reservoir, the river regime downstream of the dam experiences an altered hydrological system. The river course shows features of late maturity like braided condition, reduced flow, and decline in the wetted perimeter. Kerala has 54 dams of different dimensions meant for irrigation, hydropower, drinking water, and flood control. Besides, there are 200 small dams/check dams distributed throughout the state. Dams are constructed across the river by arresting normal flow and drainage discharge is regulated (Fig. 15.7).

Fig. 15.7: Upstream View of Malampuzha Reservoir in Palakkad District of Kerala. It is the largest reservoir in Kerala impounded on a tributary of the Bharatappuzha to cater to irrigation needs of Palakkad district and adjoining areas. Photo by G. Ravindrakumar.

Some of the reservoirs are multipurpose projects serving more than one causes. The combined area of these reservoirs is 443 km^2. In other words, as much as 443 km^2 forest area has been inundated due to human intervention. The combined storage capacity of all these dams together is 5.8 billion m^3. All major rivers in Kerala have been arrested, sometimes in more than one location. Bharathappuzha basin has six reservoirs and the Periyar basin has eleven reservoirs. The dams constructed to impound reservoirs also trap nutrients released from the catchment and destined to reach coastal waters including lagoons and coastal seas through the river system. Once the regular flow of water, sediment, and nutrients are hindered, the biogeochemical cycles operating in the catchment level are interrupted. This has a far-reaching impact on coastal processes and coastal productivity.

Reservoir Siltation

Terrain conditions surrounding these reservoirs are not uniform. Some reservoirs are located over the plateau, while some other reservoirs are in the foothill zone. The catchment areas are vulnerable and subject to a high degree of erosion and sediment yield during monsoon months. As a result, many of these reservoirs functioning as a silt trap have lost part of their capacity over the years. Available study indicates that the reservoir of Malampuzha, one of the largest reservoirs in Kerala experiences an annual average rate of sedimentation of 0.27 per cent. The Aruvikkara reservoir, the source for water supply catering to Thiruvananthapuram city has lost its capacity by 1.16 per cent per year. The sedimentation rates of selected reservoirs are given in Table 15.4.

Table 15.4: Annual Storage Loss of Selected Reservoirs in Kerala

Sl. No.	*Reservoir*	*Water spread area (Km²)*	*Year of survey*	*Agency*	*Period of records*	*Annual storage loss (per cent)*
1.	Mangalam	3.93	1985	CWRDM	29	0.30
2.	Peruvannamuzhi	10.5	1986	CWRDM	13	1.66
3.	Malampuzha	23.13	2005	KERI/NRSA	50	0.27
4.	Peechi	12.63	1985	KERI	26	0.97
5.	Neyyar	15.0	1997	CESS	38	0.71
6.	Aruvikkara	0.29	2009	KERI	37	1.16

River Sand Mining

Kerala does not have sufficient reserves of construction-grade sand. River-beds are continuously mined for sand and thus depths of rivers are increasing and river profiles are changing. It has been reported that annual sand mining from the seven rivers debouching into the Vembanad lake was 6.63 million m^3 against natural replenishment of 0.086 million m^3 in the storage zone (Padmalal *et al.*, 2004). Based on data for two locations in the Pamba river it has been estimated that annually an average amount of 4,08,800 m^3 of sand is mined against an annual replenishment of 17,734 m^3 (MSSRF, 2007). Excessive mining has several impacts like the erosion of riverbank, slumping, the lowering of channel, change in thalweg configuration, increase in turbidity, aggravated saltwater ingression in the lower reaches, fall in water table consequently fall in well water level and disturbance of flora and fauna. It is also observed that many stretches of river courses have lost all loose materials and water now flows over bedrocks.

Drainage Discharge

Drainage discharge depends on climatic factors of rainfall and temperature and catchment characteristics. When the rainfall remains largely the same, it is the catchment parameters like geology, geomorphology, soil, and land use that impact the spatial and temporal distribution of drainage discharge. Among these parameters, it is the land use that is most dynamic, and therefore, control discharge pattern. Trend analysis brings out that annual, and monsoon drainage discharge record increasing trend, whereas drainage discharge during non-monsoon period decreases in the case of most of the rivers (Table 15.5).

An increase in flow during monsoon months in spite of the reported decrease in rainfall during south-west monsoon can be attributed to excessive run-off generated due to large scale land use change that contributes to the reduction of water withholding capacity of the river catchments and excess surface run-off due to expansion of surfaced area associated with urbanization, construction of roads and other built-up areas. In one hand surface run-off is increasing and on the other hand, infiltration is declining. This leads to a reduction in water storage which results in reduced flow in the non-monsoon months and also falls in groundwater storage. River catchments showing a negative trend in all three seasons like the Pamba catchment warrants a deeper probe. However, it may be noted here that this catchment is dominated by

Table 15.5: Discharge Trend of Selected Rivers in Kerala (1998-99 to 2008-09)

River/Season	*Annual*	*Monsoon*	*Non-Monsoon*
Valapattanam	Y = 90.97X +3244.0 r = 0.32	Y = 92.66X + 3016.8 r = 0.33	Y =-1.68X +227.18 r = -0.05
Bharatapuzha	Y = 183.89x + 3227.3 r = 0.33	Y = 190.43x + 2620.9 r = 0.37	Y =-6.54X + 606.4 r =-0.01
Chalakudy	Y = 13.01X + 1757.9 r = 0.10	Y = 19.53X + 1462.7 r = 0.11	Y =-6.52X + 295.20 r =-0.16
Pamba	Y = -46.391X + 3913.5 r = -0.14	Y = -36.26X + 3097.5 r =-0.16	Y = -11.23X + 828.55 r =-0.01
Vamanapuram	Y = 2.8727X + 656.22 r = 0.05	Y = 8.4455X + 424.69 r = 0.18	Y =-5.55X + 232.45 r = -0.30

Data source: Central Water Commission (CWC), 2012, Integrated Hydrological Data Book (Non-Classified River Basins).

Plantation land use particularly rubber plantations. In the case of urban watersheds, saturation of surface run-off occurs relatively rapidly, since storage and infiltration have reduced to near zero conditions due to the spread of built-up area. Subsurface run-off is also drastically reduced and the excess rainfall augments stream flow (Nair and Chattopadhyay, 2004).

River Water Quality

There are reports of water quality deterioration across the country (CPCB, 2014). Expansion of human settlements and the use of chemical inputs for agriculture have enlarged non-point sources of pollution. All rivers in Kerala are facing water quality problems owing to the indiscriminate discharge of industrial effluents, untreated sewage, and wastewater from urban centres/major settlements, and agricultural run-off. It has been estimated that more than 8 lakh m^3 of trade effluent containing a variety of deadly pollutants are being dumped into the rivers of Kerala every day from the major industrial establishments. In addition, indiscriminate use of poisons, dynamite, etc., by unskilled fishermen and others in the upper reaches of rivers have caused serious depletion of indigenous fishes in these rivers. River water can hardly be used for domestic purposes (Table 15.6). The condition is severe for the rivers passing through urban centres. The State Pollution Control Board (SPCB) regularly monitors the quality of water in the State for many years under various programmes. Based on water the quality parameters, assessment is being done whether the water body or any stretch of a river is suitable for the designated best use or not, and if not, the aspects in the quality which warrants changes, and the remedial steps which need to be taken up to ameliorate the situation are suggested.

WATER QUALITY OF VEMBANAD ESTUARY

The Vembanad estuary is now divided into two separate water bodies lying to the north and south of Thanneermukkom barrage. There are seven rivers joining the Vembanad estuary—three rivers to the north and four rivers to the south of Thaneermukkom barrage respectively. The northern part, popularly known as Kochi estuary, with regular tidal flushing maintains its natural character. Surrounded by dense urban settlements and industrial land use, Kochi estuary serves as a receptacle for the effluents of several industrial and manufacturing units and domestic

Table 15.6: Water Quality Parameters (selected) for Selected Rivers in Kerala

Rivers	*Water Quality*					
	pH	*BOD (mg/l)*	*DO (mg/l)*	*TDS (mg/l)*	*CPCB best use **	*Remarks/problems*
Kabani (sampling year: 2006-2007 & no. of samples-21)	6.17-8.20	0.11-11.08	3.46-11.68	13.2-233.6	B (31%), C (46%) D (21%)	Quality problem, untreated domestic and municipal effluents
Bharathapuzha (2005-2010, 27)	6.67-10.17	0.03-7.0	3.40-14.6	22.9-419.2	B (17%), C (63%) D (15%)	High turbidity near sand mining sites. bacteriological contamination
Chalakudy (2009-2010, 23)	5.27-7.89	0.07-7.45	3-10.8	18.8-718.7	A (52%), C (39%), D (9%)	Sand and clay mining, excessive water withdrawal and saline water ingression in the downstream
Periyar (2005–2008, 35)	5.65-8.42	0.06-4.73	1.60-11.8	12.5-566.8	B (18%), C (57%), D (8%), E (15%)	Sand and clay mining, Saline ingression, industrial effluents along downstream, heavy metal pollution
Pamba (2007, 21)	6.15-8.60	0.13-11.8	0.26-10.23	17.6-226.2	C (45%), D (40), E (13%)	Quality problem, Cadmium and copper concentration slightly high.
Achankovil (2009, 20)	6.01-8.22	0.07-3.60	4.64-10.43	35.6-127.6	A (5%), B (20%), C (75%)	Require treatment and disinfection for drinking
Karamana (2007-2008, 20)	4.39-7.72	0.06-39.11	2.4-11.53	15.8-6533.2	C (47%), E (37%)	75 per cent of water samples are acidic. Carries urban sewage.

* CPCB (Central Pollution Control Board) best use classification: A—Drinking water source without conventional treatment but after disinfection, B—Outdoor bathing (organized), C—Drinking water source with conventional treatment followed by disinfection, D—Propagation of wildlife, fisheries and E—Irrigation, industrial cooling, controlled waste disposal. Figures in bracket indicate percentage of total samples in the Class.

Source: Chattopadhyay and Mahamaya, 2014.

sewage from Kochi and surrounding smaller townships. It is estimated that Kochi Urban Agglomeration generates 600 tons of solid waste per day of which around 60 per cent are from households and 21per cent from commercial establishments (Corporation of Kochi, 2011). The rest are from institutional, road sweeping, drain sweeping, clinical, and construction and demolition of buildings and other structures.

Kochi city alone generates 2550 million litres day^{-1} of waste-water that directly enters into the backwater untreated. Total dissolved solid content near the sewage discharge site in Kochi estuary was as high as 53750 mg l^{-1} during summer which reduces drastically during the rainy season to 160 mg l^{-1}. The sewage without proper treatment contains organic and inorganic pollutants along with pathogenic micro-organisms responsible for various water-borne diseases. The concentration of industrial nodes, along the banks of the Periyar in the Udyogamandal area, 10 km north of Kochi harbour is visible. It is estimated that nearly 260 million litres of industrial effluents discharge into the estuary daily. The wetland ecosystem is also threatened by petroleum hydrocarbon (PHC) pollution. Numerous oil tankers and other vessels plying through the waters are the major sources of pollution.

The southern part of the Thanneermukkom barrage known as the Vembanad Lake receives urban industrial effluents from Kottayam, Alappuzha, and high-density rural settlements besides, agricultural run-off from the extensive paddy fields in the greater Kuttanad area. Annual fertilizer consumption in Kuttanad is 8409 tonnes of N, 5044 tonnes of P, and 6786 tonnes of K. Pesticides/fungicides/weedicides are applied to the tune of about 500 tonnes, annually. Coconut husk retting, another human activity in the coastal tract of Kerala contributes heavily to the organic pollution load of the open water bodies. The water quality is very poor near Alappuzha town with high nitrate and ammonium concentration.

Heavy pollution load due to factory effluents, pesticides, hydrocarbons, sewage and domestic discharges, retting of coconut husks, use of on board motors in boats, etc. has stressed the wetland ecosystem irrecoverably. Incidence of large scale fish mortality due to pollution in the backwaters has become quite common. Variations in the temperature, salinity, pH, and BOD have been observed in the backwaters due to heavy pollution. The deterioration of biodiversity affects economic conditions and can have a serious social impact.

Nutrient Pollution in Vembanad Estuary and Selected Rivers

Nitrogen concentration in water is a matter of serious concern all-over the world, and it is considered as governance challenge as most of this nitrogen is human-induced and flow from agriculture, specially fertilizer, livestocks, and urban and industrial effluents. Water quality is seriously impaired in Kerala. The estimated nitrogen surplus from land in Kerala is 556kg ha^{-1}, the highest among all the states in India (Vadrevu and Badarinath, 2014). We would like to discuss the issue of nitrogen concentration in the Vembanad estuary and the rivers debouching into it. The mean concentration of total nitrogen in Kochi estuary varied from 5.82 μM/L to 93.80 μM/L (Joseph and Ouseph, 2009). The concentration varies over the seasons and towards the bottom of the estuary/lake. The Pre-monsoon (February to May) period shows higher concentration due to the restricted flow of freshwater through the rivers. With the decrease in non-monsoon flow and an increase in human activities, the concentration of nitrogen will be further elevated. An earlier study indicated a higher concentration of nitrogen in the sediments of Kochi estuary (Sankaranarayanan and Panampunnayil, 1979). The symptom of hypoxia (depletion of dissolved oxygen <2 mg/l or 30 per cent saturation) is noted in the deeper layers during monsoon months (July) both at the high tide and neap tide phases in the Kochi estuary (Martin *et al.*, 2010). Cochin mouth and surrounding coastal water is one of the 400 sites in the world showing symptoms of hypoxia due to human-induced water quality deterioration.

In the case of the Vembanad lake to the south of the Thanneermukkom barrage, the average nitrate (NO_3-N) concentration varied from 7.69 μM/L ±2.27 during the month of November to 16.85 μM/L ±14.4 during the month of May, and the highest value recorded during these two months were 13.85 μM and 49.12 μM respectively (Henkel, 2006). Long term annual average indicated that nitrate+nitrite concentration varied from 0 to 35.71 μM/L with a mean value of 8.57 μM/L (CPCB, 2014). The average dissolved nitrogen (Nitrate +Nitrite) in the downstream of rivers before joining the Vembanad lake system varied from 27.86 μM/L in Muvattupuzha to 77.14 μM/L in the Pamba river (Fig. 15.8). Considering the maximum value it may be noted that Periyar (Pathalam) and Pamba recorded dissolved nitrogen of >100 μM/L. The Pamba reported the highest maximum of 164.29 μM/L in the observation station near Chenganur. The high concentration in the Pamba river is linked to the gathering of pilgrims upstream (David *et al.,* 2016). Elevated values of nitrogen in the Pamba river and Chalakudy river are also attributed to agriculture practices, livestocks, and discharge from settlements.

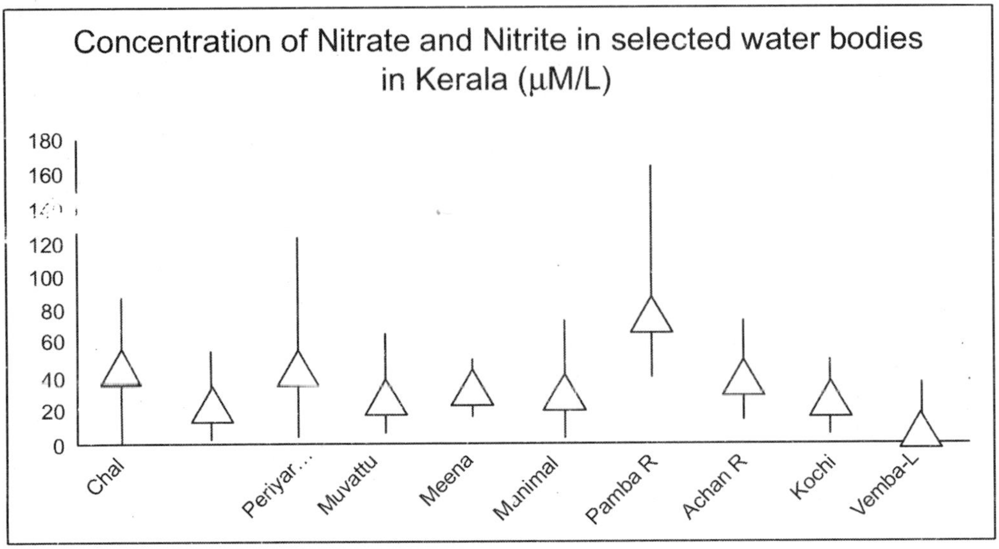

Fig. 15.8: Concentration of Nitrate and Nitrite in selected surface water bodies in Kerala. Note elevated values for the Pamba river followed by the Chalakudy and Periyar rivers.

GROUNDWATER

Dependence on groundwater is increasing over the years. As a result, the groundwater situation is undergoing perceptible change, and there are reports of depletion of groundwater level (CGWB, 2012). Census data brought out that around 62 per cent of households in Kerala use well water for drinking and other domestic use. The quality of groundwater in Kerala is by and large potable. However, there are reports of contamination. Many water-borne diseases related to the gastrointestinal system, diarrhoea, dysentery, typhoid, worm infestations, and infectious hepatitis and a high rise of fecal contamination are attributed to groundwater contamination (Harikumar and Chandran, 2013). While chemical contamination occurs in areas due to heavy industrialisation and heavy application of chemical fertilisers and pesticides, there are exceptions of *in situ* chemical contamination (iron, fluoride) due to the inherent chemical qualities of soil/ geological formations. Bacterial contamination is predominant in thickly populated areas with poor sanitation, and in the vicinity of bio-polluting industries and waterlogged areas where stagnation of surface and groundwater occurs (Chakrapani, 2014). Overpumping from coastal aquifers leads to saline intrusion in some parts of the coastal tracts.

NATURAL VEGETATION COVER

Kerala experienced drastic changes in forest vegetation cover. Natural vegetation cover is now fragmented and limited to rugged hilly areas (Fig. 15.9). According to the official statistics, Kerala has a total forest area of 11309 km^2 (2011). Of which 140 km^2 are ecologically fragile land (EFL), 7860 km^2 are forests, 1493 km^2 are under plantations, and the remaining 1816 km^2 area are allotted to various agencies, such as the Kerala State Electricity Board (KSEB) and Public Works Department (PWD) for the river valley projects, the Forest Development Corporation, the Kerala

Livestock Development Corporation, and other agencies (Government of Kerala, 2012b) (Table 15.7). Irrigation, hydropower, and water supply projects and other institutions were established within the forest boundary. Hillmen settlements now occupy 230 km^2 and it is growing over the years. Due to all these human-induced activities, forest patches are fragmented. Routes of wildlife movement, which were earlier in deep forest, are now exposed. The frequent incursion of elephants in the human habitation is a result of such interventions and loss of wildlives' niche. An earlier study based on remote sensing data and map analysis indicated that intact forest vegetation in Kerala was little over 6630 km^2 or 17 per cent of the total geographical area in 1973 (Chattopadhyay,

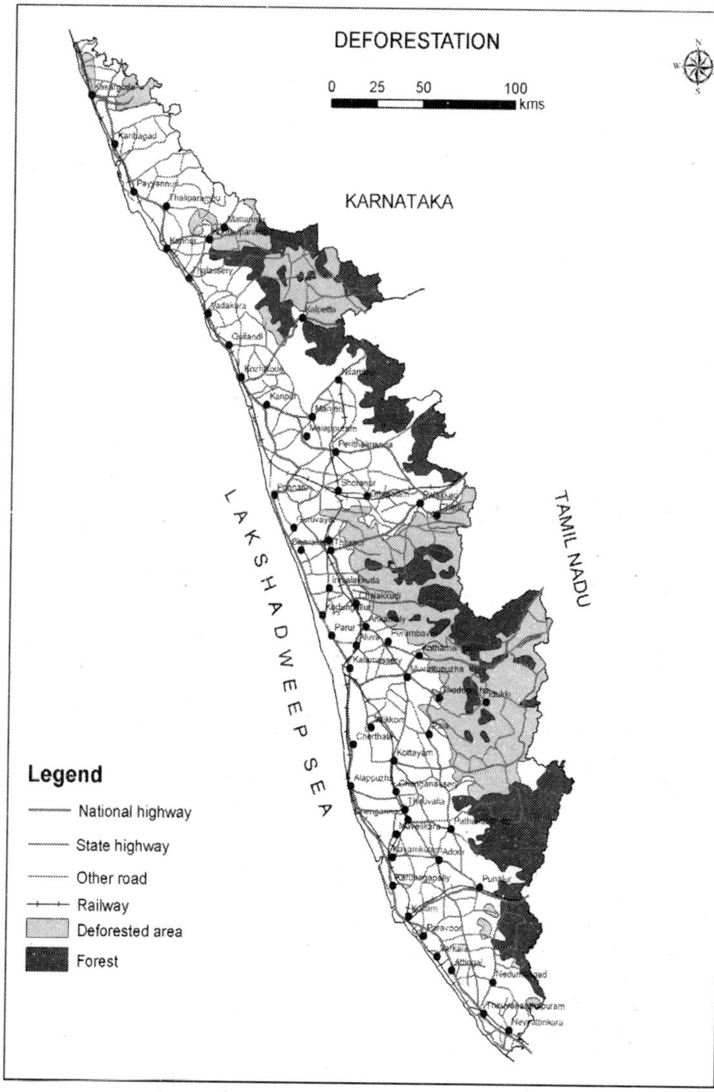

Fig. 15.9: Forest and deforested area in Kerala. Note that forest patches are fragmented and deforestation has taken place mostly along transport routes.

Data Source: Chattopadhyay, 1985.

1985). It may be recorded here that this coverage refers to the natural vegetation cover. Area reported by the forest department is the area under the forest administration and does not represent the actual vegetation cover.

Encroachment within Forest Area

Encroachment within the forest areas has a long history. Kerala is traditionally well known for spices, most of which are hill products. With an increase in demand for black pepper in the global market in the historical past, there were efforts to raise black pepper vines replacing forest vegetation. The introduction of agriculture by clearing forests began in the early nineteenth century under the patronage of the Travancore Government. The first coffee plantation was introduced by the Europeans in the year 1798, although the actual spread of coffee plantation started by 1830s in Wayanad and in the 1860s in Travancore. There were several initiatives by the Travancore State and also by the British Government to expand cultivation and settlements within the forest area providing various incentives. Waste Land Rules (1861) granting ownership rights to the tenants in the government lands and similar incentives promoted the expansion of non-forest activities within the forest area. This was intensified over the years. Even the *taungya* system developed in Burma in the middle of the 19th century was tried in Travancore in 1915. The state monopoly of cardamom cultivation came to an end by 1896 and private ownerships were allowed. The period between 1928 and 1932 witnessed a boon in encroachment and registration. During the colonial period, tea and coffee plantations sprung up within the forest area. Agriculture-forest boundary has been pushed deep inside the Western Ghats and roads were developed in various parts of the Western Ghats, thus facilitating human interventions.

Table 15.7: Use of Forest Land for Non-Forestry Purposes in Kerala (2011)

Sl. No.	*Use*	*Area*
1.	Diversion of forest land for non-forestry purposes including river valley projects	410.36 km^2
2.	Area of encroachment	45.19 km^2
3.	Encroachment regularized	25.87 km^2
4.	Settlement area	230.10 km^2
5.	Forest Plantations	1493 km^2
6.	Forest land under lease	423.39 km^2
7.	Road length within the forest area	4693.3 km
8.	Number of buildings within the forest area	2640 nos.

Source: Government of Kerala, 2012b.

Change in Natural Vegetation Cover in River Basins

Change in natural vegetation cover varies spatially and to highlight this dimension the scenario has been discussed for selected seven river basins in south Kerala. By the year 2004/05, the Pamba river basin has the largest natural vegetation cover of 856 km^2 followed by Kallada and Achankovil (Table 15.8). The Pamba basin houses some of the best forests in Kerala. The Sabarimala, the famous pilgrim centre is located deep inside the forest. Although more than 1.5 lakh people visit this holy shrine each day during the main season of 61 days, the forest area remains more or less undisturbed, perhaps due to reverence that people attach to a holy site and its surroundings. Kallada experienced the maximum reduction of 449 km^2 in a span of

37 years, with an average annual loss of little over 12 km^2. In the first phase (1967-1973) the loss was the maximum in Kallada but in the second phase, the maximum loss was marked for the Achankovil basin, where large-scale forest plantations have been established. When total catchments of the rivers are considered it emerges that the Pamba has 38 per cent of its total catchment area under natural vegetation, followed by Kallada (21%). Ittikara river basin has lost all its natural vegetation cover. All four rivers of Neyyar, Karamana, Kallada, and Pamba have been arrested for river valley projects upstream.

Table 15.8: Basinwise Change in Area under Natural Vegetation Cover in Kerala (1967 to 2004/5)

River basins	*Vegetation cover (in km^2)*			*Change in km^2 (per cent figures in bracket)*		
	1967	*1973*	*2004/05*	*1967-1973*	*1973-'04/05*	*1967-04/05*
Neyyar	89.0	80.0	65.0	– 9 (10.1)	–15 (18.4)	–24 (26.6)
Karamana	134.0	126.0	32.0	–8 (6.6)	–94 (74.8)	–102 (76.5)
Vamanapuram	182.0	110.0	16.0	–72 (39.7)	–94 (85.3)	–166 (91.1)
Ittikara	55.0	38.0	Nil	–17 (29.4)	–38 (100)	–55 (100)
Kallada	804.0	533.0	355.0	–271 (33.7)	–178 (33.4)	–449 (55.8)
Achankovil	591.0	489.0	167.0	–102 (17.3)	–322 (65.9)	–424 (71.8)
Pamba	1139.0	1058.0	856.0	–81 (7.1)	–202 (19.1)	–283 (24.8)
Total	2994.0	2434.0	1491.0	–560 (22.0)	– 943 (35.8)	–1503 (50.2)

Source: Chattopadhyay and Sajna (2009).

CAPTURING CHANGES AND KNOWLEDGE GAP

The geography of Kerala is changing at a faster rate. Changes are manifested throughout the state from the coastline to the high mountains and from settlement systems to the environment. These changes are potent to jeopardize the state's development process. Case study-based results used to illustrate the changes highlight that rainfall behaviour is erratic, soil erosion is accelerating, quality of soil is being impaired and finally, water distribution, both quantity and quality are affected. Biogeochemical cycles are interfered. Deforestation has been officially arrested, however, there are changes within the forest area. Thinning of forest and quality deterioration may continue in the future. Field investigation in various segments of Kerala indicates that the nature and intensity of changes that are taking place in Kerala vary spatially. Drivers of these changes are not always *in situ*. There are several factors including globalization that have triggered these changes. Some of the government policies have also significantly contributed to this matter.

One of the serious challenges in capturing and understanding these changes and taking ameliorating measures is the knowledge gap. Macro-level data and official statistics are hardly useful. Many of the data are conjectural. Statistics of one department rarely matches with that produced by another department. There is little interdepartmental coordination. Local-level information and actual measurements required for sectors like land, land use, and water are seldom attempted. Similarly, the social and political power structures that govern the change are little understood. It is necessary to examine all these issues closely. Understanding changes in the geography of Kerala is significant to plan for building resilient and sustainable Kerala.

CHAPTER

16

Regions, Area Development and Management

The regional dimension of development is being recognized as critical for an effective and coordinated response to address developmental problems. So far, a region used to be considered as an outcome of the deeper political-economic process, however, since the early 1980s it was asserted that region might be a fundamental basis of economic and social life 'after mass production' (Storper, 1997).

REGIONALISATION

Regionalisation is a process of identification of regions. It is a subjective exercise in the sense that the criteria for identifying regions are selected by people to serve specific purposes. The phenomena/features being described and measured may not necessarily be the product of human action, but the selection of salient details of phenomena to use in regionalisation schemes is definitely a human decision (Stransfield Jr., 1983). The utility of regionalisation is well known, and frequently there are references to regions while discussing even mundane issues like where one is going or coming from or the type of languages people are speaking. A village, a town, or a city, part of town/city or a neighbourhood, a hill, or coastal area provides a perception of a region on local scale. Boundaries are well accepted as perceived by local people, who also have a sense of belongingness and locational identity with the place. A post office, school, hospital, or any other service like a police station or fire station all cater to a definite service area/catchment, boundaries of which may or may not correspond all the time. Nevertheless, boundaries demonstrate a sense of ownership and responsibility.

Kerala is not a homogeneous single entity. There are impressive and significant variations, and traditionally, there are definite identifiable characteristics to recognise regions. The character and quality of widely different faces of Kerala can be best interpreted through regionalisation. As regions are intellectual creations, a product of human curiosity, and a means to divide an apparently complex world into subdivisions for a better understanding of the phenomena, there are different schemes to pursue regionalisation. It begins with physical identification principally based on biophysical parameters. Over the years, anthropogenic criteria figured in. The new schemes of regionalism are driven by the issues like trade, connectivity, monetary and financial coordination, and cooperation, and meeting the health, food security, and social protection concerns of an increasingly mobile population in one hand, and on the other hand, it gained importance as a buffer against global shocks and crisis at a time when global regulations and institutions have proved inadequate or ineffective (UNRC, 2011).

SIX REGIONS IN KERALA

Six regions of Kerala are determined to highlight intra-state variations of qualities of the physical and cultural landscape. We considered distinguished physical, social, cultural, and economic features for regionalisation. Each of the terrain, drainage, soil, climate, and vegetation is fundamental to the traditional, and present perception of usefulness to land use, economy, population distribution, and general appraisal of the landscape. The human population, social and economic variables, which tend to modify the natural landscape, constitute the second major component for regionalisation. The nature and intensity of economic development form another set of landscape moulders. The types of natural resource exploitation, land use system, transport routes and technologies, connectivity, industrial and commercial activities, and development challenges all are significant issues. Scale and influence of urbanisation on landscape provide yet another dimension to consider in the process of regionalisation. People's perception of the places and problems, and geopolitical reality are also taken into account while demarcating regions in Kerala. Traditionally, there are three geographic regions of Highland, Midland and Lowland often referred to in common parlance. There are also common references to Malabar, Travancore, Kochi, Palakkad, Western Ghats, coastal plain, etc., to indicate various regions in the state. We consider all these perceptions while delineating six regions in Kerala.

The regional boundaries, although appear to be precise are actually arbitrary and there are transitions from one region to another. Secondly, boundaries are not permanent. They shift and change over time. The six identified regions in Kerala are (i) Western Ghats, (ii) Malabar Uplands, (iii) Kochi-Thrissur-Palakkad corridor, (iv) Travancore-Kochi midlands, (v) Kuttanad, and (vi) Coastal zone (Fig. 16.1). Kuttanad although a part of the Coastal plain, is considered as a separate region due to its wetland condition and physiographic position at below sea level. This area requires special attention. Therefore, it can also be treated as a separate region. Brief descriptions of all these regions are given here.

1. Western Ghats

The Western Ghats, a biogeographical region, has a profound impact on the ecologic, economic, social, and cultural milieu of Kerala. It covers a little more than 40 per cent of the total area of the state and spreads over 32 talukas. Wayanad and Idukki districts fall completely under the ambit of the Western Ghats. Among the rest 12 districts in Kerala, except Alappuzha, all 11 districts have some parts falling within the Western Ghat region. The Palghat gap at an elevation of 300 m breaks the monolithic Western Ghats into north and south blocks. There are geologic, structural, and relief variations between these two blocks, nevertheless both of them can be considered as one region given their commonality. The Western Ghats in Kerala receiving more than 300 cm of rain is one of the high rainfall regions in India. The eastern part of the Western Ghats, particularly the Marayur area receives less than 50 cm rain due to its location in the rain shadow area. The Western Ghats is one of the biogeographic regions in the World heavily impacted by human activities. There was the migration of people from the lowlands and midlands of Kerala to this area since long. Tropical rainforest, forest plantation, agricultural plantation of cardamom, tea, coffee, pepper, ginger, and other spices are widely cultivated. The western parts of the Western Ghats are mostly given for rubber plantations.

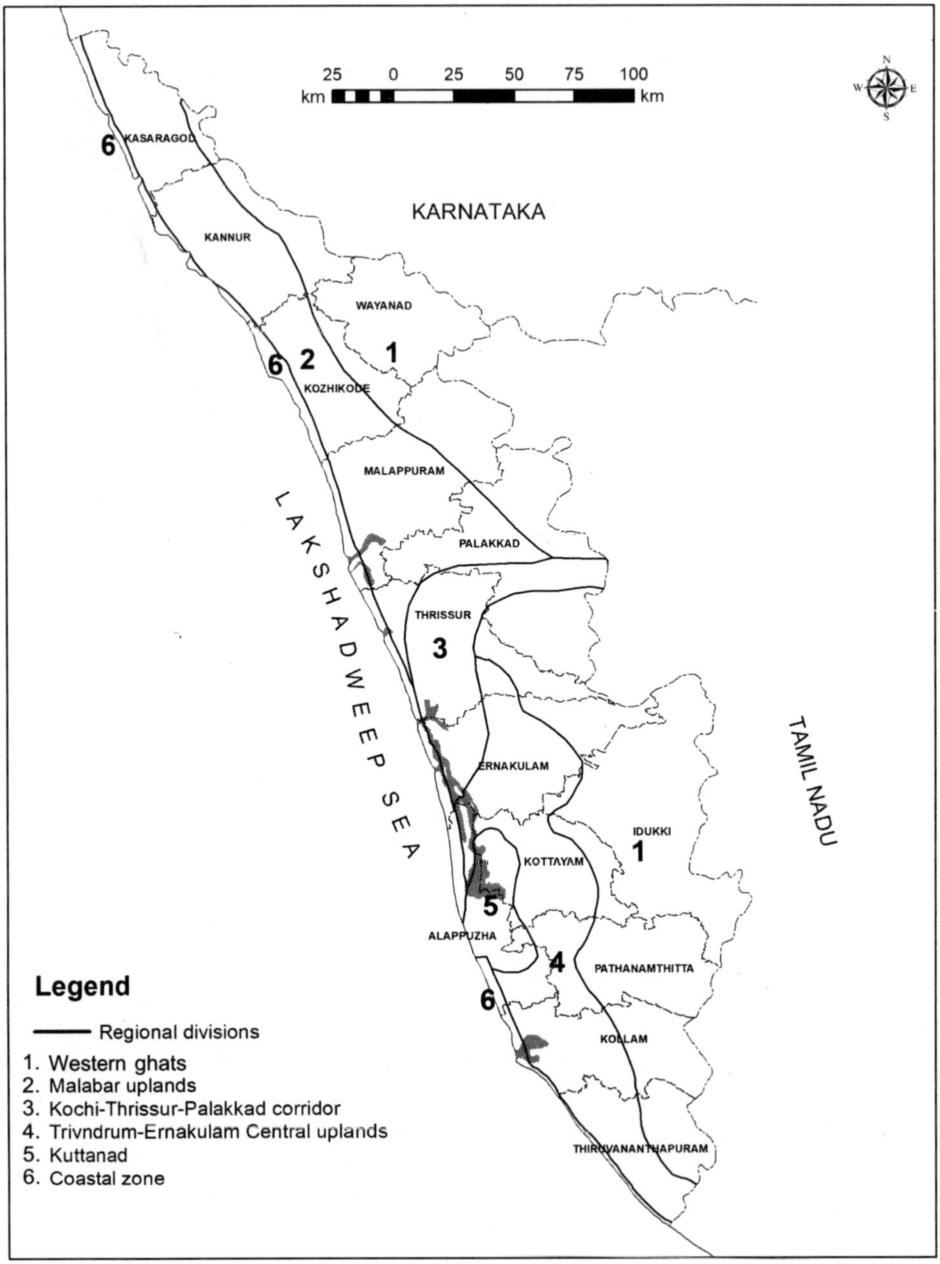

Fig. 16.1: Regions in Kerala.

There are indigenous settlements scattered all-over the Western Ghats, particularly in Wayanad, Palakkad, and Idukki districts. The Western Ghats faces several problems mainly originated from environmental degradation. Preservation of the Western Ghats, its environment and people, is a high priority activity in Kerala. Some of the issues related to the conservation of the Western Ghats have been discussed separately.

2. Malabar Uplands

The Malabar uplands extend from Palakkad district in the south to Kasaragod district in the north. The topography is lateritic and undulated with the occurrence of well-developed lateritic mesa. It is the type area for laterite. Angadipuram, where laterite was first identified and named, is located in the Malappuram district of Malabar uplands. Topographic characteristics, rainfall patterns, and soil types are different compared to the southern part. Lateritic wastelands are extensively marked particularly in Kannur and Kasaragod districts. There are highly urbanised pockets like that of Kozhikode and surroundings and Kannur-Tellicherry stretch. The population growth rate in this area is more than that of its southern counterpart. In fact, the most populous district Malappuram falls in this zone. This region is relatively lagging behind in socio-economic development, which can be traced to its colonial legacy. Malabar region receives high priority in the development planning of Kerala.

3. Kochi-Thrissur-Palakkad (KTP) Corridor

Kochi is the main natural port in this part of the country. It plays important role in the industrial development of Kerala. The hinterland of Kochi port extends beyond Kerala through the Palghat gap. The Kochi-Thrissur-Palakkad corridor is developing under this background. Although agriculture is an important occupation of people living in this area, this corridor is fast emerging as an urban-industrial-transport corridor. Palghat gap connects Kerala with Tamil Nadu plain. Railways and roadways, both from the northern and southern part of Kerala, converge in Palakkad, cross the 30 km wide east-west gap, and diverge in the rest of Peninsular India. The towns of Coimbatore and Palakkad located to the east and west of the Palghat gap respectively serve as connecting nodes. Kochi-Kalamassery-Alwaye is the most important modern industrial belt in Kerala. Kochi area is also known for receiving foreign direct investment. Thrissur houses traditional industries like clay and tiles. Palakkad has traditional industrial background and is emerging as an important modern industrial belt. This KTP corridor is playing an important role in Kerala's economy. The accessibility and efficiency of the transport system in this corridor is increasing at a faster rate. Kochi international airport, one of the four international airports in Kerala, is the busiest in the state and runs under Public Private Partnership mode. This region is part of the economically fast growing areas in Kerala. It is the industrial hub of the state with high potential of development both for traditional and modern industries.

4. Travancore-Ernakulam Midlands

Situated between the coastal zone and the Western Ghats in southern Kerala, this region spreads over the districts of Thiruvananthapuram, Kollam, Pathanamthitta, Kottayam, and parts of Ernakulam. This area is a lateritic piedmont plain characterised by undulated topography with alternate ridges and valleys. This region is densely settled and intensely used. A series of towns/urban centres have developed along foothills, which serve as market centres for the collection and distribution of goods. Rubber plantations have been extensively developed. Cashew plantations are prominent in some parts of the Kollam district. This region is experiencing rapid urbanisation both in the southern segment around Tiruvananthapuram and the northern segment around Ernakulam. Population growth is low in this region and some areas like Pathanamthitta experience negative growth of population. Economically, it is fast growing with IT and other industrial developments.

5. Kuttanad

Kuttanad is considered a problem zone in Kerala. The first report on Agro-climatic zones in Kerala identified Kuttanad as a separate zone. This area is low-lying, altitudinally positioned at below mean sea level, geologically subsiding, clayey, with an abundance of water, and criss-crossed by natural and man-made channels. This deltaic landscape is seasonally waterlogged, houses lush green paddy fields, and is dominated by primary sector of the economy. Kuttanad presents a typical ecosystem, quite different from the rest of Kerala. This area is known as the rice bowl of Kerala, although, coverage under rice has come down substantially in recent years. Cultivation of paddy, mostly in the summer season (*Puncha* cultivation), is done by dewatering of the paddy fields located 1 to 2 m below mean sea level. Recently, the Food and Agriculture Organisation (FAO) of the United Nations declared Kuttanad as Globally Important Agricultural Heritage System (GIAHS) similar to the Polder system in the Netherlands. Thanneermukkom barrage and Thottapally spillway are two major human interventions in Kuttanad. Alappuzha-Chenganasseri road is also another human intervention cutting across the core area of Kuttanad. These interventions have seriously impacting the hydrological regime. Due to the recurrence of the flood, soil constraint, water quality deterioration, economic backwardness, and livelihood problems, the Kuttanad region is a priority area for development initiatives. We provide a separate discussion on area development initiatives for Kuttanad.

6. Coastal Zone

The coastal zone is a distinct region in Kerala. It spreads over an area of 3355 km^2, which is less than 10 per cent of the total geographical area of Kerala and accommodates 29 per cent of the state's total population. The local bodies with direct sea frontage covers 1998 km^2 area. One of the most notable features of Kerala coast is the series of brackish water wetlands running parallel to the coastline and providing varieties of ecosystem services. The coastal zone is resource-rich, intensively used, densely populated, and highly urbanised. More than 70 per cent of the coastal plain is under the urban area. At the same time, the coastal zone is beset with several problems. Erosion, sea surge, loss of coastal ecosystems services, degradation of water quality, and incidence of poverty are some of the problems encountered in the coastal zone. The problems will be further compounded due to sea-level rise and related environmental consequences linked to climate change. The fishing population, one of the weaker communities, mostly lives in the coastal zone. Management of coastal zone remains a challenging task in Kerala. The issue of coastal zone management is separately discussed in this chapter.

AREA DEVELOPMENT AND MANAGEMENT

The importance of spatial planning is well recognised. It is often argued that socio-economic development should be anchored on the biophysical foundation as part of spatial planning aimed at area development and management. Six regions spelt out above may provide a basic foundation for such an exercise. Here, we discuss some of the emerging issues pertaining to three regions, namely, Coastal Zone, Kuttanad, and the Western Ghats, which are priority areas for intervention. It is suggested to adopt area-specific approach for addressing local problems and evolve development plans.

MANAGEMENT OF KERALA COAST

Kerala is one among 9 coastal states in India. The Coastal zone in Kerala is perhaps the most intensively used coastal tract in the country. People live right up to the coastline. Stretching over 590 km along the Lakshadweep sea, Kerala coastline accounts for 10.9 per cent of the total shoreline of mainland India. Kerala has a long tradition in sea trade since the early days. Musiris near Kodangular in central Kerala coast was a major emporium in the spice route that extends from the China coast in the east to the Arabian coast in the west. Coastal zone has a strong impact on the economy, social and cultural lives of Kerala. Nine districts, out of 14 districts in the state, extend up to the coastline. Kerala coast is rich in heavy mineral deposits in the Chavara area, glass sand deposits in Cherthala, and estuarine and marine fisheries. It supports a variety of habitats, like sandy beaches, spits, mudflats, lateritic and rocky cliffs, lagoons and estuaries, barrier bars, islands, a breeding ground for Olive Ridley turtle, and mangrove vegetation dotted in several small patches in different parts of the coast.

Coastal wetlands in Kerala are well known for their productivity. All west-flowing rivers (41 nos.) in Kerala debouch into the sea through lagoons and estuaries. The velocity of river water drops down substantially once the rivers enter coastal plain. As a result, the lagoons and estuaries receiving all sediments and nutrients brought down from the Western Ghats are highly productive for fishery and other aquatic lives. They also facilitate fish breeding. Large-scale human interventions are deteriorating these water bodies and are causing considerable damage to the aquatic resource base. Coastal zone is also exposed to various natural hazards including coastal erosion, sea surge, coastal flooding, and occasional tsunami. Climate change will adversely affect the Kerala coast. Sea level rise will not only inundate coastal lowlands, but also, it will affect freshwater saltwater wedge in the estuaries, and sub-surface water resources. The rise in sea surface temperature is another issue of concern.

The coastal zone is highly urbanised. The entire coastal plain from Kasaragod to Thiruvananthapuram appears to be a single urban corridor. Three main cities in Kerala, Thiruvananthapuram, Kochi, and Kozhikode are located along the coast. However, there are also pockets of high deprivation particularly among the fishing populations. In the governance front, the issue is the multiplicity of organisations involved in coastal resource management, sometimes with conflicting interests. Besides, there is a national policy governing coast and coastal water, and local self-governments engaged in local governance. All these issues warrant due attention, and coastal zone management assumes high importance in Kerala.

Coastal Regulation Zone

The introduction of Coastal Regulation Zone (CRZ) in 1991 by the Government of India may be considered as the first step to regulate activities in the coastal zone, and to bring a balance between conservation and development. There were four zones noted as CRZ-I, CRZ-II, CRZ-III, and CRZ-IV. Activities in each of these zones have been specified. Around 500 km^2 area of Kerala coastal zone falls under CRZ-I to CRZ-III (CESS, 1995). There are 2130 islands in the backwaters falling under the coastal regulation zone, where no development is permitted within

50 m from the high tide line. Coastal zone management plans have been prepared for Kerala considering CRZ norms. Recently, there were modifications of CRZ rules in 2018 to permit certain activities within 100 m distance from the high tide line in CRZ-III. It is suggested that a proper coastal zone management plan can address all issues related to the problems of environment and development in the coastal zone. Coastal Zone Management Authority (CZMA) was established by the Government of Kerala to look after coastal regulations and other development initiatives, particularly, construction of buildings. There are specified setback lines for this purpose. Violation of CRZ norms has been a very significant issue in Kerala. Three high-rise buildings, which have been constructed violating CRZ regulations in the Maradu area of Kochi have been demolished in January 2020 under the direction of the Supreme Court. The Government of Kerala has initiated the process of identification of buildings that have been constructed violating CRZ rules. There are nearly 26,000 cases of suspected coastal regulation violations in Kerala (Sudhi, 2020). CRZ is now a major governance challenge.

Coastal Zone Management: Issues and Challenges

Coastal zone is a multifunctional area and subject to competing use of resources. The traditional attempt of sectoral management has not yielded the desired results. Globally, there is a shift towards integrated coastal zone management (ICZM), which is considered a tool for sustainable development of coastal areas. Coastal zone in Kerala faces several challenges related to environment, development, and governance. We shall try to highlight some of the issues that warrant immediate attention. There is a need to address these issues and take necessary precautionary measures. Protection of shoreline and management of coastal hazards assume high significance in Kerala as people are living on the coastline itself in many segments. Assessment of vulnerability and computation of risk are important steps in this direction.

Land use zonation to allocate land for various uses is a starting point. High population density, scattered settlements, and exposure to vulnerability have called for a re-examination of the settlement systems in critical areas. This is also required for proper and effective service delivery. Rehabilitation of a group of people living in highly critical areas is one of the interventions in the coastal zone that merits consideration. It warrants settlement consolidation and identification of safe habitable places. The other sectors warranting interventions are agriculture and allied activities, inland fisheries, animal husbandry, health, sanitation, and drinking water. The growing urbanisation of coastal plain and increasing rural-urban differences is a matter of concern. Diversification of economic opportunities for rural people living in the coastal area is important. Eco-restoration and conservation, biodiversity management, and promotion of ecotourism are some other issues of immediate interest. Like land use zonation, there is a need for littoral zonation to plan for marine resource management. The promotion of artificial reef in coastal water requires detailed scientific studies. There are high priority areas along the coastal stretch like mangrove sites, breeding ground for Olive Ridley turtle (Payyoli, Kozhikode), mudbank dominated coast, cliffy coast, geosites (Varkala), forts (Anchuthengu, Bekal), and similar other places of special interest. It is important to develop management plans specifically for such areas as part of coastal zone management in Kerala.

Joint Coastal Zone Management Programme (JCZMP)

The introduction of JCZMP is necessary for the development of the Kerala coast. Under this programme the entire ICZMP of a particular segment may be entrusted to the local community with assured technical support from the Government department/research Institutes. The local communities will act as custodians of the coastal zone. The model village programme is the first step in this direction. The Gram Panchayat will be the overall co-ordinator of such initiatives. The activities under this initiative are (i) Identification of a segment of the coast, (ii) Group formation among the local communities, (iii) Organising technical support group, (iv) Preparation of plan, and (v) Execution of project activities. Training of local people in drawing coastal zone management plans and executing them as part of the panchayat plan should form a major component of Coastal Zone Management activities.

Terrain Based Management Plan for Coast

The coastal zone is not uniform. The landscape is diverse. Four dominant morphogenetic processes, namely, marine, fluvio-marine/estuarine, fluvial, and denudational are active in the coastal zone of Kerala. There are 20 terrain units that evolved under these processes (Table 16.1). The management plan for each of these units may be worked out as has been demonstrated for one of the coastal Panchayats, Neendakara, in Kollam district (Table 16.2). Eight terrain units have been identified at the micro level. Beach placer deposits, barrier bars, spit, and coastal cliffs are some of the examples of coastal assets, morphodynamic units, and sites of special interest. All these sites warrant specific management practices as indicated in Table 16.2.

Table 16.1: Morphogenetic Processes, Terrain Units and their Area Coverage in the Coastal Zone of Kerala

Morphogenetic processes	*Terrain unit*	*Area coverage (km^2)*
Marine		
Accretional	Beach	77.0
	Tidal flat/mud flat	59.0
	Spit bar	3.0
	Sand ridge	18.0
	Coastal plain-sandy	1542.0
Erosion	Coastal cliff	8.0
	Coastal plain-lateritic	116.0
Complexes	Coastal plain with palaeo strandline	90.0
Fluvio-marine/Estuarine		
	Tidal channel	3.0
	Brackish marsh	2.0
	Fluvio-lacustrine plain	99.0
Fluvial		
Accumulational	Flood plain	119.0
	Seasonally flooded area	8.0
Complexes	Transitional plain	95.0
	Alluvial plain (young)	27.0
Denudational		
Associated with relief and slope	Low rolling terrain	1586.0
	Undulated terrain	377.0
	Highly undulated terrain	14.0
	Isolated/elongated ridge	8.0
Associated with planation	Lateritic mesa	85.0
	Total	4336.0

Source: Chattopadhyay and Mahamaya, 1995.

Table 16.2: Terrain Units, Characteristics and Management Requirements in Neendakara Panchayat of Kollam District in Kerala

Terrain unit	*Characteristics*	*Suggested management activities (selected)*
Runnel	Low areas between sand ridges, dominated by finer elements, holds water	Preservation of this land unit as far as possible; restriction of land conversion; use for seasonal agriculture, water harvesting and drainage development
Bare sand	Sandy waste, without vegetation	Use for settlement, vegetation cover
Barrier bar	Sandy, densely settled with coconut plantation	Zonation for settlement consolidation and other use, creation of open space, infrastructure development, fishing facilities
Braided bar	Within the water body, sandy, vegetated, settled	Preservation of the water fronts, developing water sport area, and fishing facilities
Block reclamation	Reclaimed land mainly for agriculture	Maintaining the embankments, introduction of fish culture along with rice
Sea wall	Rocky wall along the coast	Strengthening of the wall, Maintaining fishing gaps, developing vegetative barrier
Water body within the barrier bar	Small water bodies, parts of deeper runnels	Preservation of the water bodies, pisciculture
Water body	Tidal channels with outlets on either side	Preservation of the waterline, development of vegetative barrier, pisciculture

Source: Chattopadhyay, 2010.

AREA DEVELOPMENT OF KUTTANAD

Kuttanad represents a typical ecosystem in Kerala. Between the coastline in the west and laterite landscape to the east, this basin-shaped landscape spreads over the Pamba-Manimala-Achankovil river delta adjoining the southern part of the Vembanad lake, and fan and riverine deposits of Meenachil and Muvatupuzha rivers along the eastern part of the Vembanad lake. The western part of Kuttanad is a sandy plain, which is relatively elevated. The clay belt to the east of this sandy plain is distinctly placed at a lower elevation and part of this cay belt lies 1 to 2 m below mean sea level. The deltaic and clay dominated core area of Kuttanad spreads over Chamkulam and Valianad Blocks and parts of Pulikeezh Block covering an area of 284 km^2. This core area and the surroundings, together known as Greater Kuttanad spreads over an area of 1390 km^2, and accommodates a little over 17 lakh people distributed in 103 villages falling under the Alappuzha district (62 villages), Kottayam district (33 villages), and Pathanamthitta district (8 villages).

The landscape of Kuttanad is a conjugation of bowl-shaped islands of various dimensions with a deep central part, and shallow periphery. Settlements have mostly developed along water courses occupying natural levees, and to take the advantage of movement along waterways. Much of the land area has been reclaimed from waterlogged marshy conditions, and lake beds. Therefore, island-like landscapes, known as *Padasekharams,* are encircled by embankments, which prevent water incursion into land areas. Human lives are embankment dependent. Kuttanad deserves special attention both from ecologic and economic point of view. Ecologically, it is one of the fragile zones in Kerala with the recurring incidence of flood, land degradation, water pollution, stress on agriculture, limiting soil character, and declining ecosystem services of the land and water. This area, due to its location below mean sea level, is also one of the areas in Kerala presumed to be inundated by climate change-induced sea level rise. Socio-economically, it is one of the backward regions in the state with the domination of the primary sector, limited livelihood opportunities, transport bottlenecks, inadequate social infrastructure, and incidence of poverty. A comprehensive area development plan is necessary to address all ecologic economic issues.

Strategy for Area Development of Kuttanad

The development of Kuttanad drew attention from time to time due to its peculiar landform and a host of problems mostly related to land and water. There had been several commissions/ studies to address the problems of Kuttanad and suggest appropriate action plans since the 1970s (Chattopadhyay and Sidharthan, 1985; KSPB, 2019). 'Water Balance Study of Kuttanad Region' under the Indo-Dutch Mission in 1989 is one of the most detailed analyses with a particular thrust to water management. However, recommendations of this study have not been considered for implementation due to various reasons including financial feasibility. On invitation from the Ministry of Agriculture, Government of India, the M.S. Swaminathan Research Foundation (MSSRF) proposed the 'Kuttanad Package' in 2007 (MSSRF, 2007). This package focussed on seven major themes namely, (i) Declining water spread area of Vembanad lake, (ii) Flood management, (iii) Salinity intrusion, (iv) Blocked waterways, (v) Water pollution and aquatic weeds, (vi) Loss of animals and fish biodiversity, and (vii) Health, sanitation, and drinking water. These themes are mostly area-based interventions. Some of the activities proposed under this package are being executed.

The year 2018 witnessed the most devastating flood in the recent history of Kerala. Apart from post-flood reconstruction, the Government of Kerala has undertaken various initiatives to build New Kerala (*Navakeralam*). Kuttanad being the most severely battered by flood has drawn special attention. The State Planning Board revisited all previous studies on Kuttanad and prepared a 'Special Package for Post-Flood Kuttanad' to help Government in its Rebuild Kerala initiatives (KSPB, 2019). It has recommended a new package for Kuttanad. The thrust is on water management, resilient agriculture, animal husbandry, fisheries, responsible tourism, sanitation, toilet and drinking water, and interventions for industries and power supply. The suggestions to establish the hierarchy of river and canals like primary, secondary, and tertiary drainage systems linking main watercourses to field canals, compartmentalisation of *Padasekharams*, 'room for rivers', and flood control interventions proposed under water management are noteworthy. Water retaining capacity of the drainage system and the Vembanad lake has been seriously impaired due to intense human activities and encroachments. Compartmentalisation of large *Padasekharams* warrants serious consideration. At present, the entire *Padasekharam* is affected during the flood. Once it is properly segmented considering hydrological conditions, some segments can be fully protected while flood water may be allowed in some other segments. It will also help creating room for rivers. Loss due to flood damage will be limited. It is important to restore/halt further loss of spill area to accommodate monsoon flood flow by creating room for the Pamba river and the Vembanad lake. Regulation of land use, and renovation of wetlands/abandoned river courses will help in this process.

It may also be noted that as the Kuttanad region occupies the downstream segment of five major river systems of the Muvattupuzha, Meenachil, Manimala, Pamba, and Achankovil, water management in Kuttanad may not be feasible in isolation. A comprehensive basin management plan covering the provenance of all these rivers, and contextualisation of Kuttanad within that larger spatial frame is necessary. The other sectors deserving attention for area management and development of Kuttanad are refashioning of levees/embankment, a reworking of land use, zonation of the flood plain, settlement consolidation in critical areas, and promotion of nodes for secondary and tertiary sector development. Provisioning of drinking water and greywater management should form part of settlement planning.

It may be noted that Kuttaand is not a homogeneous region. Based on soil and site characteristics Kuttanad region is traditionally classified into three segments for land management, especially for paddy cultivation. These are *Karappadam, Kari*, and *Kayal*. *Karapaddam* lands are relatively elevated, whereas the *Kari* area is organic matter rich acidic black clay soil, and *Kayal* lands are the lowest land units within Kuttanad which have been reclaimed from the lake bed. This classification prevails over the selection of rice varieties, crop practices, working hours, and wage rates. Considering greater Kuttanad and other development issues the Kuttanad region can be divided into seven landscape ecological zones, namely, Western sandy plain, Purakkad mudflat (*Kari*), Upper Kuttanad (mature delta), Lower Kuttanad (active delta), *Kayal* land, North Kuttanad (fan deposits) and Vaikom mudflat (*Kari)*. The problems and potentials of these zones are different. It is important to recognise this heterogeneity and evolve suitable development packages. Activities envisaged under the proposed Kuttanad package should be dovetailed according to these landscape ecological zones following area management and development approach.

To facilitate area management and development covering all these issues, it is necessary to build up a proper institutional mechanism for inter-departmental coordination, the involvement of local self-government organisation, other stakeholders, and local people. An apex body like Kuttanad Development Authority may be useful in this context.

CONSERVATION OF THE WESTERN GHATS

The Western Ghats along with the wet zone of Sri Lanka is now considered as one of the eight 'hottest hot spots' of biodiversity in the World (Meyers *et al.*, 2000). High topographic heterogeneity, strong rainfall gradient, diverse soil type, and vegetation have endowed the Western Ghats region with rich biodiversity which can be gauged from the known inventory of its plant and animal groups, and the levels of endemism in these taxa (Gunawardene *et al.* 2007, MOEF, 2011). The available inventory indicates that the Western Ghats houses nearly 4000 species of flowering plants or about 27 per cent of the country's total species. It is rich in lower plant groups, vertebrate and invertebrate species, amphibian fauna, reptiles, birds, and mammals. The Western Ghats region supports the largest global populations of the Asian elephant. Besides, there are a number of wild relatives of cultivated plants, including pepper, cardamom, mango, jackfruit, and plantain found here. This biological wealth has paid rich dividends to Kerala over the years. In fact, this mountainous tract was famous for its wild produce of pepper, cardamom, sandal, and ivory, all are trade items since the early part of history.

However, this biogeographic region is also one of the few biodiversity hot spots with the most elevated risks due to the high concentration of the human population and widespread anthropogenic intervention. It was estimated that natural vegetation cover has reduced from 44.5 per cent in 1905 to 17 per cent by 1973 and area with intact forest ecosystem is limited to 14-15 per cent of forest area in Kerala (Chattopadhyay, 1985). Slopes are vulnerable and prone to landslides and high soil erosion. Kerala has a 24 per cent area in the category of > 35 per cent slope, another 16 per cent area falls between 25 per cent to 35 per cent slope, and 4 per cent area is scarp slope (>100 per cent slope). Slopes of the Western Ghats are steep. Weathering is deep and intensive. Soils developed over the sloping surfaces are prone to erosion.

High-intensity rainfall accelerates soil erosion. Landslides are common during the monsoon months. Rainfall is the single most important triggering mechanism for land disturbances in the Western Ghats. The average annual rate of soil loss from the Western Ghats is around 22 tons $ha^{-1}yr^{-1}$. Sample studies indicate that the rate of erosion in some pockets is even >200 tons $ha^{-1}yr^{-1}$. The intensity of the problem can be well gauged when these figures are compared with the estimated average annual soil loss of 5 tons $ha^{-1}yr^{-1}$ in areas under dense forest (>40% Canopy). The problem of soil erosion in the Western Ghats has been recognized for a long. There are several soil conservation programmes under the Western Ghats Development Programme being executed by the Government of Kerala. However, interventions under these programmes have not yielded the desired results, and problems of the Western Ghats not only continued but also aggravated over the years.

Migration of people from lowland and midlands to highlands in different time points since the mid-eighteenth century has altered land use patterns and ecology of the Western Ghats. One of the studies indicated that in a span of 80 years (1911-1991) population growth in the highland region was 1342 per cent whereas it was only 306 per cent in the coastal plain (CESS, 1997). The dense vegetation cover of tropical rainforest with thick undergrowth used to protect the land. Human intervention has disturbed this delicate balance. At present, there are four types of agricultural typology in the Western Ghats: namely, (i) Tea, coffee, and rubber estates; (ii) Cardamom and other spices; (iii) Annual crops-based farming consisting of mainly paddy, vegetables, pulses, tuber crops; and (iv) Homestead farming. Initially, agricultural expansion took place at the expense of natural vegetation, however, now there is competition among the crops and increasing demand for agricultural land for non-agricultural use. Underlying factors range from socio-economic conditions to government policies.

The Western Ghats is experiencing large-scale change resulting in environmental stress, loss of natural vegetation, fragmentation of forests, loss of biodiversity and genetic resources, land degradation, accelerated soil erosion, landslides, rockfall, mudflow, and soil piping. There is a large-scale alteration of the landscape and altered hydrological behaviour of the rivers resulting in declining water holding capacity of the Western Ghats, elevated instantaneous flow during monsoon and reduced base flow in summer months. The impact of climate change is also evident in the mountain ecosystem. The agricultural production, particularly of homestead farming, which is an important component of the local economy in the Western Ghats is suffering due to all these changes. Besides, there is human-animal conflict, alienation of indigenous people living in the forest areas through generations, and incidence of poverty. Degradation of the Western Ghats will have far-reaching ecologic and economic consequences, and therefore conservation of the Western Ghats is a priority requirement of national importance.

Western Ghats Ecology Expert Panel Report (Gadgil Committee Report)

Concern about the Western Ghats figured at the highest level in the Government of India. The first initiative was in the early 1980s with deliberations on Eco-development of Western Ghats, and in recent years there were two panels of experts, namely Gadgil Committee—(Western Ghats Ecology Expert Panel, WGEEP, GOI, 2011) and Kasturirangan Committee (High-Level Working Group on the Western Ghats, GOI, 2013b) to look into various issues concerning the ecology of the Western Ghats and development challenges. In fact, Kasturirangan Committee was constituted to take into account the comments and suggestions advanced by different stakeholders including state governments and central ministries on the Gadgil Committee Report and to suggest an all-round and holistic approach for sustainable and equitable development while keeping in focus the preservation and conservation of ecological systems of the Western Ghats.

The Western Ghats spreads over 12 districts in Kerala. According to the WGEEP report the 150 m contour is considered as the western boundary of the Western Ghats. A little over 13,000 km^2 area in Kerala has been identified as an ecologically sensitive area, which constitutes around 34 per cent of the total area of Kerala. To work out ecologically sensitive zones (ESZ), the Panel proposed a set of attributes like (i) biological attributes (biodiversity richness, species rarity, habitat richness, productivity, estimated biological/ecological resilience, and cultural and historical significance), (ii) geo-climatic attributes (topographic features, climatic features and hazard vulnerability), and (iii) stakeholder's valuation (Gadgil *et al.*, 2011). There were limitations to consider all these variables due to the scanty availability of data. Nevertheless, ESZs were proposed under three categories of ESZ-1, ESZ-2, and ESZ-3 based on the degree of sensitivity in the ranges of the highest, high and moderate respectively. ESZ-1 will include all existing protected areas. The WGEEP suggested fine-tuning of this exercise through the participatory process starting at the Gram Sabha level and provided a guideline for sector-wise activities under all three zones. However, there were reservations to implement all the recommendations proposed in this report. Subsequently, the Government of India constituted new committees at the national and state level to the examine the applicability of this report, which had not yielded tangible results, and business as usual continued. The devastating floods and landslides in Kerala in 2018 and 2019 raised serious questions about the use of existing natural resource and environmental management practices in the Western Ghats and invoked state-wide debate on the necessity of implementing the WGEEP report.

Terrain Based Approach for Conservation and Development of the Western Ghats

The focus of the Western Ghats development programme is on conservation and regulation of human activities in ecologically sensitive areas. The Western Ghats is not a homogeneous single entity. It is diverse and spatially heterogeneous. Capturing this heterogeneity across scale is fundamental to evolve conservation and development programmes suitable for the Western Ghats. We suggest a terrain-based approach for this purpose. At the macro level, the Western Ghats can be divided into 44 river basins, which can be further subdivided down the scale. Terrain analysis can be taken up for each of these 44 river basins.

There are three main morphogenetic processes, namely, denudational, denudational-cum-accumulational, and fluvial operating in the Western Ghats, and landforms have developed accordingly. Terrain units may be identified under each of these morphogenetic processes considering lithology, slope, ruggedness, drainage density, soil type, land use, and similar other parameters. The nature of the human intervention and environmental evaluation of each of the terrain units may also be worked out as demonstrated for a case study on Wayanad Plateau (Table 16.3). These terrain units represent landscape ecological units. A micro-scale approach is necessary following this principle at the watershed level to evolve management requirements. Seven terrain units have been identified in the Thirunelli watershed in the northern part of the Wayanad plateau (Table 16.4). Watershed boundaries in the western and southern parts are mostly characterized by steep rolling terrain, moderately steep rolling terrain, and erosional scarps. All these areas are highly vulnerable and warrant protection. The central part of the watershed exhibits well-developed flood plains, mostly given for seasonal agriculture. Development potentials and limitations of each of the seven units in this watershed can be worked out considering all biophysical and socio-economic factors. These terrain units are also useful to work out land suitability, carrying capacity, and finally to demarcate zones of conservation/no development, restricted development, and development. This type of terrain-based micro-level zonation is suggested to plan for the conservation and development of the Western Ghats. The next step involves the integration of socio-economic data with terrain data to fine-tune this zonation and frame operational guidelines for Panchayats. Both top-down and bottom-up approaches are necessary.

Table 16.3: Terrain Units and their Characteristics in Wayanad Plateau of Kerala

Process, origin and lithology	*Denudational terrain on charnockites*	*Denudational hill on granites*	*Denudational terrain mainly on gneisses and gabbro/diorite*		*Denudational cum depositional terrain on gneiss*	*Fluvial depositional terrain with alluvium*
Unit name with slope (%)	Scarp slopes and hills (>100)	Isolated hills (>100)	High undulating terrain (25-35)	Low undulating terrain (10-25)	Hummocky terrain (10-15)	Flood plain, terraces and valley fills (<5)
Area (%)	30	2	20	24	15	9
Rugged ness	High to very high	Low	Medium	Low	Low	Low (Flat bottom)
Drainage density	High	Low	Medium	Low	Low	Low
Land use/land cover	Natural vegetation with patches of forest plantations	Scrub, natural vegetation and rock outcrop, few settlements	Plantation with few patches of natural vegeta-tion, few settlements	Mainly plantation with settlements	Plantation, tree crops and settlements	Rice/Seasonal crops, tree crops, settlements
Broad soil characteristics	Forest loam, shallow soil, gravelly	Shallow soil, rocky	Forest loam, Deep, gravelly	Colluvic alluvium, deep, gravelly	Colluvium, alluvium, deep	Alluvium, deep
Nature of human intervention	Low to moderate	Moderate to high	High	High	High	Very high
Environmental evaluation	Highly erosion prone, unstable slopes, landslides, rock fall	Erosion prone, rock fall	Highly erosion prone, unstable in the slope range of 35 per cent, landslides	Relatively stable slope turns unstable due to toe cutting for roads and other human activities	Stable, graded slope, unstable in places due to intense human intervention	Flood prone, seasonally water-logged, lands widely converted for non-agricultural purposes

Source: Modified after Chattopadhyay and Mahamaya, 1989.

UNDERSTANDING AND PRESERVING LOCAL GEOGRAPHY

There is re-emphasis on understanding and preserving local geography in recent years, particularly in view of the changing world, and stress on local actions/decentralisation to address societal problems and transitioning to sustainable development. It is now well accepted that local changes through spatial aggregation may turn into a major force causing large-scale environmental change even of the global dimension (Turner II *et al.,* 1990). It is, therefore, necessary to take note of the changes not only at the regional and sub-regional level but also at the local level to understand the process of change, to devise appropriate mitigation measures, and to evolve suitable governance system because origins of most of these changes are anthropocentric and require societal intervention. Unfortunately, local issues invoke little interest

Table 16.4: Terrain Units and Suggested Management Activities in Thirunelli Watershed of Wayanad Plateau in Kerala

Terrain units (with slope)	*Characteristics*	*Management activities*
Flood plain/valley fill (<5 %)	Fine alluvium, colluvic in the valley fill, high water holding capacity	Agriculture, restricted development, room for river
Low rolling terrain (5 to 10 %)	Low erosion, moderate water availability	Suitable for development, tree crops, plantation, settlements
Moderately undulating terrain (10 to 15 %)	Moderately erosion prone	Tree crops, plantations, settlements, suitable for development
Moderately steep undulating terrain (15 to 35 %)	Highly erosion prone	Tree crops, plantations, settlements, caution for land management activities, caution for landslides in upper slopes
Steep rolling terrain (> 35 %)	Unstable slope	Restricted/no development, forest and conservation, caution for landslides, avoid run out area for settlement
Erosional scarp (>100 %)	Unstable slope	No development zone, forest and conservation
Ridge (<5 %)	Thin soil, coarse grained	No development zone, forest and conservation

among geographers, although the place is a core concept in geography (Sack, 2000). It is only recently that place is being considered as a research topic (Kapur, 2016).

As described by Agnew (1987) place possesses three dimensions of (i) locale—the setting for local action and interaction, (ii) locality—arrangement of setting relative to broader scales of political-economic processes, and (iii) sense of place—the attachment and meaning associated with the locale. Dialectical and contextualised studies are necessary to recognise how social and economic forces shape locality and are in turn shaped by them. It is argued that geography still awaits a genuine turn to these concepts and only through closer engagement with place, geography can arrive at what geography itself is and might be (Maipas, 2013). Inquiry into the nature and significance of place (locality) within the ambit of geography may be traced to John Wright (1947), however, the importance of local geography as a natural starting point for the education of children was emphasised even earlier by Bowen (1911). Briault (1953) lamented about reluctance in the promotion of local geography in schools although local geography is of high importance as an essential component of a good school course. Needless to say, local geography teaches us about the place and its introduction in the school curriculum can have a far-reaching impact on society.

At present, when the notion of a place in an increasingly global, and digitally disembodied World is changing, and the post-modern society debates on rootlessness, and the idea that people no longer need places to call their own, it is significant to revisit the importance of place in self actualisation. The local issue, local custom, and local heritage help in self actualisation and nurture belongingness. Place and self are two important aspects of identity, which is pluralistic (Sen, 2006). Inquiry about place brings convergence between geography and philosophy because the "relationship between self and place is not just one of reciprocal influence (that much any ecologically sensitive account would maintain) but also, more radically, constitutive coingredience: each is essential to the being of other. In effect, there is no place without a self and no self without place" (Casey, 2001: 684).

Kerala must understand the importance of place research as part of geography, and take necessary measures to conserve local geography for strengthening her achievements and move forward to address present and future challenges. The basic premise of planning is to use environmental resources like land, water, and biomass for the benefit and interest of all sections of society. A sizable part of the land in Kerala is privately owned and therefore utilisation of these lands depends on individual decisions. Landownership is a keystone of an individual's access to resources. Much of Kerala's development is attributed to land reforms that provided a large section of people accesses to land. Taking note of the interest of larger society, government prescribed various rules and regulations to limit the use of land in certain areas like that in the coastal zone or forest. However, there are several different strata of interest hindering proper implementation of existing regulations and also the introduction of required new regulations. The changing geography of Kerala manifests the nature of alterations that are taking place in Kerala and environmental mismanagement, be it in the coastal zone, urban areas, river corridors, or the Western Ghats. Occasional incidence of natural hazards like floods, landslides, and frequent flooding in coastal lowlands and urban areas have exposed the weakness of existing environmental management practices. There is hardly any effective regulation on the use of land and water. Land use zonation for various activities is overdue in Kerala.

Hills, streams, wetlands, ponds, springs, village wood lots, sacred groves, etc., which constitute part of local geography are under severe stress. There are widespread changes, largely irreversible, in landscape configuration and land use. All these are resulting in alterations of the hydrological system and ecological dysfunctions, which, if allowed unabated, may lead to ecological collapse jeopardising the entire development process. Similarly, local customs, the local language and local culture are often being neglected thereby traditional knowledge base is withering away. Both landscape diversity and cultural diversity are being threatened due to the lincar and rcductionist approach in planning and development.

Kerala has been a trendsetter in India for her performance in various sectors of human and social development initiatives including local level planning and social mobilisation. She has demonstrated her social capability and resoluteness in tackling various problems/hazards including the latest COVID 19 Pandemic in 2020. The recent initiative, in January 2020, to demolish buildings constructed violating CRZ norms in the Kochi area by adhering directives of the Supreme Court is a good example of environmental governance. The state can lead the nation in appreciating the importance of place in ecologic, economic, and social interactions of the environment and people, and understanding and preserving local geography.

References

Agrawal, D.P., Gupta, S.K. and Sheela, K. (1970). Radiocarbon dates of Quaternary samples. *Current Science* 39: 219-222.

Agrawal, D.P., Avasia R.P. and Guzdar S. (1979). A multidisciplinary approach to the Quaternary problems in Maharashtra. In: Allachin K.R. and Chakrabarti D.K. (eds.) *Source Book of Indian Archaeology*, New Delhi: Munshiram Manoharlal Pub., pp. 189-202.

Agnew J. (1987). *Place politics: The geographical mediation of State and Society*. Boston: Allen and Unwin.

Ahmad, J., Devarajan, S., Khemani, S. and Shah S. (2005). Decentralisation and Service Delivery. World Bank Policy Research Working Paper 3603. Washington DC: The World Bank, http//econ.worldbank.org

Ajitkumar, N. and George K.K. (2009). Kerala's education system: from inclusion to exclusion. *Economic and Political Weekly*, XLIV (41) 55-61.

Ammini, P.L. and Augustine, S.K. (2003). Status of marine fisheries at Cochin fisheries harbour. Kochi: Marine Fisheries Information Service (T & E ser), ICAR, No. 176, April-June, pp. 5-8.

Andrews, S. Sr. and Kannan E. (2016). Land use under homestead in Kerala: The status of homestead cultivation from a village study. Working paper no. 369. Bangalore: Institute for Social and Economic Change.

Ananthakrishnan, R., Aralikatti, S.S. and Pathan, J.M. (1979). Some features of southwest monsoon rainfall along the West Coast of India. *Proceedings, Indian Academy of Sciences,* 88A: 177-199.

Antony, T. (2017). Depths of despair. *The New Indian Express*, Thiruvananthapuram, 10th September, 2017,

Ayyar, R.V.K. (1966). *A short history of Kerala*. Ernakulam: Pai & Co.

Azfar, O., Kähkönen, S., Lanyi, A., Meagher, P. and Rutherford, D. (1999). Decentralization, Governance and Public Services: The Impact of Institutional Arrangements—A Review of Literature. IRIS Center, Maryland: University of Maryland, College Park.

Baba, M., Thomas, K.V. and Sreekala Panicker (1997). Alternate coastal engineering strategies for Kerala. In Proceedings, Second Indian National Conference on Harbour and Ocean Engineering (INCHOE-97). Thiruvananthapuram: Centre for Earth Science Studies. Vol.2:1205-1215.

Balakrishnan, T.S. (2001). Tectonics of Western India inferred from gravity patterns and geophysical exploration. In: Gunnell, Y. and Radhakrishna, B.P. (eds.) *Sahyadri—The Great Escarpment of Indian Subcontinent.* Bangalore: Geological Society of India. Memoir 47, Vol. 1: 271-277.

Ballukrya, P.N. (2005). Tsunami of 26th December, 2004: Observations at Chennai and east coast of India. *Journal Geological Society of India*, 65 (3): 368-369

Bamford, C.G. and Robinson H. (1978). *Geography of Transport*. Estover, Plymouth: Macdonald and Evans Ltd.

Bandyopadhyay, D. (1997). People's Participation in Planning: Kerala Experiment. *Economic and Political Weekly*, 32 (40): 2450-2454.

Baum A., Fleming I. and Davidson, L.M. (1983). Natural disaster and technological catastrophe. *Environment and Behaviour*, 15(3): 333-354.

Beek, K.J., Bie K.D. and Driessen P. (eds.) (1997). Proceedings, International Conference on Geo-information for Sustainable Land Management, Enschede: International Institute of Aerospace Survey and Earth Sciences.

Bhat, P.N. Mari, and Irudaya Rajan, S. (1990). Demographic transition in Kerala revisited. *Economic and Political Weekly*, 25 (35-36): 957-80.

Bhattacharya, G.C., Almeida F., Vora, K.H. and Siddiquie, H.N. (1987). Subsurface geology of Bomaby with paleo climatic inferences interpreted from shallow seismic profiles. *Journal of Coastal Research*, 3 (4): 521-530.

Bhattacharya, S.N. and Dattatrayam, R.S. (2002). Earthquake sequence in Kerala during December 2000 and January 2001. *Current Science*, 82 (10): 1275-1278.

Boserup, E. (1965). *The Conditions of Agricultural Growth: The Economics of Agrarian Change under Population Pressure.* London: G. Allen and Unwin.

Bowen A.C. (1911). The importance and possibilities of local geography in elementary schools. *The Journal of Geography*, X (4): 109-114. Accessed online, https://doi.org/10.1080/ 00221341108985660

Briault E.W. H. (1953). The study of local geography as an integral part of the school course. *Geography*, 38 (1): 29-32. Accessed online https://www.jstor.org/stable/40563349

Brown, A. (2013). Growth and success in Kerala. *The Yale Review of International Studies*. Yris.yira.org/ essays/1150. Accessed on 02.06.2018.

Buchanan, F. (1807). *A Journey from Madras through the countries of Mysore, Kanara and Malabar.* London: East India Co.

Bulletin of Seismological Society of America (BSSA). (2010). BSSA Tip Sheet for October 2010. California, p.1.

Casinadar, R. (1987). Transformation of spatial structures in Asia. Kotadesasi in Asia, Kotadesasi in Kerala, Ph.D. research proposal.

Casey E.S. (2001). 'Between geography and philosophy: What does it mean to be in the Place-world'. *Annals of the Association of American Geographers*, 91 (4): 683-93.

Central Ground Water Board (CGWB). (2012). *Ground Water Level Scenario in India.* Ministry of Water Resources, New Delhi: Government of India.

Central Ground Water Board (CGWB). (2014). *Dynamic Ground Water Resources of Kerala, 2013*, Ministry of Water Resources, New Delhi: Government of India.

Central Pollution Control Board (CPCB). (2014). Status of Water Quality in India, 2012. Monitoring of Indian National Aquatic Resources, Series: MINARS/36/2013-14. Ministry of Environment and Forest, New Delhi: Government of India.

Central Water Commission (CWC). (2018). Study Report—Kerala Floods, 2018. Hydrology (S) Directorate. Hydrological Studies Organisation, Central Water Commission, New Delhi: Government of India.

Centre for Development Studies/United Nations 1975. *Poverty, unemployment and development policy—A case study of selected issues with reference to Kerala.* Reprinted in 2000, Thiruvananthapuram: Centre for Development Studies.

Centre for Earth Science Studies (CESS). (1984). *Resources Atlas of Kerala.* Thiruvananthapuram: Centre for Earth Science Studies.

Centre for Earth Science Studies (CESS). (1991). Panchayat level resource mapping—An approach paper. Thiruvananthapuram: Centre for Earth Science Studies.

Centre for Earth Science Studies (CESS). (1995). Coastal Regulation Zone Management Plan of Kerala. Thiruvananthapuram: Centre for Earth Science Studies.

Centre for Earth Science Studies (CESS). (1997). Analysis of the Environmental Impact of the Lowland to Highland Migration in the Western Ghats Region: Kerala. Unpublished Project Report. Thiruvananthapuram: Centre for Earth Science Studies.

Centre for Earth Science Studies (CESS). (1998). Catchment conservation/development plan for Sasthamcotta Lake, Kollam district. Unpublished Project Report. Thiruvananthapuram: Centre for Earth Science Studies.

Centre for Earth Science Studies (CESS). (2000). Panchayat level resource mapping—final technical report. Thiruvananthapuram: Centre for Earth Science Studies.

Centre for Water Resources Development and Management (CWRDM). (1989). *Tanks and Ponds in Kerala.* Kozhikode: Centre for Water Resources Development and Management.

Centre for Public Policy Research (CPPR). (2017). Public and Private Healthcare Institutions: Preference and Expenditure Pattern. Working Paper No. 1. Kochi: Centre for Public Policy Research.

Chakraborty, S.C. (1988). Extended metropolitan areas: Their relevance for understanding urban processes in India. Conference paper: Extended Metropolis in Asia. Hawaii: East-West Environment and Policy Institute, East West Centre.

Chakraborty, P., Chakraborty, L., Amarnath, H.K. and Mitra, S. (2010). Financing human development in Kerala: Issues and Challenges. Draft Report. New Delhi: National Institute of Public Financing and Policy.

Chakrapani, R. (With inputs and guidance from K. Madhavan Namboodiri) (2014). Domestic Water and Sanitation in Kerala—A Situation Analysis. Water Aid, Chalakudy puzha Samrakshana Samity and Forum for Policy Dialogue on Water Conflicts in India, Pune.

Chand Basha, S. (1997). Forests Goods and services. In Thampi K.B, Nayar N.M. and Nair C.S. (eds.) *The Natural Resources of Kerala*. Thiruvananthapuram: World Wide Fund for Nature-India, Kerala State Office, pp. 425-436.

Chandrasekhar, C.P., Ramchandran V.K. and Ramkumar, R. (2001). Issues in school education in contemporary Kerala. Paper prepared for UNICEF. New Delhi: UNICEF.

Charvak. (2000). From decentralization of planning to people's planning: Experiences of the Indian States of West Bengal and Kerala. Discussion Paper No. 21. KRPLLD, Thiruvananthapuram: Centre for Development Studies.

Chattopadhyay, S. (1985). Deforestation in parts of Western Ghats region of Kerala, India. *International Journal of Environmental Management*, 20 (3): 219-230.

Chattopadhyay, S. (1988). Urbanisation in Kerala. *Geographical Review of India*, 50 (1): 8-25.

Chattopadhyay, S. (1995). Environmental impact on urbanisation: A case of Kerala, India. *Transactions Institute of Indian Geographers*, 17 (1):15-27.

Chattopadhyay, S. (2000). Settlement system in Kerala. Paper presented in the International Congress on Kerala Studies. Thiruvananthapuram: AKG Centre for Research and Studies.

Chattopadhyay, S. (2002). Emergence of central Kerala coastal plain: a geomorphic analysis. In: Tandon, S.K. and Thakur, B. (eds.) *Recent Advances in Geomorphology, Quaternary Geology and Environmental Geosciences: Indian Case Studies,* New Delhi: Manisha Publications, pp. 287-298.

Chattopadhyay, S. (2002). GIS Analysis of Catchment Characteristics of Pamba River and its relation to Irrigation Project. CESS–PR-36-2002. Unpublished report. Thiruvananthapuram: Centre for Earth Science Studies.

Chattopadhyay, S. (2004). *Landforms in Kerala: An overview.* In Ravindra Kumar G.R. and Subhas N. (eds.) Silver Jubilee Compendium: Earth System Science and Natural Resources Management. Thiruvananthapuram: Centre for Earth Science Studies, pp. 211-228.

Chattopadhyay, S. (2010). Geomorphology for integrated coastal zone management: A theoretical approach with examples from Kerala, India. *Indian Journal of Geo-Marine Sciences*, 39 (4): 623-630.

Chattopadhyay, S. (2015). Environmental consequences of rubber plantations in Kerala. Discussion paper no. 44, National Research Programme on Plantation Development (NRPPD), Thiruvananthapuram: Centre for Development Studies.

Chattopadhyay, S. (2017). Decentralised Planning and Development: Kerala Experience. *The Indian Geographical Journal*, 92 (1): 13-28.

Chattopadhyay, S. (2019). Kerala's human development model. *Population Geography*, 41 (1): 1-21.

Chattopadhyay S. and Sidharthan S. (1985). Regional analysis of Greater Kuttanad, Kerala. Technical Report No. 43, Thiruvananthapuram: Centre for Earth Science Studies.

Chattopadhyay, S. and Mahamaya Chattopadhyay. (1989). Remote sensing application in terrain analysis: Case study in the Wayanad plateau, Kerala. *International Journal of Remote Sensing*, 10: 1075-1084.

Chattopadhyay, S. and Mahamaya Chattopadhyay. (1995). Terrain Analysis of Kerala: Concept, Method and Application. Technical Monograph No.1/95.STEC, Thiruvananthapuram: Government of Kerala.

Chattopadhyay, S. and Mahamaya Chattopadhyay. (1997). Sustainable land management in Kerala, India—A biophysical approach. In Beek, K.J., Bie, K.D. and Driessen P. (eds.) *Proceedings (in CD-ROM) of the International Conference on Geo-information for Sustainable Land Management.* Enschede: International Institute of Aerospace Survey and Earth Sciences, pp. 165-174.

Chattopadhyay, S. and Mahamaya Chattopadhyay, (2001). Remote sensing application in terrain analysis of Kerala. In: M.G. Srinivas ed., *Remote Sensing Applications,* Mumbai: Narosa Publications, pp. 345-356.

Chattopadhyay, S. and Mahamaya Chattopadhyay. (2014). Rejuvenation of Kerala rivers: Geoenvironmental setting, potentials, problems and recent initiatives. In R. Vaidyanadhan ed. *Rejuvenation of Surface Water Resources of India: Potential, Problems and Prospects.* Bangalore: Geological Society of India. Special Publication No. 3, pp. 1-12.

Chattopadhyay, S. and Mahamaya Chattopadhyay. (2020). Land resources of Kerala: An overview. In H.S. Sharma, Padmaja, S. ed. *Land Resources: Processes, Degradation and Evaluation.* NAGI Series 3, New Delhi: Concept Publishing Co. Pvt. Ltd., pp. 9-42.

Chattopadhyay, S., Sasidharan, C.K. and Mathai, J. (2004). Panchayat resource mapping—An experiment from Kerala. In Ravindra Kumar G.R. and Subhash N. (eds.) *Silver jubilee compendium—earth system science and natural resources management.* Thiruvananthapuram: Centre for Earth Science Studies, pp. 401-416.

Chattopadhyay, S., Asa Rani, L. and Sangeetha, P.V. (2005). Water quality variations as linked to land use pattern: A case study in Chalakudy river basin, Kerala. *Current Science*, 89 (12): 2163-2169.

Chattopadhyay, S. and Franke, R.W. (2006). *Striving for Sustainability—Environmental Stress and Democratic Initiatives in Kerala.* New Delhi: Concept Pub.Co.

Chattopadhyay, S. and Shravankumar, V. (2007). Marine fisheries in Kerala: A geographical analysis. *Transactions, Institute of Indian Geographers*, 29 (2):119-130.

Chattopadhyay, S. and Suresh Kumar, S. (2007). Fractal dimensions of selected coastal water bodies in Kerala—A case study. *Indian Journal of Marine Sciences—Special volume on 'Fractals in Marine Sciences'*, 36 (2): 162-166.

Chattopadhyay, S. and Sajna, A.N. (2009). Landuse-landcover change and its impact on biophysical system: a case study of Pamba-Agasthyamalai hill tract. Unpublished Project Report. Thiruvananthapuram: Centre for Earth Science Studies.

Chattopadhyay, S., Shravankumar, V., Sarvanan, E., Sukumar, B., Roy Chowdhury, R., Mohandas, K. and Narayana Kuttty, N. (1984). Morphology and landuse of the Mitraniketan campus—A micro level study. CESS T.R. No. 41/1984, Thiruvananthapuram: Centre for Earth Science Studies.

Chattopadhyay, S., Krishnakumar, P. and Rajalekshmi, K. (1999). Panchayat resource mapping to panchayat level planning in Kerala: an analytical study. Discussion Paper No. 14, KRPLLD, Thiruvananthapuram: Centre for Development Studies.

Chandrasekhar, U.M. and Sankar, S. (1998). Ecology and management of scared groves in Kerala, India. *Forest Ecology and Management,* 112 (1998) 165-177.

Chundamannil, M. (1993). History of forest management in Kerala. KFRI Research Report No. 89. Peechi: Kerala Forest Research Institute.

Constain, J.K., Bollinger, G.A. and Speer, J.A. (1987). Hydroseismicity—A hypothesis for the role of water in the generation of intraplate seismicity. *Geology*, 15 (7) 618-521. DOI: http://doi.org/10.1130/0091-7613(1987)15<618:HHFTRO>2.0.CO;2

Coque-Delhuille, B. (1991). The Long term geomorphic evolution of the English south-west massif (UK). *Z. Geomorphologie*, 35 (1): 65-84.

Corporation of Kochi. (2011). City Sanitation Plan for Kochi—Draft report, Vol. 1, Main report, Kochi: Corporation of Kochi.

corona.mygov.in Accessed on 9th May, 2020

covid19.who.int Accessed on 9th May, 2020

Crutzen P.J. and Stoermer E.F. (2000). The Anthropocene. IGBP Global Change Newsletter, 41:17-18.

David, S.E., Chattopadhyay, M., Chattopadhyay, S. and Jennerjahn, T.C. (2015). Impact of human interventions on nutrient biogeochemistry in the Pamba river, Kerala, India. *Science of the Total Environment,* 541: 1420-1430.

Davis, W.M. (1899). The geographical cycle. *Geographical Journal*, 2: 481-504.

Department of Town and Country Planning (DTCP). (2012). State Urbanisation Report, Kerala, Unpublished report. Thiruvananthapuram: Government of Kerala.

Dikshit, K.R. (1981). The Western Ghats: A geomorphic overview. In L.R, Singh ed. *New Perspectives in Geography.* Allahabad: Thinkers Library, pp. 1-25.

Dilip, T.R. (2008). Role of Private Hospitals in Kerala: An Exploration, Working Paper No. 400, Thiruvananthapuram: Centre for Development Studies.

Dinesh, R., Srinivasan, V., Hamza, S. and Anandaraj, M. (2014). Massive phosphorus accumulation in soils: Kerala's continuing conundrum. *Current Science*, 106: 343-344

Dreze, J. and Sen, A. (1989): *Hunger and Public Action.* World Institute for Development Economics Research (WIDER), Oxford: Clarendon Press.

Dube, S.C. (1958). *India's Changing Villages*. London: Routledge, Kegan and Paul.

Dutta, A.K. and Parai, A. (1999). Defining urban corridors: Case studies of four mega cities of India. In Chapman, G.P., Dutta, A.K. and Bradnock, R.W. (eds.) *Urban Growth and Development in Asia, Vol. II: Living in Large Cities,* Studies in Development Geography. London: School of Oriental and African Studies. pp. 93-110.

Ekbal, B. and Isaac Thomas T.M. (1988). *Science for Social Revolution. The Experience of Kerala Sasthra Sahitya Parishat (KSSP).* Thiruvananthapuram: KSSP.

Evans, P.B. (ed). (2002). *Livable Cities? Urban struggle for Livelihood and Sustainability.* Berkley: University of California Press.

Food and Agriculture Organisation (FAO), (1984). Intensive multiple-use forest management in Kerala. FAO Forestry paper No. 53. Rome: FAO.

Food and Agriculture Organisation, (FAO) (1985). Towards improved multilevel planning for agriculture and rural development in Asia and Pacific. FAO Economic and Social Development Paper 52. Rome: FAO.

——(1995). Planning for Sustainable Use of Land Resources—Towards: A New Approach. FAO Land and Water Bulletin-2. Rome: FAO, United Nations Organisation (UNO).

——(1999). Future of our land—Facing the Challenges. Land and Water Development Division, Rome: FAO in collaboration with United Nations Environment Programme.

Formento-Trigilio, M.L. and Pazzaglia, F.J. (1998). Tectonic geomorphology of the sierra Nacimiento: Traditional and new techniques in assessing long term landscape evolution in the southern Rocky Mountains. *Journal of Geology*, 106: 433-453.

Franke R.W. and Chasin B. (1989). *Kerala: Development through Radical Reform.* San Francisco: Food First.

Franke R.W. (1992). Land reforms versus inequality in Nadur village, Kerala. *Journal of Anthropological Research*, 48 (2): 81-116.

Franke R.W. (1996). *Life is a little better: Redistribution as a development strategy in Nadur village, Kerala.* Delhi: Promilla.

Franke R.W. (2008). Local Planning: Kerala Experiment. In Chris Spannos ed. *Real Utopia: Participatory Society for the 21st Century*. Oakland CA: AK Press, pp. 130-135.

Friedmann, J. and Weaver, C. (1979). *Territory and Function: The Evolution of Regional Planning.* Berkeley: University of California Press.

Gadgil D.R. (1967). District Development Planning. Gokhale Institute of Politics and Economics, Poona. R.R. Kale memorial lecture series. Bombay: Asia Publishing House.

Gadgil, M., Daniels, R.J.R., Ganeshaiah, K.N., Prasad, S.N, Murthy, M.S.R., Jha, C.S, Ramesh, B.R. and Subramaniam, K.A. (2011). Mapping ecologically sensitive, significant and salient areas of Western Ghats: proposed protocol and methodology. *Current Science.* 100(2): 175-182.

Gangadharan, K. (2003). Utilisation of Health Services in Urban Kerala—A socio-economic Study. Unpublished Ph. D. thesis, Department of Economics, Pondicherry University, Mahe.

Gasper, C. (2003). Vocational education in the context of globalisation and technological change in Kerala. Paper presented in the Journal of Vocational Education and Training's Fifth International Conference on Research Policy and Practice in Vocational Education and Training, London: University of Greenwich, 16-18 July.

Gaussen, H., Legris, P. and Mehir Homji, V.M. (1986). *International map of vegetation and environmental conditions—India. Sheet-Cape Comorin*. Pondicherry: French Institute.

Ghosh, A. (1988). Popular participation and decentralized planning in West Bengal. *Economic and Political Weekly*, 23 (3): 66-68.

Greffe, X. (2004). *Local Governance and Partnerships*. Paris Université. Sorbonne: Paris.

George K.K. (1998). Historical roots of Kerala model and its present crisis. *Bulletin of Concerned Asian Scholars*, 30 (4) 35-52.

George K.K. (2011). Kerala economy—Growth, Structure, Strength and Weakness. Working Paper No. 25. Kochi: Centre for Socio-Economic and Environmental Studies.

George, P.S. and Nair, K.N. (1990). *Livestock economy of Kerala*. Thiruvananthapuram: Centre for Development Studies.

George, P.S. and Chattopadhyay, S. (2001). Population and land use in Kerala. In *Growing Populations, Changing Landscapes: Studies from India, China, and the United States*. National Academy of Sciences, Washington DC: National Academy Press.

Geological Survey of India (GSI) (2005). *Geology and Mineral resources of Kerala*. Thiruvananthapuram: Government of India.

Gopalan, U.K. (2002). Backwaters of Kerala—An endangered wetland ecosystem. In: *Wetland Conservation and Management in Kerala. Compendium volume*. 14th Kerala Science Congress, State Committee on Science Technology and Environment, Thiruvananthapuram: Government of Kerala, pp. 23-29.

Gopinath, D. (2009). Decentralisation in India: Towards 'localism' or 'regionalism'? *Commonwealth Journal of Local Governance*, 4:101-112. http://epress.lib.uts.edu.au/ojs/index.php/cjlg

Gopokumar, C.S. (2011). Impacts of climate variability on agriculture in Kerala. Unpublished Ph. D. thesis. Faculty of Marine Sciences, Cochin University of Science and Technology, Kochi.

Government of India. (1951). *Census of India 1951*. New Delhi: Office of the Registrar General.

Government of India. (2001). *Census of India 2001: Migration tables, Kerala*. Thiruvananthapuram: Office of the Registrar General.

Government of India. (2011). *Census of India 2011: Population totals, Kerala*. Thiruvananthapuram: Office of the Registrar General.

Government of India. (1993). The Constitution (Seventy-fourth Amendment) Act, 1992 and The Constitution (Seventy-third Amendment) Act, 1992. Ministry of Law and Justice. New Delhi: Government of India Press.

Government of India. (1995). *Achievements under education for all: A Status Paper*. Department of Education. Ministry of Human Resource Development, New Delhi: Government of India, pp. 38-39.

Government of India (2011). Disaster Management in India. New Delhi: Ministry of Home Affairs.

Government of India (2011). *Report of the Western Ghats Ecology Expert Panel (WGEEP)*. Submitted to the Ministry of Environment and Forests, New Delhi: Government of India.

Government of India. (2013a). *All-India Survey on Higher Education, 2011-12 (Provisional)*. Department of Higher Education. New Delhi: Ministry of Human Resource Development.

Government of India. (2013b). *Report of the High Level Working Group on Western Ghats, Vol. 1*. Submitted to the Ministry of Environment and Forest, New Delhi: Government of India.

Government of India (2017a). *Millennium Development Goals—Final Country Report India*. Central Statistical Office. New Delhi: Ministry of Statistics and Programme Implementation. www.mospi.gov.in

Government of India (2017b). *Health and Family Welfare Statistics in India, 2017*. National Health Mission. New Delhi: Ministry of Health and Family Welfare. www. Mohfw.nic.in

Government of India. (2018). *Agricultural situation*. Ministry of Agriculture and Farmers Welfare, New Delhi: Government of India.

Government of India. (2019). *Provisional estimates of annual national income*. Ministry of Statistics and Programme implementation. New Delhi: Government of India.

Government of Kerala. (1958). *Report of the Administrative Reforms Committee*. Thiruvananthapuram: Government of Kerala Press.

Government of Kerala. (1974). *Water Resources of Kerala*. Thiruvananthapuram: Public Works Department.

Government of Kerala. (1974). *Report of the Committee on Agro-Climatic Zones and Cropping Patterns*. Thiruvananthapuram: Department of Agriculture.

Government of Kerala. (1978). *Soils of Kerala*. Soil survey branch, Soil Survey Staff, Thiruvananthapuram: Department of Agriculture.

Government of Kerala. (1981). *Economic Review, 1980.* Thiruvananthapuram: State Planning Board.
Government of Kerala. (1991). *Economic Review, 1990.* Thiruvananthapuram: State Planning Board.
Government of Kerala. (1992). *Economic Review, 1991.* Thiruvananthapuram: State Planning Board.
Government of Kerala. (1993). *Economic Review, 1992.* Thiruvananthapuram: State Planning Board.
Government of Kerala. (1996). *Agricultural Statistics, 1995-96.* Thiruvananthapuram: Department of Economics and Statistics.
Government of Kerala (1999). *Natural Resource Management: Community Based Action Research Programme for Kerala.* Project Profile. Thiruvananthapuram: State Planning Board.
Government of Kerala. (2001). *Agricultural Statistics, 2000-01.* Thiruvananthapuram: Department of Economics and Statistics.
Government of Kerala. (2002). *Economic Review, 2001.* Thiruvananthapuram: State Planning Board.
Government of Kerala. (2004). *Economic Review, 2003.* Thiruvananthapuram: State Planning Board.
Government of Kerala. (2005). Economic Review, *2004.* Thiruvananthapuram: State Planning Board
Government of Kerala. (2006a). *Agricultural Statistics, 2005-06.* Thiruvananthapuram: Department of Economics and Statistics.
Government of Kerala. (2006b). *Economic Review, 2005.* Thiruvananathapuram: State Planning Board.
Government of Kerala. (2006c). *Human Development Report, 2005.* Thiruvananthapuram: State Planning Board.
Government of Kerala. (2008). *Forest Statistics, 2008.* Thiruvananthapuram: Forest and Wildlife Department.
Government of Kerala. (2010). *Economic Review, 2009.* Thiruvananathapuram: State Planning Board.
Government of Kerala. (2011). *Agricultural Statistics, 2010-11.* Thiruvananthapuram: Department of Economics and Statistics.
Government of Kerala. (2012a). *Economic Review, 2011.* Thiruvananthapuram: State Planning Board.
Government of Kerala. (2012b). *Forest Statistics.* Thiruvananthapuram: Kerala Forest Department.
Government of Kerala. (2013a). *Economic Review, 2012.* Thiruvananthapuram: State Planning Board.
Government of Kerala. (2013b). *Facts and Figures of Agriculture in Kerala.* Thiruvananthapuram: Department of Agriculture.
Government of Kerala. (2014a). *Perspective Plan—Kerala 2030. Volume IV Social Sustainability.* Thiruvananthapuram: State Planning Board and National Council of Applied Economic Research.
Government of Kerala. (2014b). *Economic Review, 2013.* Thiruvananthapuram: State Planning Board.
Government of Kerala. (2014c). *Kerala State Action Plan on Climate Change.* Thiruvananthapuram: Department of Environment and Climate Change.
Government of Kerala. (2014d). *Evaluation study on soil conservation in Kerala, 2011-12.* Thiruvananthapuram: Department of Economics and Statistics.
Government of Kerala. (2015a): Report of Nineteenth Quinquennial Livestock Census 2012. Thiruvananthapuram:. Directorate of Animal Husbandry.
Government of Kerala. (2015b). *Economic Review, 2014.* Thiruvananthapuram: State Planning Board.
Government of Kerala. (2016). *Report on Health in Kerala, NSS 71 Round.* Thiruvananthapuram: Department of Economics and Statistics, NSS Division.
Government of Kerala. (2017a). *Agricultural Statistics, 2016-17.* Thiruvananthapuram: Department of Economics and Statistics.
Government of Kerala. (2017b). *Report of the Working Group on Animal Husbandry and Dairy Development including Veterinary Research.* Thirteenth Five Year Plan (2017-2022). Agricultural Division, Thiruvananthapuram: Kerala State Planning Board.
Government of Kerala. (2017c). *Economic Review, 2016*, Thiruvananthapuram: State Planning Board.
Government of Kerala. (2017d). *Forest Statistics, 2016.* Forest Department. Thiruvananthapuram: Forest Department.
Government of Kerala. (2018). *Economic Review 2017,* Vol. 1 & 2. Thiruvananthapuram: State Planning Board.
Government of Kerala. (2019). *Economic Review 2018,* Vol. 1 & 2. Thiruvananthapuram: State Planning Board.
Ground Water Department (GWD), Central Ground Water Board (CGWB). (2014). *Dynamic Ground Water Resources of Kerala (2013).* Thiruvananthapuram: Government of Kerala.

Gunawardene, N.R, Daniels, A.E.D., Gunatilleke, I.A. U.N, Gunatilleke, C.V.S., Karunakaran, P.V., Nayak, K.G., Prasad, S., Puyravaud P., Ramesh, B.R., Subramanian, K.A. and Vasanthy, G. (2007). A brief overview of the Western Ghats—Sri Lanka biodiversity hotspot. *Current Science,* 93: 1567-1572.

Gunnell, Y. (2001). Interaction between geological structure and global tectonics in multistoreyed landscape development: a denudation model for the south Indian shield. In: Gunnel, Y. and Radhakrishna, B.P. (eds.) *Sahyadri: The Great Escarpment of the Indian sub-continent, Memoir 47 (1),* Bangalore: Geological Society of India, pp. 599-644.

Gunnell, Y. and Radhakrishna, B.P. (2001). Editorial Introduction. In: Gunnel, Y. and Radhakrishna (eds.) *Sahyadri: The Great Escarpment of the Indian sub-continent, Memoir 47 (1),* Bangalore: Geological Society of India, pp. 89-117.

Gunnell, Y. and Fleitout, L. (2001). The morphotectonic evolution of the western ghats, India: An exploration of denudation driven flexural models. In: Gunnel, Y. and Radhakrishna (eds.) *Sahyadri: The Great Escarpment of the Indian sub-continent, Memoir 47 (1)*, Bangalore: Geological Society of India, pp. 581-598.

Harikumar, P.S. and Chandran, K.M. (2013). Bacteriological contamination of groundwater due to onsite sanitation problems in Kerala state: a case study, *International Journal of Life Sciences Biotechnology and Pharma Research,* 2 (3): 190-202.

Hedge, N.G. (2000). Community Development in India: An Overview. Presented at the Asian Productivity Organization, Tokyo, Japan seminar on Comparative Study on Planning Process of Community Development: Component of People's Participation. Ulaanbaatar, Mongolia. August 21-26, 2000.

Heller, P. (2005). Reinventing public power in the age of globalisation. Decentralisation and the transformation of movement politics in Kerala. In Ray R. and Katzenstein M.F. (eds.) *Social Movements in India—Poverty, Power and Politics*. New York: Rowman & Littlefield Publishers Inc., pp. 79-106.

Heller, P. (2008). Local democracy and development in comparative perspective. In van Donk, M., Swilling, M., Pieterse, E. and Parnell, S. (eds.) *Consolidating Developmental Local Government—Lessons from South African Experience.* Cape Town: An Isandla Institute Book Project, pp. 153-174.

Heller, P. (2020). A virus, social democracy, and dividends for Kerala. Thiruvananthapuram: *The Hindu*, 18th April, 2020.

Henkel S. (2006). Biogeochemical studies of small mountainous rivers in south Kerala (India). Diploma Thesis, Geosciences Department, Bremen: University of Bremen.

Ibrahim, P. (1978). The development of transport facilities in Kerala: A historical review. *Social Scientist.* 6(8) : 34-48.

India Brand Equity Foundation (IBEF), undated. *Indian States—Economy and Business, Kerala*. Confederation of Indian Industry. www.ibef.org

Indian Council for Agricultural Research (ICAR). (1991). *Agroclimatic Zone Specific Research,* National Agricultural Research Project, ICAR, New Delhi: Government of India.

Independent Evaluation Group (IEG) Public Sector Evaluation. (2013). *Project Performance Assessment Report, Republic of India, Kerala Rural Water Supply and Environmental Sanitation Project "JALANIDHI" (Credit 3431 IN) and Maharashtra Rural Water Supply and Sanitation Project "JALSWARAJYA"(Credit 3821 IN),* Repo No. 78786, Document of the World Bank.

India Meteorological Department (IMD). (2017). *History of cyclones*. Thiruvananthapuram. www.imdtvm.gov.in/index.php

Indo-Dutch Mission, (1989). *Kuttanad Water Balance Study: Final Report, Vol. IV.* Thiruvananthapuram: Government of Kerala.

Ingles OG, (1968). Soil chemistry relevant to the engineering behavior of soils. In Lee I.K. ed. *Soil Mechanics.* London: Butterworth.

International Geographical Union Commission on Geographical Education. (2015). *International Declaration on Research in Geography Education*, pp. 1-17.

Isaac Thomas, T.M. and Franke R.W. (2000). *Local Democracy and Development*, New Delhi: Left Word.

Jafar, K. (2014). A note on Decentralised Planning Initiative: possible lessons from the Kerala experience. MPRA Paper No. 65710, Munich Personal RePEc Archive, http://mpra.ub.uni-muenchen.de/65710.

Jarrett, H.R. (1977). *Tropical Geography*. Plymouth: Macdonald and Evans Ltd.

Jayasankar, B. (2000). Bamboo Processing Industry in Kerala: A Study of an Unorganised household sector in

Adimaly Panchayat (Idukki district). Discussion paper No. 26. Kerala Research programme on Local Level Development. Thiruvananthapuram: Centre for Development Studies.

Jaydev, A. and Ha, H. (2015). Land reforms in Kerala: An aid to ensure sustainable development. In Ha, H. (ed) *Land and Disaster Management Strategies in Asia.* New Delhi: Springer India, pp. 15-28. Doi: 10. 1007/978-81-322-1976-7_2.

Jeffrey, R. (1992). *Policies, women and well being: How Kerala become a 'Model'.* London: The McMillan Press.

Jha, S.N. (1999). Introduction. In Jha, S.N. and Mathur, P.C. (eds.) *Decentralisation and Local Politics: Readings in Indian Government and Politics.* London: Sage, pp. 13-44.

John, J. (2009). A study on Kudumbashree Project—A poverty eradication programme in Kerala: Performance, Impact and Lessons for other States. Study sponsored by Planning Commission of India, New Delhi: Kerala Development Society.

John, J. (2014). Homestead farming in Kerala: A multi-faceted land-use system. *Review of Agrarian Studies,* 4(1): 80-94.

Johnson, C. (2003). Decentralisation in India: Poverty, Politics and Panchayati Raj. Working Paper No. 199. London: Overseas Development Institute.

Jose, T.K. (undated): Kudumbashree—An Innovative Poverty Reduction Strategy. Thiruvananthapuram: Government of Kerala.

Joseph, P.V. (2009). Meteorological disasters. In Proceedings of the National Workshop on Natural Hazards, Disaster Mitigation and Management. Thiruvananthapuram: Centre for Earth Science Studies, pp. 19-21.

Joseph, P.V. and Simon, A. (2005). Weakening trend of the south-west monsoon current through peninsular India from 1950 to the present. *Current Science,* 89 (4): 687-694.

Joseph, S. and Ouseph, P.P. (2010). Assessment of nutrients using multivariate statistical techniques in estuarine system and its management implications: a case study from Cochin Estuary, India. *Water and Environment Journal,* 24: 126-132. DOI: 10.1111/j.1747-6593.2008.00163x.

Jurlina Alibegovic, D., Slijepcevic, S. and Kordej-de Villa, Z. (2014). Regional development and decentralization—Two options to overcome lack of funding. *Transylvanian Review of Administrative Sciences*, 43 E/2014: 146-167.

Kabir, M. (2010). On the periphery: Muslims and the Kerala model. In Ravi Raman K. ed. *Development, Democracy and the State—Critiquing the Kerala Model of Development.* London: Routledge.

Kalirajan, K. and Otsuka, K. (2010). Decentralisation in India: Outcome and Opportunities. ASARC Working paper 2010/14. Canberra: Australian National University.

Kalpka, P., Halas, M. and Tonev, P. (2013). Functional regions: Concept and types. *Sborník pøíspìvkù XVI. Mezinárodní kolokvium o regionálních vìdách* 19–21 (6): 94-101. DOI: 10.5817/CZ.MUNI.P210 6257 2013 11.

Kannan, K.P. (undated). People's planning—Kerala's dilemma. *Seminar, web edition.* 1-10.

Kannan, K.P. and Pillai, V.N. (2004). *Development as Freedom: An Interpretation of the 'Kerala Model'.* Thiruvananthapuram: Centre for Development Studies.

Kannan, K.P., Thankappan, K.R., Kutty, V.R. and Aravindan, K.P. (1991). *Health and Development in Rural Kerala.* Thiruvananthapuram: Kerala Sasthra Sahitya Parishat (KSSP).

Kapur, A. (2008). On Disasters in India. New Delhi: Cambridge.

Kapur, S. (2010). Vulnerable India: A Geographical Study of Disasters. New Delhi: Sage.

Kapur, A. (2016). *Made Only in India: Goods with Geographical Indications.* New Delhi: Routledge, Taylor and Francis Group.

Karunakaran, C. and Sinha-Roy, S. (1981). Laterite profile development linked with polycyclic geomorphic surfaces in South Kerala. In: *Lateritisation Processes,* Geological Survey of India. Calcutta: Oxford and IBH Pub. Co., pp. 221-231.

Karunaratne, G. (1976). The failure of community development programme in India. *Community Development Journal*, 11 (2): 95-119. http://doi.org/10.1093/cdj/11.2.95

Katz, M.B. (1978). Tectonic evolution of the Archean granulite facies belt of Sri Lanka—south India. *Journal Geological Society of India*, 19 (2): 185-205.

Katz, M.B. (1982). Lineament analysis of Landsat imagery applied to mineral exploration. In Laming D.J.C.

and Gibbs A.K. (eds.) *Mineral Exploration Techniques in Tropical Rainforest Areas.* AGID Report 7. pp. 157-166.

Kerala Industrial and Technical Consultancy Organisation (KITCO). (2003). Survey on Status of Coir Industry in Kerala. Thiruvananthapuram: Government of Kerala.

Kerala State Council for Science Technology and Environment (KSCSTE). (2007a). *State of Environment, Kerala.* Thiruvananthapuram: Government of Kerala.

Kerala State Council for Science, Technology and Environment (KSCSTE). (2019). Report of the committee to examine the causes of repeated extreme heavy rainfall events, subsequent floods and landslides and to recommend appropriate policy responses. Submitted to Kerala State Planning Board, Thiruvananthapuram.

Kerala State Disaster Management Authority (KSDMA) (2010). Kerala State Disaster Management Plan Profile, Department of Revenue and Disaster Management. Thiruvananthapuram: Government of Kerala.

Kerala State Drought Monitoring Cell (KSDMC) (2016). Drought Situation Assessment Report, 2016 (1st June to 30th September, 2016). Kerala State Disaster Management Authority, Department of Revenue and Disaster Management. Thiruvananthapuram: Government of Kerala.

Kerala Statistical Institute (KSI). (1994). Study on Conversion of Paddy land in Kerala, 1992-94. Project report, Thiruvananthapuram: KSI.

Kerala Sasthra Sahitya Parishat (KSSP). (1991). A Report on the Intensive Campaign for eradication of Illiteracy in Ernakulam District. Thiruvananthapuram: KSSP.

Kerala State Planning Board (KSPB). (2019). A Special Package for Post-Flood Kuttanad. Thiruvananthapuram: Government of Kerala.

King, L.C. (1953). Canons of landscape evolution. *Geological Society of America Bulletin,* 64: 721-752.

Kirk, D. (1996). Demographic transition theory. *Population Studies*, 50 (3): 361-387.

Kodath, P. (Undated). Globalisation and higher education in Kerala: Access, equity and quality. Report of a Study sponsored by the Sir Ratan Tata Trust. Thiruvananthapuram: Centre for Development Studies.

Kreimer, A. and Munnasinghe, M. (1990). Managing Natural Disaster and the Environment: Selected materials from the colloquium on Environment and Natural Disaster Management. Environmental Policy and Research Division, Washington DC.: World Bank.

Krishnan, T.N. (1985). Health Statistics in Kerala State. In Halstead, S.B., Walsh, J.A. and Warren, K.S. (eds.) *Good Health at Low Cost.* New York: The Rockefeller Foundation.

Krishnakumar, K.N., Rao, G.S.L.H.V.P. and Gopakumar, C.S. (2009). Rainfall trends in twentieth century over Kerala, India. *Atmospheric Environment,* 43: 1940-1944.

Krishnaswamy, V.S. (1981). The Deccan volcanic episode related tectonism and geothermal manifestations. In Subba Rao, K.V. and Sukheswala, R.N. (eds.) *Deccan Volcanism and Related Basalt Provinces in Other Parts of the World. Memoir, 3.* Bangalore: Geological Society of India, pp. 1-17.

Kunte, P.D. (1995). On some aspects of barrier islands of the west coast, India. *Journal of Coastal Research,* 11 (6): 508-515.

Kuriakose, S.L., Sankar, G. and Muraleedharan, C. (2009). History of landslide susceptibility and a chronology of landslide-prone areas in the Western Ghats of Kerala, India. *Environmental Geology,* 75 (7): 1553-1568.

Kurien, J. (1995). The Kerala Model: Its Central Tendency and the Outlier. *Social Scientist,* 23 (1/3): 70-90.

Kurien, N.P., Prakash, T.N., Baba, M. and Nirupama, N. (2006). Observations on tsunami impact on the coast of Kerala, India. *Marine Geodesy,* 29 (2): 135-145.

Lekshmi, S., Mohanta G.P., Revikumar K.G. and Manna P.K. (2014). Development and emerging issues in public and private health care systems of Kerala. *International Journal of Pharmacy and Pharmaceutical Sciences,* 6 (2) 92-98.

Lewis, J.P. (1997). *India's Political Economy: Governance and Reform.* Delhi: Oxford University Press.

Lin, G.C.S. (2001). Evolving spatial form of urban-rural interaction in the Pearl river delta, China. *The Professional Geographer,* 53 (1): 56-70.

Logon William. (1887) (1995). *Malabar–volumes I & II.* New Delhi: Educational Services.

Mahamaya Chattopadhyay, (1984). Morphometric analysis of the Valapattanam drainage basin, Cannanore district, Kerala. Technical Report 42. Thiruvananthapuram: Centre for Earth Science Studies.

Mahamaya Chattopadhyay, (1998). Geomorphic analysis for sustainable landuse development: case study of a south Kerala drainage basin. In: Chattopadhyay S. and Sharma H.S. (eds.) *Sustainable Development: Issues and Case Studies,* New Delhi: Concept Pub. Co., pp. 220–225.

Mahamaya Chattopadhyay and Sakunthala, C. (1986). Land form morphology and its impact on landuse pattern in the Kuppam River Basin, Kerala. Technical Report No. 45. Thiruvananthapuram: Centre for Earth Science Studies.

Mahesh Chand. (2002). Strategies for better transportation management in Kerala. Paper presented in the Annual management Convention. Thiruvananthapuram: National Transport Planning and Research Centre.

Mahadevan, R., Chandramohan, P. and Van Holland, G. (2005). Hydrodynamics of tsunami. Paper presented in the brainstorming session on tsunami mitigation strategies. Tiruchirapally, pp. 25-26.

Maipas J. (2013). *Thinking Topographically: Place, Space and Geography*. http://jeffmalpas.com/wp. pdf accessed

Mandal. G.S. and Mohapatra, M. (2010). Cyclone Hazard Prone Districts in India—A Report. National Disaster Management Authority, New Delhi: Government of India.

Manimaran, G. and Roy Chacko, P.T. (2006). Indian Ocean Tsunami: A historical perspective. *Journal Geological Society of India*, 68 (7): 163-166.

Mallik, T.K. and Suchindan, G.K. (1984). Some sedimentological aspects of Vembanad lake, Kerala, West coast of India. *Indian Journal of Marine Science*, 13: 159-163.

Marshall, T. (2015). *Prisoners of Geography*. London: Elliott and Thompson Ltd.

Mathai, J. (2009). District level natural hazard zonation maps for Kerala State. In Proceedings, National Workshop on Natural Hazards, Disaster Mitigation and Management. Thiruvananthapuram: Centre for Earth Science Studies, pp. 1-18.

Mc Farlane, M.J. (1981). Morphological mapping in laterite areas and its relevance to the location of economic minerals in laterite. In *Lateritisation Processes*, Calcutta: Oxford & IBH Pub., pp. 308-317.

Mc. Gee, T.G. (1989). Urbanisasi or Kotadesai? Evolving patterns of urbanisation in Asia. In Frank G. Costa *et al.* (eds.) *Urbanisation in Asia*. Honolulu: University of Hawaii. pp. 93-108.

——(1991). The emergence of desakota regions in Asia: Expanding a hypothesis. In N. Ginsburg *et al.* (eds.) *The Extended Metropolis: Settlement Transition in Asia*, Honolulu: University of Hawaii Press, pp. 3-25.

M.S. Swaminathan Research Foundation (MSSRF). (2007). Measures to Mitigate Agrarian Distress in Alappuzha and Kuttanad Wetland Ecosystem. MSSRF Commission Report. New Delhi: Government of India.

Menon, K.P.P. (1924). *History of Kerala, Vol. 1.* T.K. Krishna Menon (ed.) Ernakulam: Cochin Govt. Press.

Menon Sreedhara, A. (ed.) (1965). *Kerala District Gazetteers*, Ernakulam. Thiruvananthapuram: Government of Kerala.

Menon Sreedhara A. (1967) (2007). *A survey of Kerala History.* Kottayam: D.C. Books.

Menon Sreedhara A. (1978). *Cultural Heritage of Kerala*. Kottayam: DC Books.

Menon Sreedhara A. (1997). Kerala—Land and People. In Thanpi, K.B., Nayar, N.M. and Nair, C.S. (Eds.) *The Natural Resources of Kerala.* Thiruvananthapuram: World Wide Fund for Nature-India, Kerala State Office, pp. 1-10.

Menon, C.A. (1995) (First print in 1910). *The Cochin State Manual.* Thiruvananthapuram: Government of Kerala.

Menon, V.S. (1997). Sacred groves. In Thanpi, K.B., Nayar, N.M. and Nair, C.S. (Eds.) *The Natural Resources of Kerala.* Thiruvananthapuram: World Wide Fund for Nature-India, Kerala State Office. pp. 159-167.

Mridul, E., Chattopadhyay, S. and Elamon, J. (2000). *Development Perspective: First Step in Formulation of a District Plan-Case study of Thiruvananthapuram District.* Thiruvananthapuram: State Planning Board.

Misra, R.P. and Sundaram, K.V. (1980). *Multilevel Planning and Integrated Rural Development*. New Delhi: Heritage Publishers.

Misra, U.A. and Irudaya Rajan, S. (2018). Internal migration. Draft Thematic Paper 2, International Labour Organisation. Thiruvananthapuram: Centre for Developmant Studies.

Mukerjee, R.K. (1938). *Food Planning for Four Hundred Millions.* London: Macmillan, p. 20.

Murali Das, S., Mohan Kumar, G. and Sampath, S. (2009). Investigations into mechanisms of involvement of objects and personnel in lightning disasters. *Journal of Lightning Research,* 1(1): 36-51.

Myers, Norman (1999). The next green revolution: Its environmental underpinnings. *Current Science,* 76 (4): 507-513.

Myers, N., Mittermeier, RA, Mittermeier, CG, da Fonseca GAB, and Kent, J. (2000). Biodiversity hotspot for conservation priorities. *Nature,* 403: 853-858

Nababe, K. (2003). The health care system in Kerala, its past accomplishments and new challenges, *Journal of the National Institute of Public Health,* 52 (2) 140-145.

Nair, C.T.S. and Muraleedharan, P.K. (1983). Rural Institutions for the Development of Appropriate Forestry Enterprises: A Case Study of Traditional Reed Industries in Kerala. Research Report No. 18. Peechi: Kerala Forest Research Institute.

Nair, G.P.R. (1989). Universilisation of primary education in Kerala. In P.R. Panchmukhi (ed.) *Studies in Educational Reforms in India. Vol II.* New Dehi: Himalaya Publishing Co.

Nair, K.M., and Padmalal, D. (2004). Quanternary geology and geomorphic evolution of South Kerala Sedimentary basin, West Coast of India. In Ravindra Kumar, G.R. and Subhash, N. (eds.) Silver Jubilee Compendium. Earth System Science and Natural Resources Manangement: Centre for Earth Science Studies, pp. 62-81.

Nair, K.M., Sajikumar, S. and Padmalal, D. (1998). Silting up of a mega lagoon along Kerala Coast. National Seminar on Coastal Evolution, Processes and Products. Thiruvananthapuram: Centre for Earth Science Studies. Abstract Volume, p. 12.

Nair, K.M. (1976). Stratigraphic analysis of Kerala basin. Symposium on Geology and Geomorphology of Kerala (Abst.). Thiruvananthapuram: Geological Survey of India, pp. 9-10.

Nair, K.M. and Rao, M.R. (1980). Stratigraphic analysis of Kerala basin. In *Geology and Gemorphology of Kerala. Special Publication No. 5.* Thiruvananthapuram: Geological survey of India, pp.1-8.

Nair, K.N. and Chattopadhyay, S. (2005). *Water Resources of Kerala: Issues and Case Studies.* Thiruvananthapuram: Centre for Development Studies.

Nair, M.M. (1987). Coastal geomorphology of Kerala. *Journal, Geological Society of India,* 29: 450-458.

Nair, M.M. (1990). Structural trend line patterns and lineaments of the Western Ghats, South of 13° latitude. *Journal Geological Society of India,* 35 (1): 99-105.

Nair, N.J.K., Reghunathan Pillai, K. and Nalina Kumar (1987). Environmental Geomorphic Atlas, Kerala Coastal Zone. Final report submitted to the Department of Environment, New Delhi: Government of India.

Nair, R.R. and Hashimi, N.H. (1980). Holocene Climate inferences from the sediments of the Western continental shelf. *Proceedings of the Indian Academy of Sciences (Earth & Planetary Science),* 84: 299-306.

Nair, M.A. and Sreedharan, C. (1986) Agroforestry farming systems in the homesteads of Kerala, southern India. *Agroforest System* 4, 339–363. https://doi.org/10.1007/BF00048107

Nandakumar, G., Rajan, K.N. and Chellappan, K. (2001). Is the deep sea prawn fishery of Kerala sustainable? *Marine Fisheries Information Service (T & E ser),* ICAR, No. 170, October-December.

Narayanswamy and Chattopadhyay, S. (1996). Lateritic landscape in Cannanore district in Kerala: A geomorphic analysis. *Annals, National Association of Geographers, India,* 16(1):1-8.

Nathan, K.K. (2000). Characteristics of drought in Kerala. *Drought Network News* (1994-2001). Linclon: University of Nebraska, pp. 3-5.

National Academy of Sciences, NSA. (1997). Committee on Geosciences, Environment and Resources. Rediscovering geography: New Relevance for Science and Society. Washington D.C., National Academy of Sciences.

National Sample Survey Organisation, NSSO, (2014). 71 Round (January to June, 2014), New Delhi: Government of India.

National Transport Planning and Research Centre (NATPAC). (1989). Inland Water Transport and Development of Inland Waterways in Kerala—Basic Approach. Discussion paper for high level committee on inland water transport. Thiruvananthapuram: Government of Kerala.

Nayar, B.R. (1960). Community development programme—its political impact. *The Economic Weekly,* 12 (39): 1401-1410.

Nayar, T.V.R. and Nayar, N.M. (1997). Tuler Crops. In Thampi, K.B., Nayar, M.M. and Nair, C.S. (Eds.). The Natural Resources of Kerala, Thiruananthapuram: World wide Fund for Nature—India, Kerala State office, pp. 263-273.

Nicolson, D.H., Suresh, C.R. and Manilal, K.S. (1988). *An interpretation of Van Rheede's Hortus Malabaricus.* Koenigstein (Germany): Koeltz Scientific Books.

Notermans, C., Nugteren, A. and Sunny, S. (2016). The changing landscape of sacred groves in Kerala (India): A critical view on the role of religion in nature conservation. *Religions*, 7 (38) 1-14. www.mdpi.com/journal/religions doi:10.3390/rel7040038.

Nukala, R.B. and Mutz, D. (2015). Strategic approach for sustainable land use in an emerging country—Case of India. Paper presented at the "2015 World Bank Conference on Land and Poverty", The World Bank, Washington D.C.

O' Kelly, M. and Byran, D. (1996). Agricultural location theory: Von Thunen's contribution to economic geography. *Progress in Human Geography*, 20 (4): 457-475.

Oommen, M.A. (2004). Deepening Decentralised Governance in Rural India: Lessons from the People's Plan Initiative of Kerala. Working paper No. 11. Kochi: Centre for Socio-economic & Environmental Studies.

Padmanabhan, N. (1990). Poor performance of private sector in Kerala. *Economic and Political Weekly,* 25 (37): 15 September.

Panikar, P.G.K. and Soman, C.R. (1984). *Health Status of Kerala—The Paradox of Economic Backwardness and Health Development.* Thiruvananthapuram: Centre for Development Studies.

Padmalal, D., Arun, P.R., Maya, K. and Ramachandran, K.K. (2004). Mining of some natural resources of greater Kochi region: Problems and prospects. In: Ravindrakumar, G.R. and Subhash N. (Eds.) *Earth System Science and Natural Resource Management, Silver Jubilee Compendium,* Trivandrum: Centre for Earth Science Studies, pp. 347-348.

Pannell, C.W. (1974). Development and the middle city in Taiwan. *Growth and Change*, 5(1): 21-29.

Parayil, G. (1996). The 'Kerala model' of development: Development and sustainability in the Third World. *Third World Quarterly*, 17 (5) 941-957.

Parayil, G. (ed). (2000). *The Kerala Model of Development: Perspectives on Development and Sustainability.* London: Zed Press.

Patnaik, P. (2004). A theoretical note on Kerala style decentralised planning. *The Marxist*, 20 (1): 1-11.

Pike, A., Rodríguez-Pose, A. and Tomaney, J. (2006). *Local and Regional Development.* London: Routledge.

Pillai, V.N. (2008). Infrastructure, Growth and Human Development in Kerala. Munich Personal RePEc Archive. Online at http://mpra.ub.uni-muenchen.de/7017/MPRA Paper No. 7017.

Pillai, M.P. (1994). Performance of State sector enterprises in Kerala. In Prakesh, B.A. ed. *Kerala's Economy—Performance, Problems and Prospects.* New Delhi: Sage Publications, pp. 259-278.

Planning Commission of India. 2008a. *Kerala Development Report.* Government of India, New Delhi: Academic Foundation.

Planning Commission of India. 2008b. *Manual for Integrated District Planning.* New Delhi: Government of India.

Prabhakaran, P.V., Balakrishna Pillai, P. and Ramachandran, K.M. (1991). Climatic classification of Kerala. *Mausam*, 43 (4):434-36.

Prasad, M.N. (2003). Transport development—A rational approach. *Kerala Calling*: October 2003, pp. 16 17.

Pushpangadan, P. and Nair, K.N. (1997): Medicinal plants. In Thanpi, K.B., Nayar, N.M. and Nair, C.S. (Eds.) *The Natural Resources of Kerala.* Thiruvananthapuram: World Wide Fund for Nature-India, Kerala State Office, pp. 130-139.

Quarantelli, E.L. (1987). What should we study? Questions and suggestions for researchers about the concept of disaster. *International Journal of Mass Emergencies and Disaster*, 5 (1): 7-32.

Quarantelli, E.L. (1991). Disaster research centre: University of Delaware. *Disasters*, 15(3): 274-277.

Radha Devi and Ajit Kumar. (2011). Population Pressure on Land in Kerala. Working Paper No. 24. Kochi: Centre for Socio-Economic and Environmental Studies.

Radhakrishna, B.P. (2001). The Western Ghats of the Indian Peninsula. In: Gunnell Y. and Radhakrisha, B.P. (eds.) *Sahyadri: The Great Escarpment of the Indian sub-continent. Memoir*, 47 (1), Bangalore: Geological Society of India, pp. 133-144.

Raj, K.N. (1971). Planning from below with reference to district development and state planning—A note. *Economic and Political Weekly*, 6 (30-32) 24th July.

Raj Rachna, Maurya, D.M. and Chamyal, L.S. (1998). Late Quaternary sea level changes in Western India: Evidences from lower Mahi valley. *Current Sciences*, 79: 910-914.

Raj, P.P.N. and Azeez, P.A. (2011). Temperature rise in the Bharathapuzha river basin, Southern India, *Current Science,* 101(4): 492.

Rajan, K.R. (1991). Mass Literacy: The Ernakulam Experiment. Cochin: Ernakulam District Literacy Society.

Rajendran, C.P. (1982). Morphometric studies of the Pamba river basin in Kerala State. CESS T.R.No. 15/1982. Thiruvananthapuram: Centre for Earth Science Studies.

Rajendran, C.P. (1987). Systematic Stratigraphic Study of Cenozoic Sediments of Kerala with Special reference to Warkalli and Quilon Formations. Unpub. Ph.D. Thesis, Kochi: Cochin University of Science & Technology.

Rajendran, C.P., Rajagopal, G. and Narayanswamy. (1989). Quaternary geology of Kerala: Evidences from radiocarbon dates. *Journal Geological Society of India,* 33: 218-222.

Rajendran, C.P. and Rajendran, K. (1996). Low-moderate seismicity in the vicinity of Palghat Gap, south India and its implications. *Current Science,* 70: 304-307.

Rajendran, C.P., John, B., Kesavan, S. and Rajendran, K. (2009). Reassessing the earthquake hazard in Kerala based on the historical and current seismicity. *Journal Geological Society of India*, 73(6): 785-802.

Rajendran, K., Rajendran, C.P., Kesavan, S. and Naveen, R. (2012). Recent microtremors near the Idukki reservoir, Kerala, South India. *Current Science*, 102 (10): 1446-1451.

Ramankutty, V. and Panikar, P.G.K. (1995). Impact of fiscal crisis on the public sector health care system in Kerala. Thiruvananthapuram: Achuta Menon Centre for Health Science Studies.

Ramankutty, V. (2000). Historical analysis of the development of health care facilities in Kerala State, India. *Health Policy and Planning*, 15 (1) 103-109.

Ramaswamy, N.S. (1998). Rural transport in India. *Current Science*, 75 (8): 801-804.

Rammohan, K. (1999). Technological Change in Kerala Industry—Lessons from Coir Yarn. Discussion paper no. 4. Kerala Research Programme for Local Level Planning Development. Thiruvananthapuram: Centre for Development Studies.

Rao, G.S. L.H. V.P. (2010). Climate change adaptation initiatives in Kerala (India) under humid tropics. Power point presentation. Trissur: Kerala Agricultural University.

Rao, G.S. L.H. V. P., Rao, A.V. R.K., Krishnakumar, K.N. and Gopokumar, C.S. (2009). Impact of climate change on food and plantation crops in the humid tropics of India. ISPRS Archives XXXVIII-8/W3 Workshop Proceedings: Impact of Climate Change on Agriculture, pp. 127-130.

Rao, P.S. (1978). Some aspects of structure and tectonics of the Kerala region and related mineralisation. In Proceedings, Symposium on Tectonics & Mineralisation of South East Asia and Far East. Misc.Pub.no 34, part 3. Kolkata: Geological Survey of India, pp. 51-64.

Rao, V.P., Veerayya, M., Thamban, M. and Wagle, B.G. (1996). Evidences of late Quaternary neotectonic activity and sea level changes along the Western continental margin of India. *Current Science,* 71 (3): 213-219.

Ravindrakumar, G.R. (1986). The petrology and geochemistry of massif anorthosites and gabbros of the Bavali fault zone, North Kerala. *Proceedings, Indian Academy of Sciences (Earth & Planetary Science),* 95: 117 130.

Ravi Raman, K (ed), (2010). *Development Democracy and the State—Critiquing the Kerala Model of Development.* London: Routledge.

Rodriguez-Pose, A. (2008). Decentralisation and local and regional development. CAF Working paper No. 2008/04.

Sack, R. (2000). Review of the fate of place: A Philosophical history. *Progress in Human Geography,* 24: 136-37.

Sajeev, T.S. and Alex, C.J. (2017). Mapping of granite quarries in Kerala, India—A critical mapping initiative. Paper presented during the *Erudite Lecture Series of Prof. Madhav Gaadgil at School of Legal Studies,* Kochi: Cochin University of Science and Technology.

Salim, A.A. (2002). *Educational Development in India: The Kerala Experience since 1800.* New Delhi: Anmol Publication Pvt. Ltd.

Sampath, S. and Vinayak, P.V.S.S.K. (1989). Rainfall in Kerala. Unpublished report. Thiruvananthapuram: Centre for Earth Science Studies.

Sanitha, V.P. and Singla, N. (2016). Structural transformation of Kerala's economy: Is there any role of agricultural sector. *Journal of Regional Development and Planning,* 5(2): 45-58.

Sankar, G. (2005). Investigation of the land subsidence in Chattivayal locality of Cherupuzha Grama Panchayath, Taliparamba taluk, Kannur district, Unpublished report. Thiruvanathapuram: Centre for Earth Science Studies.

Sankar, G. and Tomson, J.K. (2017). Land disturbances due to soil piping in the Western Ghats. Field Excursion Guide, Thiruvanathapuram: National Centre for Earth Science Studies.

Sankaranarayanan, K.C. and Bhai, M.M. (1994): Industrial development of Kerala—Problems and prospects. In B.A. Prakesh (ed.) *Kerala's Economy—Performance, Problems and Prospects.* Sage Publications, New Delhi, pp. 298-315.

Sankaranarayanan, V.S. and Panampunnayil, S.V. (1979). Studies on organic carbon, nitrogen and phosphorus in sediments of Kochi backwaters. *Indian Journal of Marine Sciences,* 8 (1): 27-30.

Sasidhar, V.K. and Lakshmi, S. (1997). Homestead farming. In Thanpi, K.B., Nayar, N.M. and Nair, C.S. (Eds.) *The Natural Resources of Kerala.* Thiruvananthapuram: World Wide Fund for Nature-India, Kerala State Office, pp. 365-368.

Sathish Chandran, T.R. (2000). Karnataka. In G. Mathew (ed.), *Status of Panchayati Raj in the States and Union Territories of India.* Institute of Social Sciences, New Delhi: Concept Publishing Company.

Scaria, S. (2010). Changes in land relations: The political economy of land reforms in a Kerala village. *Economic and Political Weekly,* XLV (26, 27): 191-198.

Schleicher-Tappeser, R., Lukesch, R., Strati, F., Thierstein, A. and Sweeney, G. (1997). Sustainable regional development: An integrative concept. Contribution to the Conference "Science for a Sustainable Society" Roskilde, www.eures.de

Sen, A. (2006). *Identity and Violence—The Illusion of Destiny,* New Delhi: Allen Lane An imprint of Penguin Books.

Seethi, K.M. (2017). 'Muthanga 19/2'—A milestone in land struggle in Kerala. http://countercurrents. Org. accessed on 02.06.2018.

Sharma, V.K. and Asutosh, D.K. (2012). Disaster management in India. *Yojana,* March, 2012, pp. 30-36.

Shewale, M.P. and Shravan Kumar. (2005). Changing Faces of Nature—Climatological Features of Drought Incidences in India. Meteorological Monograph, Climatology no. 21/2005. Pune: India Meteorological Department, p. 22.

Shravankumar, V., Chattopadhyay, S. and Nair, N.J.K. (1988). Bio climatic characteristics of Kerala. *Indian Geographical Journal,* 63 (1):80-89.

Simon, A. and Mohankumar, K. (2004). Spatial variability and rainfall characteristics of Kerala. *Proc. Indian Academy of Sciences. (Earth Planet.Sci),* 113 (2): 211-221.

Singh, S.V., Kripalani, R.H. and Sikka, D.R. (1992). Interannual variability of the Madden-Julian oscillations in Indian summer monsoon rainfall. *Journal of Climate,* 5: 973-978.

Singh, N., Krishnakumar, K. and Soman, M.K. (1989). Some features of the periods contributing percentages of rainfall to annual total in Kerala, India. *Theoretical and Applied Climatology,* 39:160-170.

Sivadas, S. (1991). How Ernakulam become fully literate district of India. Bharat Gyan Vigyan Samithi (BGVS), India Unit for Cooperation with UNICEF and WFP, Paris: UNICEF.

Soil Survey Organisation, (2007). *Bench Mark Soils of Kerala.* Department of Agriculture (Soil Conservation Unit), Thiruvananthapuram: Government of Kerala.

Soman, M.K., Krishnakumar, K. and Singh, N. (1988). Decreasing and trend in the rainfall of Kerala. *Current Science,* 57 (1): 7-12.

Soman, K. (1997). *Geology of Kerala.* Bangalore: Geological Society of India.

Soman, K. (2002). *Geology of Kerala.* Bangalore: Geological Society of India.

Sreekumar, T.T. (1993). Urban processes in Kerala. CDS Occasional paper series. Thiruvananthapuram: Centre for Development Studies (CDS).

Stansfield Jr., C.A. (1983). *A Geography of New Jersey.* New Jersey: Rutgers University Press.

State Planning Board. (1990). Report of the Task Force on State Public Sector Industries. Eighth Five Year Plan, (mimeo). Thiruvananathapuram: Government of Kerala.

Storper, M. (1997). *The Regional World: Territorial Development in a Global Economy.* New York: The Guilford Press.

Subramanian, K.R. (1987). Evolution of the Western Ghats, India—a simple model. *Journal Geological Society of India*, 29: 446–449.

Subrahmanyan, K.R. (1998). Relative fall in sea level in parts of southern Karnataka coast. *Current Science,* 75 (7): 727-729.

Subrahmanian, K.K. (1994): Some facets of the manufacturing industry in Kerala. In Prakesh, B.A. ed. *Kerala's Economy—Performance, Problems and Prospects.* New Delhi: Sage Publications, pp. 237-258.

Subramanyan, A.R. and Murthy, N.S. (1982). Climatc zones of Kerala state and their water balances. *Deccan Geographer,* 20 (2):285-293.

Subrahmanian, K.K. and Pillai, M.P. (1986). Kerala's industrial backwardness: Exploration of alternative hypotheses. *Economic and Political Weekly*, 21 (14) 5 April.

Subramanian, K.S. and Muraleedharan, M.P. (1985). Origin of the Palghat gap in south India—A synthesis. *Journal Geological Society of India*, 26 (1): 28-37.

Sudhi, K.S. (2020). Crisis along the coast. *The Hindu*, February 1, 2020, Thiruvananthapuram.

Sundaram, K.V. (1983). Multilevel planning: The Indian experience. In Chatterjee L. and Nijkamp P. (eds.) *Urban and Regional Policy Analysis in Developing Countries,* London: Grower Publishing Company.

Sundaram, K.V. (2017). Multiple level planning—Concept, principles, and its application to sector planning. In Thakur B., Sharma, H.S., Misra S., Chattopadhyay S. and Singh S. (eds.) *Regional Development—Theory and Practice, Vol. 1: Concept of Regional Development*, New Delhi: Concept Publishing Co., pp. 162-183.

Sunny, K.P. (1990). Uneven development of Kerala's transport system—Some economic aspects. *Economic and Political Weekly*, 25 (37): 2093-2096.

Suryanarayan, M.H. (2008). Morbidity profiles of Kerala and all-India—An economic perspective. Working Paper No. 2008-007. Mumbai: Indira Gandhi Institute for Development Research.

Taaffe, E.J, Morrel, R.L. and Gould, P.R. (1973). Transport in underdeveloped countries. In B.S. Hoyle ed. *Transport and Economic Development*, London: McMillan.

Tandon, R. and Mohanty, R. (2003). *Does Civil Society Matter?* New Delhi: Sage Publishers.

Tello, M.D. (2010). From national to local economic development: theoretical issues. *Cepal Review,* 102: 49-65.

Thampi, P.K., Mathai, J., Sankar, G. and Sidharthan, S. (1998). Evaluation study in terms of landslide mitigation in parts of Western Ghats, Kerala. Research Report submitted to the Ministry of Agriculture, Government of India, Thiruvananthapuram: Centre for Earth Science Studies, p. 100.

Thara, K.G. and Soman K. (1993). Evidence for an isothermal depression path of granulites of the Palghat Gap and its implications on the origin of the Gap. Proceedings, V. Kerala Science Congress. Thiruvananthapuram: Kerala State Council for Science, Technology and Environment, pp. 288-292.

Tharakan, Michael, P.K. (1984). Socio-economic factors in educational development: The case of Nineteenth Century Travancore, *Economic & Political Weekly,* November 7-17.

Tharakan, Michael, P.K. (1990). The Ernakulam District Total Literacy Programme. Report of the Evaluation. Thiruvananthapuram: Centre for Development Studies,

The Hindu, 2020a. Daily News Paper, dt. 15th March, Thiruvananthapuram.

The Hindu, 2020b. Daily News Paper dt. 9th May, Thiruvananthapuram.

The Hindu, 2020c. Daily News Paper dt. 6th May, Thiruvananthapuram.

Thummarukudy, M. (2017). Kerala is among the most dangerous places in terms of drowning. *The New Indian Express*, Thiruvananthapuram, 10th September, 2017.

Tilak, J.B.G. (2001). Higher Education and Development in Kerala. Working Paper No. 5. Kochi: Centre for Socio-Economic and Environmental Studies.

Tornquist, O. and Tharakan, M.P.K. (1995). *Next Left? Democratisation and Attempts to Renew the Radical Political Development Project: The Case of Kerala*, Copenhagen: NIAS Books.

Tornquist, O. (1999). Of new popular politics of development: The Kerala experience. In Parayil G (ed): *The Kerala Model—An Anthology*, Zed Books, pp. 1-28.

Tornquist, O. (2001). Movement, politics and development: The case of Kerala. *Social Scientist*, 29 (11-12): 57-87.

Torrisi, G., Pike, A., Tomaney, J. and Tselios, V. (2011). Defining and measuring decentralisation: a critical review. Munich Personal RePEc Archive, (MPRA) Paper No. 51441. http://mpra.ub.uni-muenchen.de/51441.

Turner II, B.L., Kasperson, R.E., Meyer, W.B., Dow, K.M., Golding, D., Kasperson, J.X., Mitchell, R.C. and Ratick, S.J. (1990). Two types of global environmental change—definitional and spatial scale issues in their human dimensions. *Global Environmental Change,* 1 (1):14-22.

Ullman, E.L. (1973). *Geography as Spatial Interaction. Studies in Regional development—Cities and Transportation.* Washington DC: University of Washington Press.

United Nations Environment Programme (undated). Water in The Transition to Green Economy—A UNEP Brief. www.unep.org/greeneconomy

United Nations Development Programme (UNDP) (undated). Decentralisation in India: Challenges and Opportunities. Discussion Paper Series No. 1. Human Development Resource Centre, UNDP. New Delhi.

United Nations Regional Commission (UNRC). (2011). The Regional Dimension of Development and the UN System. A study sponsored by the UN Regional Commissions, New York: United Nations.

Unnikrishnan, A.S. and Shankar, D. (2007). Are sea-level rise trends along north Indian Ocean coasts consistent with global estimates? *Global Planet Change,* 57: 301-307.

Unnikrishnan, A.S., Nidheesh, A.G. and Lengaigne, M. (2015). Sea-level rise trends off the Indian coasts during the last two decades. *Current Science,* 108 (5): 966-971.

Vadrevu, K.P. and Badarinath, K.V.S. (2014). Agriculture and the nitrogen problem in India—Environmental implications. In Sutton, M.A., Mason, A.E., Sheppard, L.J., Sverdrup, H., Haeuber, R. and Hicks, H.K. (eds.) *Nitrogen deposition, Critical loads and Biodiversity.* Springer, New York. pp. 439-446.

Vaidyanadhan, R. (1967). An outline of the geomorphic history of India, south of 18° N latitude. *Proceedings of the Seminar on Geomorphological Studies in India, 1965,* Centre for Advanced Studies, University of Saugar, Sagar, pp. 121-130.

Vaidyanadhan, R. (1971). Evolution of the drainage in Cauvery in south India. *Journal Geological Society of India,* 12(1): 14-23.

Vaidyanadhan, R. (1990). Morphology of east coast modern deltas. *Proceedings of the Field Seminar on Recent Delta,* Geology Department, Visakhapatnam: Andhra University, pp. 31-35.

Valdiya, K.S. and Rajagopalan, G. (2000). Large Palaeo lakes in Kaveri basin in Mysore plateau, Late Quaternary fault reactivation. *Current Science,* 78 (9): 1138-1142.

Van Campo, E. (1986). Monsoon fluctuations in two 20,000 yr. B.P. oxygen-isotope pollen records off south west India. *Quaternary Research,* 26: 376-388.

Varadarajan, K. and Nair, K.M. (1978). Stratigraphy and structure of Kerala Tertiary basin. *Journal, Geological Society of India,* 19 (2): 217-220.

Varadarajan K. and Balakrishnan, M.K. (1980). Kerala coast — a Landsat's view. *Geology and Geomorphology of Kerala. Special publication no. 5,* Thiruvananthapuram: Geological Survey of India, pp. 67-68.

Vermon, R. (2001). The new Kerala model: Lessons for sustainable model. *World Development,* 29(4) 601 617.

Verstappen, H. Th. (1987). Geomorphic studies on Sri Lanka with special emphasis on the north west coast. *ITC Journal,* 1 (1): 1-17.

Vorosmarty, C.J,. D. Lettenmaier, C. Leveque, M. Meybeck, C. Pahl-wostl, J. Alcamo, W. Cosgrove, H. Grassi, H. Hoff, R. Kambat, F. Lansigan, R. Lawford, R. Naiman (2004). Humans transforming the global water system. *EOS, Transactions, American Geophysical Union,* 85 (48) 509, 513-514.

White, F.G. (1978). Natural hazards and Third World—A reply. *Human Ecology*, 6(2): 229-231.

White, S. and Gasser, M. (2001). *Local Economic Development: A Tool For Supporting Locally Owned and Managed Development Processes that Foster the Global Promotion of Decent Work.* Geneva: Job Creation and Enterprise Development Department, ILO.

Widiastuti, I. (2013). The living culture and typo-morphology of vernacular houses in Kerala. *Journal of the International Society for the Study of Vernacular Settlements,* 2 (4): 41-53.

Williams, J.F. (1999). Taiwan's urban system in transition. In Chapman, G.P., Dutta, A.K. and Bradnock, R.W. (eds.) *Urban Growth and Development in Asia, Vol. II, Living in Large Cities*, SOAS Studies in Development Geography, London: University of London, pp. 24-41.

World Bank, (2000). *Overview of Rural Decentralisation in India: Volume II—Approaches to Rural Decentralisation in Seven States.* Washington DC: World Bank.

World Commission on Environment and Development (WCED) (1987). *Our Common Future*. New Delhi: Oxford.

World Travel and Tourism Council (WTTC) (2019). Travel and Tourism Economic Impact, (2019), World. London: WTTC.

Wright, J.K. (1947). Terrae Incognitae: The place of the imagination in geography. *Annals of the Association of the American Geographers*, 37 (1): 1-15. Published online: 16 Jun, 2008.

Zachariah, K.C. and Irudaya Rajan, S. (1997). *Kerala's Demographic Transition—Determinants and Consequence,* New Delhi: Sage Publications.

Zachariah, K.C. and Irudaya Rajan, S. (2018). *Emigration from Kerala: End of an Era.* Kochi: RedInk, Nalanda Books.

Ziadat F., Bunning, S. and De Pauw, E. (2017). Land Resource Planning for Sustainable Land Management. FAO Land and Water Division Working Paper 14. Rome: Food and Agriculture Organisation, United Nations Organisation.

www.keralapolice.org

www.metoffice.goc.uk/learning/library

www.thrissorkerala.com/census-2011/sex ratio

Index